Rhythms of Life

Thyroid Hormone
&
The Origin of Species

Susan J. Crockford Ph.D.

see also ***www.rhythmsoflife.ca***
comments & questions welcome

Note for Librarians: A cataloguing record for this book is available from Library and Archives Canada at www.collectionscanada.ca/amicus/index-e.html
ISBN 1-4120-6124-5

Printed in Victoria, BC, Canada. Printed on paper with minimum 30% recycled fibre. Trafford's print shop runs on "green energy" from solar, wind and other environmentally-friendly power sources.

Offices in Canada, USA, Ireland and UK
This book was published *on-demand* in cooperation with Trafford Publishing. On-demand publishing is a unique process and service of making a book available for retail sale to the public taking advantage of on-demand manufacturing and Internet marketing. On-demand publishing includes promotions, retail sales, manufacturing, order fulfilment, accounting and collecting royalties on behalf of the author.

Book sales for North America and international:
Trafford Publishing, 6E–2333 Government St.,
Victoria, BC V8T 4P4 CANADA
phone 250 383 6864 (toll-free 1 888 232 4444)
fax 250 383 6804; email to orders@trafford.com
Book sales in Europe:
Trafford Publishing (UK) Limited, 9 Park End Street, 2nd Floor
Oxford, UK OX1 1HH UNITED KINGDOM
phone 44 (0)1865 722 113 (local rate 0845 230 9601)
facsimile 44 (0)1865 722 868; info.uk@trafford.com
Order online at:
trafford.com/05-1025

10 9 8 7 6 5 4 3

Contents

Figures

Boxes

TABLES

Abbreviations

AA	arachidonic acid
ACTH	adrenocorticotropin (aka *adrenal cortical stimulating hormone*)
Bmp-4	bone morphogenetic protein-4 (a gene)
BMR	basic metabolic rate
BP	before present
BPA	bisphenol-A (a biochemical)
ca. (also, c)	circa (approximately)
Ca^{2+}-ATPase	an enzyme that moves calcium across cell membranes, especially in skeletal muscle (part of the *calcium pump*)
CNS	central nervous system
CRH	corticotropin-releasing hormone (aka *corticotropin*)
D2	deiodinase type II (an enzyme)
DHA	docosahexaenoic acid
EGF	epidermal growth factor
Fgf8	fibroblast growth factor 8 (a gene)
FSH	follicle stimulating hormone (aka *follitropin*)
GH	growth hormone (same as STH)
GHRH	growth hormone-releasing hormone
GnRH	gonadotropin-releasing hormone
IDD	iodine deficiency disorders
IGF-I	insulin-like growth factor-I
LH	luteinizing hormone (aka *lutropin*)
MSH	melanocyte-stimulating hormone (aka ⊠-*melantropin*)
mtDNA	mitochondrial DNA
mU/l	milli-international units per litre
mya	million years ago
Na^{+}/K^{+}-ATPase	an enzyme that moves sodium and potassium across cell membranes, especially in nerves (part of the *sodium pump*)
nDNA	nuclear DNA
PCBs	polychlorinated biphenyls (a class of biochemicals)
PFCs	perfluorinated chemicals (a class of biochemicals)
PRL	*prolactin*
SCN	suprachiasmatic nucleus (located in the hypothalamus)
SH	steroid hormones
Shh	sonic hedgehog (a gene)
SSRIs	selective serotonin reuptake inhibitors
StAR	steroidogenic acute regulatory protein
STH	*somatotropin* (same as growth hormone, GH)
T_3	*triiodothyronine*
T_4	thyroxine (aka *tetraiodothyronine*)
TBG	thyroxine-binding globulin
TH	thyroid hormone
TR	thyroid hormone receptor
TRE	thyroid responsive element
TRH	thyrotropin releasing hormone
TSH	thyroid stimulating hormone
WHO	World Health Organization
ya	years ago
µl	microlitre

Acknowledgements

The road to the place where this book could be written has taken years to traverse. At times it's been lonely but I have not been alone. Friends, family and colleagues (many of whom I've never met in person) carried much of the load. In countless ways and at different times, they made it possible for me to finish what I started and I am grateful to them all. For practical help and support in preparing this volume, special thanks go to my editor Ruth Dyck and cover designer Sharon Belley. Last but not least, I am indebted to my children, Jesse and Laura McMillan, whose faith and good humor never failed through the entire journey (including the embarrassing "obsession with spotted animals" phase): I could not have done it without you.

Prologue

Why Yet Another Book About Evolution?

False facts are highly injurious to the progress of science for they often endure long; but false hypotheses [theories] do little harm, as everyone takes a salutary pleasure in proving their falseness; and when this is done, one path toward error is closed and the road to truth is often at the same time opened.

Charles Darwin (preface to *The Origin of Species*)

My introduction to evolution came early, just after I'd started elementary school. I was six years old and hung quietly in the background while my mother helped my sister illustrate the story of evolution on a 90-foot roll of kitchen shelf paper for her school science project. The scroll depicted the descent of living organisms from a single common ancestor, with each geological era plotted to scale, and the whole thing lavishly decorated with drawings and photos cut from an ancient encyclopedia. The finished product wrapped impressively around three walls of the fourth grade classroom, and there, near the very end of the roll, were humans and their ancestors—clearly occupants of an insignificant sliver of evolutionary time. The dramatic message wasn't lost on my sister's classmates—or on me.

At home over the next couple of years, evolution became the focus of many discussions over dinner and the impetus for family camping weekends to the shores of Lake Erie in southern Ontario. There we spent long summer days chipping trilobites and other fossil creatures out of the brittle shale, and evenings admiring our finds and discussing their evolutionary significance. Over one of the intervening winters of that period, I developed a fascination with wolves and sled dogs.

Jack London's adventure story of the Canadian North, *White Fang*, not only became my favourite book but defined my image of what a real dog should be. No matter that we lived in suburban Toronto—not the Yukon—or that I was barely ten years old: I campaigned heavily for a sled dog of my own.

My father relented after about a year and bought me a registered Alaskan malamute, the largest and strongest of the sled breeds. That wolf-grey female was the first of half a dozen malamutes who've shared my life over the years, and she remains the most memorable. Chaka had astonishing wolf-like moments, including a remarkably expressive howl. The combination of wild and civilized nature typical of this breed captivated me both intellectually and emotionally, and I've rarely been without one since.

Years later, malamute number three—a massive black-and-white male—was the first family dog for my own two children. When my son was about three, we saw a chihuahua being led down the street, prompting the familiar pointing and "Wha's tha?" His response to my answer left me dumbfounded. The disbelief and hurt on his face was as easy to read as a stop sign: "A dog? I'm not *that* gullible. I know what a *dog* looks like!" My glib reply was much easier than a real answer: "You're right," I said, "it's a pussycat."

Figure P.1 The animal my children understood a dog to be (here with daughter Laura)

It never occurred to me then to explain the concept of dog breeds to a three-year-old. It also didn't occur to me just how significant the incident was, biologically speaking. However, years later I realized that what I've come to think of as "the chihuahua conflict" actually says all there is to be said about what makes dogs special. Think about it—the term *dog* describes an animal with almost endless variation, so when we want to be really clear about what a particular animal looks and acts like, we use breed names. Most breeds of domestic animals not only have a distinctive, consistent appearance but often a characteristic temperament as well. In other words, for domesticates, the name of the *breed* describes animals of relatively uniform appearance and behaviour in the same way as the *species* name does for wild animals (such as the grey wolf, *Canis lupus*).

When you think of it this way, breeds of domestic animals are similar in many respects to species of wild animals, because when individuals of the same breed mate, the offspring resemble their parents—and each other—to a large degree. Granted, all individuals (even so-called identical twins and higher multiples) are never absolutely the same—there is always some amount of difference, or *variation*, among them. Within a litter of puppies for instance, these differences among individuals can be physical (like contrasting hair colour or size) or behavioural (like the variations in temperament that make one pup bold and another shy). Other distinctions may exist that won't be obvious until later, like differences in resistance to disease. In spite of such variation, however, the result of humans carefully controlling which individuals are allowed to mate is that breeds of animals stay more or less the same, generation after generation, in the same way as species of animals do.

This control by humans over breeding is an essential and significant difference, of course (as we will see later). If dogs are allowed to mate indiscriminately, the breed distinctions disappear. Breeds remain similar to species only as long as our influence persists. As soon as people relinquish control over the mating habits of a breed, or cease to provide opportunities for mating within the breed, it becomes extinct—as well and truly extinct as any wild species can become. This has been the fate of many local varieties of dogs, including some European breeds that were lost during the chaos of the first and second world wars. Many local aboriginal varieties in the South Pacific, like the Hawaiian *poi-dog* and the New Zealand *kuri*, were lost through interbreeding with the European dogs that early settlers and missionaries brought with them. The one true native breed that appears to have existed in the Americas fared a similar fate, but for another reason. The Northwest Coast Salish wool dog (illustrated next page), bred for its long woolly hair, which was shorn and woven into highly-valued blankets on simple wooden looms, became extinct because of a sudden and irrevocable loss of effort in maintaining control over its breeding—within a few decades after Europeans arrived with their colourful Hudson's Bay trade blankets, the wool dog was no longer a breed distinct from other local dogs.

I suppose few people would ever think of breeds of dogs as species–equivalents. But after many years of study, I've come to the conclusion that understanding precisely why there are such similarities between breeds of domestic animals and wild species is paramount to understanding evolution. Charles Darwin came tantalizingly close to a similar conclusion more than one hundred years ago and I believe it's time to revisit this line of thinking. I suggest that in today's world, dogs—not fruit flies—may be the best models for studying and explaining evolution.

Fruit flies have become a popular subject for evolutionary studies because they can produce many generations of offspring in a year, and so the results of experiments that mimic natural selection can be seen quickly. But they're poor models when it comes time to explain how the process actually works to people who aren't scientists, for reasons that only begin with the fact that you can barely see a fruit fly without a microscope. More importantly, it requires a huge leap of faith for people to believe that what works for insects could transform apes into humans and leaps of faith, in my opinion, are not what science should be about.

Figure P.2 Extinct breeds are as forever lost as extinct species. Shown here, a sketch of a reconstructed Salish wool dog of the Central Northwest Coast of North America, a unique native (aboriginal) breed that became extinct so fast during the early years of European contact that no specimens were ever collected—it's known only from archaeological skeletal remains and historical descriptions. Artist Cameron J. Pye (see refs. 145, 153, 365).

For many people, the evolutionary processes of speciation and natural selection are much easier to comprehend, and to apply to our own evolutionary history, when domestic animals are the models. While Darwin would probably have chosen pigeons, I think dogs are a better choice for the twenty-first century. Most people—regardless of where in the world they live or what they do for a living—know something about the range of form and behaviour that dogs exhibit.

However, in order to use dogs—or any domestic animal, for that matter—as a model to explain evolution, some misconceptions have to be addressed. We've all been taught that dogs became domesticated because of human manipulation and design, that someone very long ago got the bright idea that a wolf might be a useful hunting assistant and so captured a few young pups to raise. Taming of these captive wolves is said to have begat dogs. Having had success with dogs, so the story goes, the same method was applied to wild sheep, goats and cattle—and lo, there was livestock.

In reality, this scenario of domestication is little more than a widely accepted myth. It's a story concocted by anthropologists to fit assumptions of human behaviour and historical events, but there's no concrete evidence whatsoever to support it. Although it's rarely challenged outright, a few people—myself included—have long been suspicious of this traditional definition of domestication, for reasons that I'll go into in more detail later. My own scepticism doubled, however, when a genetic research project on dogs and wolves I was involved in during the early 1990's didn't generate the results it should have if this concept of domestication was correct. This lack of correspondence didn't matter to the research project or to my graduate school supervisor, but it mattered to me: if domestication didn't happen the way we'd all been taught, how exactly *did* dogs come to be?

I lost patience with assumptions masquerading as fact and went looking for some scientific answers that made sense. As I will show in the chapters that follow, dog domestication is fully explainable as a natural evolutionary process, a true *speciation event* that did not require intentional human interference. The same can be said for all of the other major domestic animals (this doesn't mean humans *couldn't* have deliberately initiated domestication events, it just means they didn't *have* to). Although a few of my colleagues had made this argument before, it wasn't an idea that really caught on because none of them were able to suggest a plausible alternative explanation for how and why the changes we see in domestication actually came about.

I was able to come up with just such an alternative explanation for domestication because I took a very straightforward approach. I began by asking one simple question: what had to happen to a wolf, in strictly biological terms, for it to become a dog? My preliminary dip into the scientific literature, in January of 1994, suggested that vital information on the topic already existed: I didn't need to do research experiments myself because they had already been done. As often happens in science, however, no one had realized the broader significance of these experimental results—they saw only the immediate applications to their own work or to other problems of a similar nature.

I was hooked. I ended up reading volumes of material, searching the literature on embryology, reproduction, animal behaviour, physiology, genetics and endocrinology. I probably spent the equivalent of a Hawaiian vacation making photocopies of journal articles and book chapters. I talked incessantly with colleagues, visiting scientists, friends and family as I struggled to piece it all together. Although I knew full well that the whole thing might be a pointless intellectual exercise, I also realized that the seeds of this inquiry had been planted more than thirty years ago, at the time my childhood fascination with wolves and wolf-like dog breeds began. I felt I owed it to myself to follow the investigation as far as it would go and truly considered it a matter of personal indulgence.

However, within the year I realized I had not only an answer to my initial question but one that went much further. When I asked how a wolf could turn into a dog without deliberate human intervention, I came up with an explanation for how *any* animal could transform into something else—and it was clear to me that if my answer was valid for domestication, the concept simply had to work for evolutionary change in all other animals as well. My model not only explained *why* animals change from one form to another when they do, but also *how*—in intuitively comprehensible terms. In other words, I was able to turn my model for domestication into a general evolutionary *theory* that describes the invisible biological mechanism responsible for evolutionary change. Although in everyday use the word *theory* often serves as a synonym for *speculation,* in science a theory must be a general explanation that fits many situations—and must also be based on verifiable facts and testable by scientific methods.

My theory explains precisely how evolutionary processes could have transformed wolves into dogs *and* created new species of fish, birds and mammals (including primates and our immediate ancestors). With slight modification, it also works for plants and invertebrate animals, like insects. It's also scientifically testable: that is, it should eventually be possible to prove by experimentation that the concept is wrong—if indeed it isn't correct. It was, I'd like to think, the kind of answer Charles Darwin dreamed of finding.

Most people know that Darwin's theories about evolution revolutionized the way people thought about the world, but few realize the one significant thing he *didn't* do: he never actually delivered what the title of his famous book seemed to promise—a precise explanation for the origin of species. Modern biologists have not done much better despite decades of research by thousands of competent scientists. Ornithologist Ernst Mayr, who died last year at the age of one hundred, actively applied himself to this particular dilemma without success for well over sixty years (see William Provine's 2005 essay and refs. 504-511); Stephen Jay Gould, palaeontologist and author of many popular books and essays on evolution, did the same—although his approach was different and he didn't live as long (275-278). While both made significant contributions to knowledge about how evolution proceeds in theory and in practice, neither resolved the question of how species arise in precise biological terms.

As it turns out, both Gould and Mayr were far too deeply entrenched in the field of evolutionary theory to contribute a really novel solution to the species problem. Tom Kuhn explained this years ago in his classic 1970 volume, *The Structure of Scientific Revolutions* (updated recently by Robert Root-Bernstein in a livelier treatment called *Discovering: Inventing and Solving Problems at the Frontiers of Scientific Knowledge,* 1989): historically, solutions to seemingly intractable problems in science have always come from either outside a field or the edges of it, not from someone ensconced in the mainstream. Gould may have felt himself peripheral, but in the end he wasn't nearly close enough to the edge.

History shows not only that revolutionary concepts and inventions tend to come from outside a discipline but that discovery itself is only half the battle. The other half is realizing what you've found—and then letting people know about it. Often, for a very new idea, this last requirement is a real challenge. Hence this book, in this format. (And in regards to the format, I've chosen some particular conventions: references cited are listed by number *and* author/date in the bibliography; there is a short list of references geared to nonspecialist readers at the end of each chapter (designated *recommended readings*); as much as possible, complex details that may be of interest only to specialist readers are confined to shaded boxes.)

My theory for the role of thyroid hormone in evolution is destined to have a revolutionary impact, not just on the field of biology but on the practice of both human and animal medicine. At first glance, I might seem an improbable person to have come up with such a far-reaching concept—and my low-tech/low budget approach unlikely to have resulted in an important discovery of any kind. However, I had some important factors in my favour. First of all, I had the right educational background: an early childhood exposure to the concept of evolution, a bachelor's degree in comparative zoology and a graduate school upgrade in molecular genetics. All of this helped prepare me for reading masses of scientific papers on a diversity of subjects.

I had also chosen the right career. Having fortuitously taken all the right undergraduate courses, I jumped early on at an offer to enter the burgeoning field of *archaeozoology*, the identification and interpretation of animal bones recovered from archaeological sites. The process of building a career in this discipline had a huge impact on my evolutionary thinking. Archaeozoology depends on the fact that species, and their bones, are constant over time. The same thing is true for Stephen Gould's field of palaeontology: it's just that the time scale in archaeology is shorter and the bones are still bone (not turned to rock, as they are in fossils).

In archaeozoology, an ancient bone or bone fragment is compared with the same bone from a number of modern animals until a match is found. I became intimately acquainted with the skeletal anatomy of modern animals as I turned scavenged carcasses into the labelled skeletons that made up our reference collection of bones. What became patently clear was that despite slight individual variation, virtually every bone in the skeleton of a distinct species (not just the skull or teeth) is always the same shape regardless of when it was collected. A black-tailed deer thigh bone (femur) from five or ten thousand years ago might be slightly larger or smaller, but it's the same shape as one found today—and it's always distinguishable from the femur of a mountain goat. The femur of a white-tailed deer will be more similar to that of a black-tailed deer than to that of a mountain goat, but it will be distinguishable nevertheless. This basic premise is true for every kind of animal there is: fish, birds, both marine and terrestrial mammals, reptiles and amphibians.

I didn't need to accept an intellectual concept of what a species was and what it meant: I'd handled the proof of it—day after day—for twenty years. Archaeozoology provided me with a critical evolutionary insight that bridged the infamous gap between the view of life that a palaeontologist gets (biased because it has no living component) and the one that a field biologist gets (biased because it lacks an historical component). This unique perspective put me in a position to assess the question of how species arise from a totally new perspective.

In the end, I believe the main reason I was able to come up with an alternative explanation for domestication that made sense was simply that I asked the right question—at the right time. As it turned out, this was all that was really needed for the development of an entirely novel concept of how evolution works.

The timing component was really crucial: when you're trying to advance a new theory that contradicts what is currently accepted, no matter how well supported or eloquently presented, people need to be ready to hear what you have to say or you'll be talking to the wind. I found to my relief that many members of the scientific community were already questioning the reigning theoretical model of evolutionary biology (the accepted *paradigm*), which hinges on the primacy of the gene. Some might say such a challenge is long overdue. Certainly, it's clear that despite decades of research and billions of dollars in funding, biologists still have little more than the vaguest of explanations for how genes transform one species into another. They can list the DNA code for the entire human genome, but can't say which genes create the essential differences between ourselves and chimpanzees—or between ourselves and carrots, for that matter. In short, the current evolutionary paradigm has outlived its usefulness, and many biologists know this in their hearts even if they're not ready to admit it publicly. This isn't meant to imply that genes and their mutations aren't important, just that they are not necessarily the biological mechanism that drives the bulk of evolutionary change.

So why yet another book about evolution? What we need, and what I'm offering, is an opportunity for geneticists and evolutionary biologists to take a few steps backward—a chance to look at the problem again from a new, less "geno-centric" perspective. More than a few of my colleagues have already happily embraced the prospect. My theory requires thinking about how species change in a somewhat different fashion than most of us have been taught, but it doesn't contradict Darwin's basic tenets—it simply comes at the problem from a slightly different direction. My theory presents a new way of thinking that provides an intuitively understandable solution for some things we didn't understand before. It also resolves an increasing number of conundrums that have only come to light with recent advances in molecular genetics. The really unexpected bonus is that the concept also has profound implications for human and animal health. As a consequence, my answer to the question of how species arise actually makes evolution *personal.*

The battery of experimental tests of my theory have barely begun, which might prompt some of my colleagues to suggest this publication is premature. But I've come to realize that whether or not the theory is upheld, in totality or in part, it represents such an important new way of thinking about life that its value transcends absolute validation. Charles Darwin saw this about his own work, and the comment he made is equally true here: "False facts are highly injurious to the progress of science for they often endure long; but false hypotheses [theories] do little harm, as everyone takes a salutary pleasure in proving their falseness; and when this is done, one path toward error is closed and the road to truth is often at the same time opened."

I've written several peer-reviewed journal articles and book chapters on this topic, as well as a Ph.D. dissertation, in scientific formats appropriate for scholarly consumption (146-151). Encouraging responses to these publications from my colleagues assure me of the validity and usefulness of my theory. But non-scientists and scholars in unrelated fields were asking for a more accessible treatment, something more pragmatic and with less jargon. I hope this fits the bill. My aim with this book is to make evolution as comprehensible and personal for you as it's become for me. By the end of the story, you will really understand how evolution works, even if you thought you understood it before. I guarantee it will change the way you look at life.

Recommended reading

Kuhn, T.S. 1970. *The Structure of Scientific Revolutions, Second Edition.* University of Chicago Press, Chicago.

Provine, W.B. 2005. Ernst Mayr, a retrospective. *TRENDS in Ecology and Evolution* 20(8):411-413.

Root-Bernstein, R. S. 1989. *Discovering: Inventing and Solving Problems at the Frontiers of Scientific Knowledge.* Harvard University Press, Cambridge.

Chapter 1

Darwin, Dogs & Dilemmas

When breeds extremely different (as the grey-hound and bull-dog) are crossed, are their offspring equally prolific, as those from between nearer varieties (such as from the grey-hound and shepherd-dog)?

Charles Darwin (1840: Question #8)

Darwin: what you need to know

What drove Charles Darwin to make such a significant contribution to our understanding of evolution? Although much has been written on the subject, to my mind only two things are really important: he cared passionately about his work and he saw the significance of plant and animal breeding as a model for how evolution worked.

Charles was born in the small country town of Shrewsbury, England in 1809—the fifth child of six and the second son (167, 391). Although his father's plan was that both sons follow in his footsteps and become physicians, Charles was such a dedicated naturalist that he could give his full attention to nothing else. He neglected his University of Edinburgh medical studies so badly that his father relented and sent him to Cambridge to study theology instead (while it may seem incongruous to us now, in the 1800's the study of religion was considered an appropriate educational route for those interested in the natural sciences—many went on to became ordained ministers).

At the age of twenty-two, Darwin graduated with a B.A. from Cambridge and signed on almost immediately as naturalist for the *H.M.S. Beagle* and "gentleman companion" to her captain—fitting employment, given his theological training. The *Beagle* had been commissioned to survey the coast of South America, a journey expected to take several years.

The voyage of the *Beagle* ended up consuming the five years following Darwin's university studies, taking him to numerous locations around South and Central America, with stops at New Zealand and Australia on the way home. Those years had an enormous impact on Charles. The specimens he collected and the observations he made on that trip proved instrumental to the development of his ideas on how the natural world operated.

Back in England, he settled down to work. "Work," for Charles, entailed a disciplined regimen of keeping up with incoming and outgoing correspondence, reading, cataloguing collections of specimens, conducting experiments and writing up his notes on plant and animal life. He never had to hold down a formal job because he had supporting income from an inheritance. He nevertheless worked hard throughout his life, and wrote a prodigious number of articles and books on various topics, including an impressive but virtually unheard of two-volume treatise on barnacles.

Aside from a few short forays into the countryside, Darwin lived the rest of his life in London and nearby Kent. At the age of thirty, he married his cousin Emma, and although often quite ill with an ailment that no one could name, he was fit and fertile enough to sire ten children. There were no further expeditions to study the world's diversity of animals such as he had made aboard the *Beagle*. He seemed content to stay at home and study the diversity that surrounded him there.

Darwin saw quite early that domestic animals were actually very good subjects for studying evolution. He devoted a huge amount of time to the study of domestic animals and plants, both prior to and after the publication of his best known book, *On The Origin of Species* (henceforth, the *Origin*), in 1859. Darwin peppered friends, relatives, acquaintances and complete strangers alike with a barrage of questions about what happens when certain kinds of animals mate and produce offspring. In 1840, almost twenty years before the *Origin* was published (but after the *Beagle* voyage), he circulated a pamphlet directed at farmers and other animal breeders called "Questions About the Breeding of Animals."

In this questionnaire, he posed twenty-one questions about the progeny that result from particular breed crosses. Several of these are actually quite long *lists* of questions, a veritable inquisition that would have taken pages to answer; one of the few *short* questions he asked heads this chapter. Darwin was driven to find answers for the things in nature that perplexed him, and ironically perhaps, many aspects of domestic animals fit that category.

Darwin and dogs

Darwin once described his interest in the crossing of various domestic animals as his "prime hobby." Pigeons in particular fascinated him, and he owned and bred many varieties. He also kept, over the years, several different breeds of dogs (he especially liked terriers) and even attended a few dog shows. Such first-hand experience with the variations in size, shape and behaviour that exist between different animal breeds, together with the information supplied by farmers and breeders, contributed enormously to Darwin's thoughts on evolution.

At a time when little was known about such things, Darwin learned a lot about the inheritance of various characteristics by studying breed crosses. He didn't know about genes and chromosomes, of course, since they hadn't yet been discovered (our current understanding about the underlying genetic basis of inheritance didn't become part of general accepted knowledge until about 1930). Darwin could see, however, that certain characteristics of both plants and animals must be passed along from generation to generation—in fact, he insisted this must be so even though he couldn't figure out exactly how it happened. He made his best guess (a process he called *pangenesis*) but it turned out to be quite wrong.

What Darwin did right, however, was to suggest that what humans do when "improving" domestic breeds (breeding together, generation after generation, only those individuals that possess particular desired characteristics) must mirror what happens to wild species in nature. In other words, he suggested that deliberate *selective breeding* by humans can slowly change established breeds—to suit our own needs or desires—in essentially the same way as some force of nature must slowly shape wild species to suit the habitats in which they live. The "force of nature" that shapes wild species is what Darwin called *natural selection.* This concept became key to Darwin's thinking on evolution: that a gradual adaptation of species to local environmental conditions via natural selection could be extrapolated to explain gradual change from one species to another over geological time. In this way, all living things could be connected, via *descent with modification*, to a common ancestor.

Darwin spent many pages of the *Origin* on the concept of common descent and listed a large number of examples. He was, after all, trying to advance an unusual and very new idea to people of his own time, many of whom were perplexed by the patterns in distribution and similarities in anatomy among the amazing diversity of plants and animals, both those found at home and those brought back to Europe by intrepid world travellers. All this could now be explained as a consequence of descent from a common ancestor. And although Darwin's concept of evolution hinged on the idea that new species are continuously generated—that new species arise and also increase in number over time—he could not come up with an explanation for this phenomenon that swayed his peers. He was convinced that *natural selection* described the powers that prevail in nature that parallel the selective breeding of domestic animals and plants by humans; he saw the similarities between the two and directed much of his energy into documenting as many examples as he could. But the "descent with modification" argument was the only one that was widely accepted at the time—his theories on speciation and natural selection were heartily disputed by many (505-507).

Eleven years after the *Origin* came out, Darwin published his observations and thoughts regarding domestic breeds and breed crosses (also with copious examples) in a two-volume set called *The Variation of Plants and Animals Under Domestication.* The book begins with a discussion of dog origins. At the time, there was little convincing evidence one way or another as to whether dogs descended from wolves or jackals (Darwin thought probably both). Everyone seemed to have a different opinion on whether the domestication of the dog happened once or several times—again, little in the way of real evidence for either was available at the time. Darwin dealt with these nebulous origin issues relatively quickly and moved on to the topic that really interested him: selective breeding. The bulk of Darwin's discussion on dogs and other domesticates concerns how selective breeding can improve breeds over time, what happens in future generations when you cross various kinds, and speculations about what might control the characteristics that are clearly being passed from one generation to the next.

Darwin thought it unlikely, for example, that people thousands of years ago would have practiced the same kind of selective breeding techniques that had become common in the 1800's. Instead, he was probably the first to suggest that *unconscious selection* may have been the only kind of direction imposed by humans in the early years of dog evolution. By unconscious selection, he meant things like making a special effort to feed dogs that were particularly good sentinels, even during hard times, or by taking on migration voyages only those dogs which were easy to get along with. By doing these things, people would have gradually increased the number of dogs in each generation that were agreeable in temperament and otherwise useful to the group. Such results could have been achieved without any real plan and certainly didn't require an understanding of the genetic basis of behaviour: changes occurred because of human selection, but selection was inadvertent rather than deliberate. Deliberate selective breeding, such as Darwin engaged in with his fancy pigeon varieties, was obviously another thing altogether.

What is abundantly clear from these volumes is that Darwin was not after an explanation for how domestic animals came to be in the first place but rather how they changed afterward as a result of human selection. By leaving out the initial transformation events necessary for drawing parallels between wild and domestic species, Darwin left a gap that's only now ready to be filled.

Darwin's dilemma

Six years after the books on domesticates were published, Darwin finally turned his attention to human evolution. In *The Descent of Man*, published in 1871, he tackled in detail what the public perceived as the most controversial aspect of his theories: that humans too had evolved through a long series of speciation events. If ancient primates eventually became modern humans, it meant that we had been subjected to the same forces of natural selection as all other beasts. This was a bitter pill for many of his contemporaries to swallow, and many gagged.

In addition, a number of Darwin's colleagues disagreed with the idea that species could change via natural selection, an objection due largely to a general misunderstanding about how characteristics are passed from one generation to the next (because no one at the time knew about genes and genetic mutations). It wasn't until well after the publication of Theodosius Dobzhansky's *Genetics and the Origin of Species* in 1937 that many long-standing objections to Darwin's principles of natural selection were resolved to any kind of general satisfaction.

By this time, Darwin's theories were in need of formal updating. So, in the early 1940's, a small group of scientists collaborated on putting together a unified concept that came to be called the *Modern Evolutionary Synthesis*. This synthesis (first presented by Ernst Mayr in 1942, in *Systematics and the Origin of Species*, and updated at regular intervals) was a well-organized effort that merged the doctrines of theoretical population genetics with Darwinian principles. With the addition of information about how genes control inherited characteristics and change over time via chance mutation, natural selection became a widely accepted evolutionary concept (623). But as Mayr (ref. 510:7) pointed out more than fifty years later, "one of the great ironies in the history of biology [is] that Darwin failed to settle the problem of speciation in the volume entitled *On the Origin of Species*." The "species problem" needed resolving but Mayr and his colleagues felt themselves able to unravel Darwin's confused thoughts on how new species arise and, by adding their new-found understanding of genetics, described what they perceived as the essence of the process.

According to the *Modern Synthesis* (504-508), speciation goes something like this: interbreeding groups of animals—*populations*—are composed of individuals who all have slightly different genes. Speciation occurs when a population becomes divided, either because a small number of colonizers (founders) leaves the ancestral group or because the ancestral group becomes subdivided into two or more parts by some new barrier (like a river that changes its path, or an ice-age glacier). Under either circumstance, each divided portion of the original population has a slightly different mix of the original set of genes. Each split portion adapts gradually to its new surroundings via reproductive advantage—meaning that individuals with the combination of genes that best fit the new habitat leave the most offspring. But, as the new surroundings for each group are different, and because each group will by chance be composed of a different set of individual genetic variants, "adaptation" will have different consequences in each splinter group. Eventually, a splinter group can become so different genetically from the ancestral group that they are no longer able to interbreed. At this point, Mayr and colleagues would declare the two populations to be new species.

The "architects" of the *Modern Synthesis* were very pleased with themselves, and for awhile, so were most of their peers. The generations that followed expended considerable effort debating fine points and documenting examples (e.g., 257, 487, 571, 627, 739).

What many eventually came to realize, however, is that the *Modern Synthesis* doesn't actually give a satisfactory explanation for precisely how this process works in a way that corresponds with the hard facts of evolution. For example, if the process depends entirely on chance (i.e., chance mixing of individuals, chance mutation of genes), why do we see so many of the same kinds of changes—in fact, similar *suites* of changes—in vastly different lineages of plants and animals over extended periods of geological time (e.g., 216, 351, 464, 473-478, 838)? How can there be such strong patterning to evolutionary change if the process depends so completely on random genetic mutation? Even today, this is as much a puzzle as it was in Darwin's time (e.g., 35, 36, 262, 277, 278, 337, 482-484, 580).

In addition, a challenge to the prevailing dogma that evolutionary change is *gradual* arose with a vengeance in the 1970's. Palaeontologists Niles Eldredge and Stephen Jay Gould insisted that the fossil record of most organisms did not indicate a gradual steady change from one form to another, but instead seemed to show that short bursts of rapid changes in form were most often followed by long periods without change, a pattern they called *punctuated equilibrium* (201, 275, 278-280). Facts supporting the rapid change hypothesis have accumulated over the years, and recently, the fields of molecular genetics and developmental biology have heaped ever more evidence on the pile (e.g., 520, 720). Today, most scientists would say the evidence for rapid change is incontrovertible (278, 111, 572, 802).

Rapid changes in form are now know to be quite common—in fact, rapid change may actually be the most common means of speciation (e.g., 381, 396, 474, 475). The swift and highly coordinated alterations we see in this type of speciation (where behaviour, body shape and reproductive traits all change together) appear to be the result of slight shifts in growth rates or changes to the time when certain stages of development begin or end. However, biologists still don't have any clear idea of how such developmental changes are initiated or implemented (e.g., 35, 111, 297, 323, 465, 481, 535, 545, 622). They've come frustratingly close, but can't provide a comprehensive answer, for the most part because they're still fixated on the idea that the solution will come through decoding genes and DNA (e.g., 23, 172, 246, 257, 336, 386, 393, 414, 430, 479, 553, 554, 589, 590, 611, 612, 616, 627, 712, 731, 761, 813, 836). An understated admission made by Ernst Mayr almost twenty years ago is unfortunately still true today:

> *[I]n spite of all the advances of genetics, we are still almost entirely ignorant as to what happens genetically during speciation. (Mayr 1988:208, ref. 507)*

And so, despite all we continue to learn about genes, the origin of species remains the enigma of evolutionary biology. In short, the *Modern Synthesis* needs revision.

Resolving evolution's enigma

Even though it's been known for quite some time that there aren't nearly enough genes to account for all of the traits that change in evolution, most researchers refuse to abandon the concept that a gene exists for every characteristic that can be seen or measured, although they may concede that some traits require input from more than one gene. Although I'm certainly not the first to suggest that this "one trait, one gene" concept is the wrong approach (e.g., 35, 111, 278, 300), even I can see why geneticists are loath to give it up: it has, in fact, resolved many other questions and advanced our understanding of biology in many important ways (e.g., 333, 695). But significant discoveries have been made during the past twenty years that have yet to be incorporated into evolutionary thinking; these include, especially, the multi-faceted roles that hormones play in the regulation of growth, behaviour and gene function (e.g., 177, 196, 216, 223, 228, 265, 273, 299, 317, 331, 348, 399, 629, 647, 801, 802, 842).

In my opinion, the story of evolution as it's currently presented is missing several essential components that my theory addresses:

1) It lacks a demonstrable link between genes, individuals and the environment that explains how populations can change in response to changing environments over evolutionary time (the concept that genes are solely responsible for controlling evolutionarily significant traits is lacking an explicit statement of this connection).
2) It lacks an explanation for the fact that there are not enough genes available to account for the number of traits that actually change during evolution (and its corollary, the fact that even between closely related species, whole suites of traits change simultaneously).
3) It lacks the identification of a precise mechanism capable of allowing species to adapt rapidly enough to changing environmental conditions that the process is effective over the short term ("in ecological time").
4) It lacks an explicit explanation for why some species show more evolutionary flexibility ("adaptability") than others.

As I explain in the pages that follow, determining how and why wolves became dogs—that question Darwin skipped over so quickly—is not only essential for understanding evolution in general but also critical to any real advancement in modern medical diagnosis and treatment techniques. I suggest that only when we understand precisely how humans developed into the animal species we have become can we expect new and significant insights into some of the debilitating health problems that continue to plague us: high cholesterol, heart dysfunction, depression, obesity, infertility, birth defects, diabetes, schizophrenia and learning disabilities, among many others. We can't escape the biological processes that rule our bodies because they are the same processes that rule evolution; if escape isn't possible, understanding is the next best thing.

Recommended reading

Darwin, F. (ed.) 1958. *The Autobiography of Charles Darwin and Selected Letters.* Dover Publications, New York.

Keynes, R. Darwin (ed.) 1988. *Charles Darwin's Beagle Diary.* Cambridge University Press, Cambridge.

Mayr, E. 1991. *One Long Argument: Charles Darwin and the Genesis of Modern Evolutionary Thought.* Harvard University Press, Cambridge.

Mayr, E. 2001. *What Evolution Is.* Basic Books, New York.

Pigliucci, M. 2005. Expanding evolution: a broader view of inheritance puts pressure on the neo-darwinian synthesis. *Nature* 435:565-566. [a book review]

Rennie, J. 2002. Fifteen answers to creationist nonsense. *Scientific American* (July):78-85.

Schwartz, J.H. 1999. *Sudden Origins: Fossils, Genes, and the Emergence of Species*. J. Wiley and Sons, New York.

Chapter 2

Rethinking Domestication Dogma

Humans seeking a steady source of food, hides and wool, and companionship, have tamed everything from wolves to turkeys to guinea pigs. Learning when and why each of the more than two dozen domesticated animals was brought under human rule has been a continuing quest for archaeologists.

Heather Pringle (*Science* 1998:1448)

The traditional story of domestication

All of us have been taught that dogs and other domestic animals were deliberately created by people. We've been told that about 14,000 years ago, someone got the bright idea that a wolf would be a useful assistant for hunting and so a couple of newborn pups were captured and brought back to the village. Tamed and confined over many generations, those wolves became the ancestors of all modern dogs. A few thousand years later, the same process was repeated with sheep, goats, cattle and pigs, and voilà—we had livestock.

All definitions of domestication either state explicitly or imply that humans made a conscious choice to tame certain wild animals for their own use or benefit, requiring that some animals be deliberately removed from the wild to form the domestic population (165, 604, 617, 735). Taming and deliberate removal of animals from the wild are considered the two essential components of domestication.

Beyond this point, however, anthropologists can't seem to agree. Most find it exceedingly difficult to define precisely what a domestic animal *is*, and virtually all definitions vary to some degree. One is deceptively simple: domesticates are animals whose breeding is, or can be, controlled by humans (617). Another lists five criteria that can be used to define fully domestic animals (ref. 352:20):

1) The animal is valued and there are clear purposes for which it is kept.

2) The animal's breeding is subject to human control.
3) The animal's survival depends, whether voluntarily or not, on man.
4) The animal's behaviour (i.e., psychology) has changed as a result of domestication.
5) Morphological characteristics that appear in individuals of the domestic species occur rarely if at all in the wild.

A third view defines a domestic animal as one that has been bred in captivity, for purposes of economic profit, by a community that maintains a mastery over its breeding, organization of territory and food—and insists also that animals cannot be domesticated unless they are owned (128, 129). In other words, the dilemma is whether we should define domestic animals by their cultural relationships to people or by their biological characteristics or both.

Because opinions vary widely on the degree of emphasis that biological differences between domesticates and their ancestors should have in any definition of domestication, lists of what constitute "real" domestic animals also vary considerably (e.g., 139). Some authors include reindeer but not zoo or circus animals, although the latter often exist for generations under total human control; some call Asian elephants *semi-domestic*, while others list them as true domesticates; rats, mice and pigeons are excluded by some (e.g., 128, 352, 846) but not by others (e.g., 62, 314, 499, 500, 559, 641, 649, 655, 735). It can get very confusing indeed, compounded by inconsistency in applying scientific names to domesticates (255, 263).

It's also difficult to reconcile the concept that taming of deliberately acquired wild animals is required for domestication when you consider that the biological differences between wild and domestic forms are virtually identical for all animals, at least where the ancestors are known (including dogs, pigs, cattle, sheep, goats, water buffalo and mithan; in Chapter 5, I explain why the changes associated with domestication in horses and cats appear to be an exception, but are not). The parallel differences among domesticates are apparent regardless of whether the animals are carnivores, herbivores or omnivores, and regardless of the area of the world in which domestication occurred. Quite simply, the similar biological changes that have taken place amongst all domesticates have to make you wonder how an explanation for domestication that depends on deliberate intent could have the same result for wolves as for sheep and pigs.

How domestic animals differ from their ancestors

All of the major domestic animals studied in detail so far differ from their wild counterparts in the same ways: they're smaller overall (at least they were initially, even if they aren't any longer), they have wider skulls that make their snouts appear shorter (and, for some animals, smaller horns), they become pregnant more often or produce larger litters per pregnancy, they're much more docile in behaviour (have markedly lessened fearfulness combined with decreased aggression), and finally, they possess different dominant coat colours (with the mottled greys and browns of wild forms replaced by solid shades of yellow or red), as well as a marked increase in the incidence of whitespotting, or *piebald* patterns (74, 128, 129, 142, 170, 206, 207, 325, 517, 562-564, 639, 735, 738, 762). A summary of the common physical and physiological changes that have occurred in all animals as a result of domestication is listed below:

- overall body size reduction
- widening of the skull (making the face appear shorter), accompanied in the early stages by tooth crowding and later by tooth size reduction
- reduction of horn size (in animals with horns)
- increased incidence of *piebald* coat colours (solid with white spots or all white with a few spots)
- increases in solid coat colours (red, yellow and black)
- increased docility/lessened fearfulness
- larger litter sizes or more frequent pregnancies
- lowered age of sexual maturity
- changes in timing of pregnancy (increases in frequency or shifts in its seasonal occurrence)

Barking up the wrong tree?

If domestic animals all change in the same ways biologically, why would taming as a domestication approach that worked on wolves also be successful for, and have the same effect on, goats, cattle and pigs, given the differences in behaviour and food requirements of these animals? And why would the method of taming always be so similar as to have the same result when the cultures of the people involved were so different? Would people in Europe really have taken precisely the same approach to taming wild cattle—and have it result in exactly the same biological changes—as people working with pigs in southeast Asia or llamas in South America?

Although such questions nagged at me, as a zoologist trained within the strictures of the scientific method, the way that domestication is inevitably portrayed made me especially wary. In marked contrast to a concept as widely accepted as evolution or natural selection, for example, the traditional explanation for how domestic animals came to be is always presented not as a theory, but as fact (even if this is not explicitly stated). You can confirm this yourself by looking up the word *domestication* in any dictionary: *taming* and *domestication* are often presented as synonymous. The fact that domestication is considered a deliberate act is implied by the intentionality inherent in the word *taming*, and the lack of any reference to *assumption*, *hypothesis* or *theory* implies that this definition is not open to question because it is a known scientific fact.

Domestication is also routinely touted as one of the first significant instances of human innovation, a prime example of our unique ability to manipulate our environment to better suit our needs: first we learned to control fire, then to subjugate wild animals.

For example, biologist Jared Diamond's *Guns, Germs and Steel* (187) is an insightful look at the impact that domestic animals and plant crops (that is, deliberate food production) had on the development and history of human societies. Diamond considers the domestication of animals to be one of the key innovations that separated "have" from "have not" societies around the globe.

Domestication, when viewed from this perspective, is always about *us*: what we did to the animals to bring them within our sphere of control. The animals were but hapless pawns in a well-orchestrated game we devised. And because domestication is viewed as a direct consequence of human behaviour, domestic animals are invariably studied within the discipline of Anthropology, the study of human cultures—rather than within Zoology, the study of animals and their evolutionary history.

This anthropological interpretation of how domestic animals came to be had bothered me for many years. To my mind, a number of other questions begged for an answer. Why, if domestication was a deliberate act, did wolves come first—thousands of years before livestock animals? What would have motivated ancient people to tame wolves *before* they applied the same process to animals they could eat? Wild sheep, goats, pigs and cattle had been sources of meat and skins for millennia before wolves became domestic: why not begin any attempts at domestication with potential food animals? More significantly perhaps, having figured out *how* to create a domestic animal from experience with wolves, why wait three to five thousand years to move on to other animals, and then tackle only one species at a time (i.e., first goats, then sheep, pigs, cattle, horses, etc.) rather than all (or at least several) together?

My scepticism for this traditional view of domestication increased dramatically when the results of genetic research I was involved in during the early 1990's did not reflect the relationship between dogs and wolves that it should have—if the traditional interpretation of dog evolution was correct. Until then, the most widely held view of dog history proposed that different wolf subspecies from distinct geographic regions gave rise to a few specific dog "stock" types (Figure 2.1). This model explained the similarities among spitz-type breeds (such as malamutes and Eskimo dogs) as being a consequence of their descent from an Arctic wolf common ancestor. Similarly, this scenario suggested a descent of sighthounds (salukis, greyhounds, etc.) from Middle Eastern wolves, and mastiffs from European wolves. Deliberate and accidental crosses between these basic physical types—coupled with deliberate selection for desired traits of other kinds, like small size and particular coat types and colours—is said to have created, over many generations, the plethora of breeds we know today.

Figure 2.1 The most common explanation for modern variation among dogs is that each of these four distinct physical types or "stock breeds" descended from a different geographic subspecies of ancient wolf (for example, the malamute and similar breeds are said to descend from the North American Arctic wolf). Adapted from Bruce Fogle's Encyclopedia of the Dog (1995).

What our genetic study (using *mitochrondrial* DNA) told us was something quite different. The results did not reveal any genetic markers that identified ancient "stock" breed types with any one geographic subspecies of living wolf. Subsequent studies by other researchers confirmed this finding for dog breeds (and for breeds of other major domestic animals as well, including cattle, sheep, goats, and pigs) (see ref. 98 and Table 5.2 in Chapter 5). It appears, then, that any wolf subspecies can generate the diversity of body types we see in dogs today.

What this means is that a robust mastiff-type dog is just as likely to descend from a middle Eastern wolf as from a European one, and that an Alaskan malamute is no more closely related to a wolf than is a spaniel. It means that the long-accepted story of breed development in dogs—as a complex mix of several basic body types associated with specific geographic regions—can't be correct. More importantly, it made me wonder what other aspects of what we "know" about the origin of dogs were also incorrect.

It also made me seriously question, for the first time, what hard evidence actually existed for the deliberate nature of domestication as it's traditionally defined. Is there any evidence that animals deliberately captured from the wild and held in captivity by people are the source of domestic stock? Or for the notion that the earliest dogs were used as hunting assistants, as would be expected if dogs were deliberately domesticated for the express purpose of assisting with human hunting?

In my quest to make sense of it all, I read both the classic literature on domestication theory and reports on the archaeological record of the relevant prehistoric period. I soon found out the truth: there has never been any direct evidence (strong *or* weak) to support the traditional explanation for domestication as a deliberate act of human innovation.

It seems that because we have had such great success over recent centuries in manipulating animal breeds, and creating new ones, scholars simply assumed that the initial transformation of domestic animals must have involved a conscious decision to proceed in that direction. (As archaeozoologist Darcy Morey has pointed out (531-533), most of us seem unable to separate intellectually the intense symbiotic relationship we now have with domestic animals from the process that created them in the first place.) Scholars were even more convinced of this view of domestication as human innovation when they understood that the mere possession of domestic animals and plants precipitated enormous cultural change and progress in ancient human societies—Jared Diamond may have summed up this equation recently for lay audiences, but anthropologists and historians have known it for decades.

Against such a backdrop of anthropological and historical rationale, it seems entirely reasonable to us—perhaps even essential—that the creation of domestic livestock and crops must have been part of a well-calculated plan. Surely such a dramatic cultural change had to have been both intentional and premeditated. Ancient humans had been hunting big game for over 200,000 years by the time domestic animals came along. Why would our ancestors have altered their lifestyle so dramatically—revolutionized everything—without a plan?

We understand intuitively that one of the reasons we evolved and progressed in a way unique from other animals is because of our superior brain. Brain power enabled us to learn to control fire and then to make it ourselves. The light and heat thus created not only made our lives more comfortable but permitted us to live in climates that would otherwise have been intolerable. No other animal has been able to do such a thing. Our ability to control and use fire to our advantage immediately set us apart from other animals.

Why would our *Homo sapiens* ancestors not eventually have come to realize that controlling the animals they depended on for food and clothing could be similarly helpful? Instead of traipsing miles and miles over the landscape in all kinds of weather in search of food to kill, why not bring the food home alive and then collect it as you need it out of your own back yard? This seems such a logical connection for our ancestors to have made that when anthropologists tell us that domestic animals and plants are end products of a deliberate and calculated plan, we accept it without question. It is, to most of us, such a plausible explanation that no others need be considered.

However, this view of domestication must be exposed for what it is: an idea constructed entirely from assumptions. There are simply no facts to support the story—and because there's no way to test if it's true or not, it can't even be called a theory or hypothesis. I came away from my reading with the realization that the traditional explanation for domestication is nothing more than a widely accepted myth. Frankly, a "widely accepted myth" is known correctly by another word: *dogma*. Dogma is the word we use to describe ideas adopted without question, accepted on faith. Dogma is therefore a concept most often used to describe religious beliefs, where it's considered inappropriate and futile to ask for proof.

It was clear to me that few scholars, in any field, have ever really questioned the concept of domestication as an act of human innovation. Take this excerpt from a recent article, part of a special section reporting advances in archaeology, in the prestigious scientific journal *Science*. In this article, the author (ref. 605:1448) is reporting on interviews with a number of researchers who have challenged long-standing assumptions regarding animal domestication. However, she begins her discussion with the statement that I've repeated at the head of this chapter: "Humans seeking a steady source of food, hides and wool, and companionship, have tamed everything from wolves to turkeys to guinea pigs. Learning when and why each of the more than two dozen domesticated animals was brought under human rule has been a continuing quest for archaeologists."

This statement actually sums up quite nicely the traditional definition of domestication: that domestication began with taming and involved a deliberate act of domination over animals that resulted from a specific need or goal. It also tells us that the dogma of animal domestication itself remained virtually unchallenged by archaeologists in the year 1998: it's certainly not identified as one of the "long-standing assumptions" being confronted. The rest of the article confirms this impression—the most radical objection to existing beliefs that the author has to report is an investigation of the notion that pigs (rather than goats or sheep) were the very first species of farmed livestock.

Especially hard for me to accept was the fact that virtually all of my colleagues in biology, most of them quite rigid in their refusal to acknowledge any newly presented idea until proven to their full satisfaction, seemed quite happy to accept dogmatic rhetoric about an event as important as domestication. Biologists are as accepting of this idea as the anthropologists who promoted it in the first place and this is abundantly clear from the language used in reports of genetic studies on the origins of various domestic animals. Clearly, hoodwinking has been accomplished on a major, global scale. And so, at the risk of belabouring the point, I state it again because it's so critical: the traditional definition of domestication we have been taught to accept as fact is no more than a deeply entrenched myth.

Although I'm not the first to suggest such a heretical idea, I may be the first to do more than complain. More than ten years ago, science writer Steve Budiansky (79) described many of the objections to the traditional explanation of domestication (including a few I haven't mentioned already) in the first half of his book, *The Covenant of the Wild: Why Animals Chose Domestication.* For example, if taming animals really does lead to domestication, why did the Egyptians have so little success with domesticating the many species they captured and tamed? The diversity of animals they attempted to domesticate is impressive: monkeys, baboons; antelope, gazelles, ibex, zebras, leopards, hyenas and cheetahs, among others (360).

The Egyptians *seemed* to be taking the right approach. In principle, one of the essential components of creating a domesticated form is that you must start with more than one animal—at bare minimum, a single breeding pair of tamed wild individuals. These animals must be kept alive and healthy long enough to become sexually mature (which may take years). Most importantly, each individual must find the mate you've chosen for them acceptable, and females must be capable of producing living offspring and raising them to maturity so that the taming sequence can continue into the next generation. When you think about all that we've learned about animal behaviour and husbandry in recent years, it becomes clear that this is a very tall order indeed.

Although the Egyptians clearly had the resources and motivation to provide the essential components of domestication, they still couldn't make the concept work: while they managed to tame and keep alive small numbers of sexually mature animals, few would mate in captivity and fewer still produced offspring. In the end, no new domesticated forms were actually created.

Is it reasonable to suppose that people who lived thousands of years before the Egyptians, people who were only just beginning to learn how to live together in permanent communities themselves, would have done better? Even with the concrete and metal fencing materials available today, tamed wild animals can be very difficult to keep confined. After they hit sexual maturity, most tamed wild animals become extremely difficult to manage and are often dangerous—simply feeding and keeping them healthy become major chores. How do we suppose that people 10,000 years ago managed the same job?

Some people might argue that by Egyptian times (about 5,000 years ago), all of the animals capable of being domesticated had already changed—that certain biological qualities of the species chosen by the Egyptians (not social enough, without dominance hierarchies, etc.) thwarted their success.

However, even in modern times, tamed wolves have been kept in captivity for generations without any of them transforming into an animal we would call a dog. This suggests that the basic premise of domestication is false—that taming doesn't lead to domestication. Which also means we're not understanding how the process works at all.

Steve Budiansky did a good job of criticizing the traditional story for how and why domestication occurred but wasn't able to suggest a viable alternative explanation. A few researchers, also bent on debunking the domestication myth, have similarly skirted around this issue (including biologists Ray and Laura Coppinger in their recent book on dog evolution and behaviour, ref. 136): when pushed to describe the actual process of transformation, they suggest the literary equivalent of "something happens." It's heartening that more and more scholars are now willing to abandon the dogma and acknowledge that some wild animals may simply have chosen to live with us and changed as a consequence (552). However, without an explanation for precisely how and why such changes occurred, we are no further ahead than we were with the myth. "Something happened" is simply not good enough.

Rethinking the dogma

If people didn't deliberately create domestic animals, how did they come to be? This question is not an insignificant intellectual dilemma to be skipped over lightly as if it were an inconvenient puddle in the road: it's a problem of vital importance. Jared Diamond quite rightly identifies the relationship that arose between humans and domestic animals as an association that changed the course of history. While few would dispute this, anthropologists have yet to acknowledge that improving our understanding of changes in human cultures can't be achieved without a plausible explanation for the process that created that distinct class of animals in the first place.

But more importantly, the question is significant on another, much broader level. I believe that settling the issue of how wild animals became domestic is even more critical to the field of biology, where the number one question is how *any* species of animal can transform into another. Resolving the dilemma of how wild animals became domestic, if not by deliberate design, may well teach us something really important about how all new species come to be, and thus, about how evolution itself actually works.

It seemed most likely to me, as it has to others, that domestication was actually a natural evolutionary event. This conclusion is based primarily on the fact that the same kinds of differences that exist between domestic animals and their wild ancestors also occur between many other closely related wild animal species. My research began in earnest when I asked one simple question: what, in strictly biological terms, had to happen to a wolf for it to become a dog? What parts of its body had to change, and what biological systems controlled those parts? I ignored the assumption that humans did anything deliberate to bring about the initial change from wild to domestic form and searched the scientific literature for information that made biological sense.

I concluded, after months of reading, that beginning about 14,000 years ago, some animals simply chose to live among our ancestors, and changed physically and behaviourally as a consequence. In essence, the animals *domesticated themselves* by choosing to associate with early human settlements. Only after these animals had transformed naturally from a wild to a domestic form—a process that probably took much less than one *hundred* years rather than thousands—did humans become actively involved in their management and development. Proximity and the amenable nature of those early domesticates encouraged humans to take an active role in these symbiotic relationships—management and ownership of domestic animals eventually transformed history. But it's my firm belief that it was the animals that started it all, not us, and that the process must have been extraordinarily fast. We can get a sense of just how quickly domestication could have proceeded from looking at the results of a fascinating selection experiment on foxes that began in Russia more than sixty years ago.

Playing with nature

We know that a rapid transformation from wild to domestic forms is possible in part because of reports on an intriguing domestication experiment undertaken by geneticist Dmitri Belyaev, whose work I came across early in my investigation (50-52). In an experiment that began in Siberia in the late 1950's, Belyaev began with a large population of silver foxes (a naturally occurring form of the red fox, *Vulpes vulpes,* that's black with white-tipped hairs and relatively common in Alaska and northern Canada). Silver foxes were being farmed commercially for their fur as an alternative to trapping wild animals. The farmed silver foxes retained all of the characteristics of wild foxes: they had one reproductive cycle per year and an annual fall moult, and they were generally timid around people.

In fact, the behaviour of these captive foxes was so wild-like that keeping them healthy and reproducing at economically viable levels was both difficult and unpleasant. The farms were intensely noisy places, as many of the animals shrieked and yelped when anyone approached the cages, even for daily feeding. Sometimes the foxes reached such a state of panic that they injured themselves; some females became so stressed they couldn't reproduce or, if they did, ate their own pups. Belyaev wondered if selecting among these captive foxes for those with less excitable temperaments might generate an animal that was easier to keep. In other words, he was attempting to change the behaviour of the foxes so that they would be easier to farm commercially, nothing more.

Belyaev and his fellow researchers assessed a randomly selected subset of foxes from several fur farms (465 animals altogether). The selection method was simple and straightforward: a gloved hand was extended into each animal's cage. Those foxes which attacked, ran, cowered or bit were excluded. Only those showing a quiet, tolerant curiosity (without fear or aggression) were chosen. These curious foxes, about 10% of the total, demonstrated noticeably less fearful behaviour toward humans than the others (in other words, they were more *stress tolerant*, as I prefer to call it). These stress tolerant animals were selected to form a small experimental population.

Interestingly, when farm breeding records were checked, it was noticed that the females in this less fearful population that had previously given birth had all been early breeders: all the breeding-aged females selected for their stress tolerant behaviour had consistently ovulated within the first week or two of the six-to-eight-week fox breeding season. This correlation suggested that a pre-existing link existed between the behavioural response to stress (the "gloved-hand intruder") and the timing of ovulation in these foxes, although the researchers didn't understand what this might mean.

Nevertheless, the experiment continued. After several generations of mating stress tolerant, early breeding females with stress tolerant males and selecting among each year's litter of pups for tolerant behaviour only, the time of ovulation (necessary for breeding) as well as the annual moult for many females shifted forward by several months (to October or November, from early February).

As the experiment continued, ovulation and hair moult receded further still, until several females were ovulating twice a year (two full reproductive cycles). After twenty generations (twenty years), some females were able to produce two litters a year. Completely unexpectedly, however, a number of other novel traits suddenly appeared, including a curled tail, flopped-over ears, a pattern of distinctive brown markings that the researchers called "brown piebald," and a classic white piebald pattern of black-and-white markings similar to that seen in a typical Border collie. The animals from this last generation had smaller adrenal glands and hormonal changes of several kinds; subsequent analysis of selected foxes found small changes had also occurred in the shape of the skull (753, 756).

Belyaev described the foxes with these novel physical and physiological traits as having "doglike" behaviour: they barked and were quite unafraid of people—they would even crawl into his lap when he sat in a large enclosure with them. Such behaviours were "doglike" because they were *juvenilized* traits (more on this in the next section), which confirmed Belyaev's initial assumption that such selection might lead to domestication. And although the experiment was a failure as far as the fox farms were concerned (the farmers certainly *did not* want black-and-white-spotted foxes), Belyaev recognized that the results were scientifically important and wrote them up for publication.

Although follow-up studies on the selected foxes continued into the mid-1980's, the investigators were still unable to explain their results. Genetic analyses confirmed that the changes in the selected foxes were not caused by spontaneous mutations in any known gene, or in fact by any known biological mechanism (53, 54, 568, 602, 752, 754, 757, 758). Additional experiments undertaken later—by other scientists in other countries, and involving red foxes (in the US) and fallow deer (in Germany)—gave similar kinds of results (325, 382).

Belyaev had concluded that something in the selection for tolerant behaviour toward humans was not only causing juvenilization of physical form and behaviour, but was somehow "disrupting" the normally constrained timing of sexual reproduction—he just couldn't explain how or why. However, the fact that none of the characteristics found among the selected foxes had been present in the original population suggests that all might be *inevitable consequences* of the selection for stress tolerant behaviour. We now know slightly more about this process, but not much.

The juvenilization of domestic animals

Some of the distinctive differences between wild ancestors and their domestic descendants listed earlier in this chapter have recently been shown to result from changes in growth rates and the timing of sexual maturation. This is a general evolutionary process known as *heterochrony*—the particular process is called *paedomorphosis,* which can be thought of more simply as a process of juvenilization (e.g., 136-138, 400, 401, 474-478, 482-484). For example, dogs have been shown to be juvenile in relation to wolves, based on several studies of behaviour and growth (138, 274, 305, 531-533, 790-792). This means that the most obvious differences between the wolf and dog result from the dog maturing at a stage of growth similar to that of a juvenile wolf in both form and behaviour. These juvenilized traits seem to have been caused by a predictable change in growth rate that's implemented during the first few weeks after birth, as shown by the changes illustrated in Figure 2.2 (790-792).

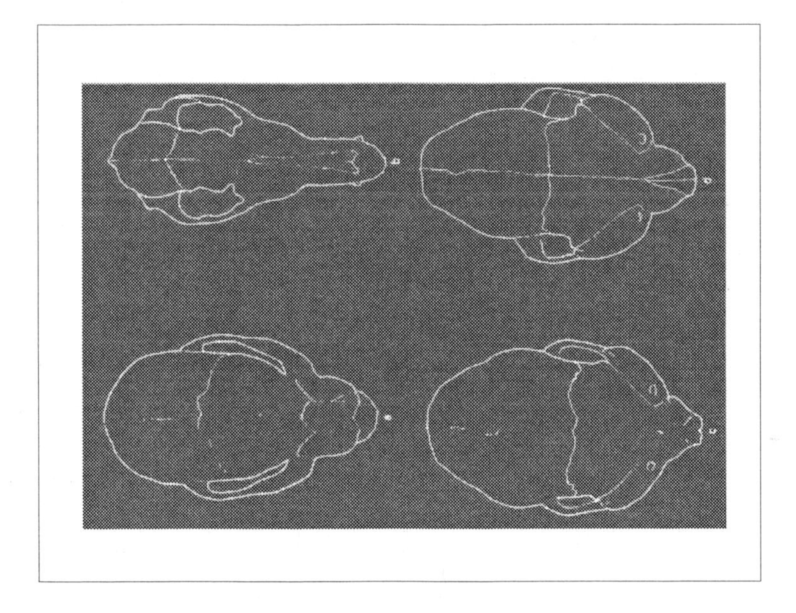

Figure 2.2 Differences in skull shape between adult (top row) and newborn, in the wolf (left) and wildcat (right). (From Wayne 1986:256, figure 9, with permission.)

Because such juvenilized traits are inherited, they're assumed to have a genetic component—although no such "growth-controlling" gene or genes have yet been found. More confounding still is the fact that the other traits seen in early domesticates (such as increases in litter size, docile behaviour, whitespotting, and frequency of reproduction)—as Belyaev discovered—appear not to be controlled by separate genes but to be associated with, and inevitable *consequences* of, the changes in growth rates themselves. In other words, growth rate is the *only* characteristic that really changes—all the rest simply go along for the ride.

Such dramatic experiments as Belyaev undertook, unfortunately, have taught us relatively little about how the process of domestication via juvenilization actually works—only that it can happen very fast. Even today, after another twenty years of study, as outlined by Ludmillia Trut in a 1999 *American Scientist* article (755), the investigators working on the Russian fox project have been unable to explain why they got the results they did.

However, by challenging the assumption of domestication as a deliberate act and asking a new, biologically-based question, I've come up with a plausible and testable explanation for those results. My review of the literature suggests that selecting for a single behavioural characteristic, as Belyaev did with his foxes, could trigger a suite of well-coordinated physiological and morphological traits within a few generations because there is a common control mechanism for these traits. This control mechanism is hormonal, and thyroid hormone is the pivotal agent. If we can figure out how and why thyroid hormone might direct the changes associated with domestication, we should gain significant insights into how speciation proceeds under other circumstances, and thus come to understand more precisely how evolution actually operates.

Protodomestication: a speciation event

In considering domestication from an evolutionary perspective, I suggest that the process is actually composed of two distinct parts. The first stage (which I refer to as *protodomestication)* is a natural speciation process whereby certain wild ancestors generate descendants with modified biological features, a rapid process associated with the natural colonization of human-dominated habitats. *Classic domestication* is the term I suggest be used to describe the gradual processes of conscious and unconscious human selection (working in concert with natural selection) that can modify any captive or commensal population, whether those animals are products of prior protodomestication or individuals deliberately removed from the wild.

Although these definitions contrast sharply with the traditional view of domestication (which collapses the two stages together for all animals), I argue that many animals—especially dogs—have undergone both protodomestication and classic domestication while a few may be products of classic domestication only. Although the terms *protodomestication* and *classic domestication* may ultimately be replaced with other terminology, they suit the purposes of this discussion, which intends only to distinguish between rapid speciation events and the gradual adaptation that comes afterward. I maintain that the two processes are so distinct that they require different terms of reference.

Figure 2.3 Human settlement changes landscapes, whether intentional or not (photo by Leland Donald)

I define protodomestication as a colonization process that occurred within human-dominated, or *anthropogenic,* environments. *Anthropogenic environment* is a term used to describe a localized set of environmental conditions created by the impact of human populations (Figure 2.3)—in this case, the conditions resulting from the formation of permanent settlements that began at the end of the last ice age (733-735). Settlements need not have been absolutely permanent to have created anthropogenic effects. For example, Magdelanean horse hunters that lived in Germany and the Czech Republic about 14,000 years ago (547) may have established large camps in the same location every year that were occupied over several seasons, since the wild horses they harvested were non-migratory animals. While not strictly permanent, such repeatedly occupied multi-season settlements may occasionally have attracted wolf colonizers.

An anthropogenic environment is also a habitat dominated by the continuous presence or proximity of people. Colonization of these new habitats was a choice animals could make, whether driven exclusively by the advantages offered (such as easily available food) or in concert with pressures exerted by changing conditions in their natural habitat (of which we know very little). Individual animals with the highest tolerance to the associated stresses of these new environments provided the source population that ultimately became "domesticated." Figure 2.4 illustrates this colonization model.

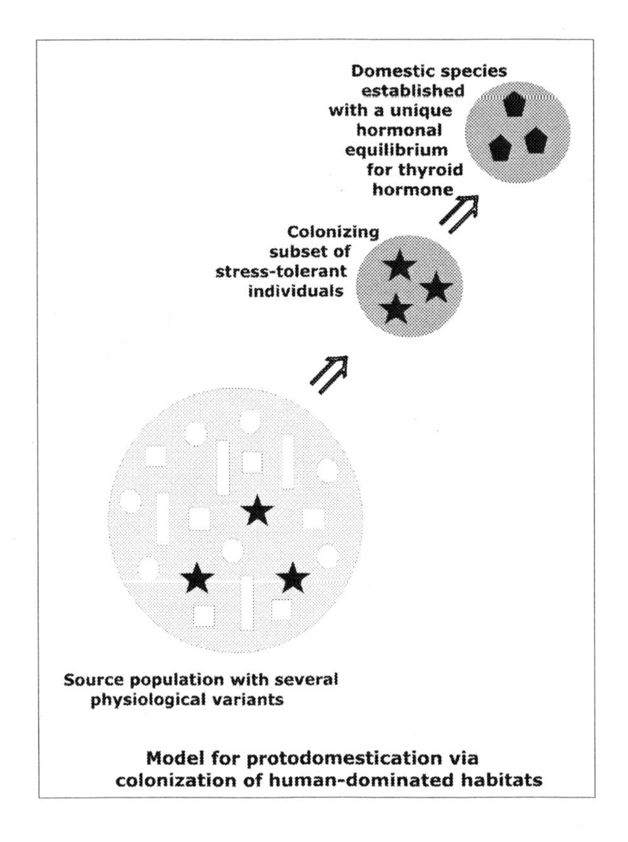

Figure 2.4 A generalized model for protodomestication via colonization of anthropogenic habitats by stress tolerant individuals of wild animal populations

Individual variation in behavioural responses to stress is the key to explaining how wolves could have evolved rapidly into dogs without direct and deliberate human interference. There's an intimate connection between the adrenal stress-response system that mediates such behaviour and thyroid hormone metabolism, which I'll go into in more detail in the next two chapters. This physiological connection suggests that a hormonal mechanism exists that could account for all of the physical and physiological changes documented in all domesticates—because thyroid hormone also mediates the control of fetal and postnatal growth, pigment production in hair, and reproductive timing. It must be said that a physiological basis for protodomestication has been sought for at least forty years (e.g., 325), so mine is not a totally new approach. However, thyroid hormone has never been specifically implicated before now. My proposal that thyroid hormone metabolism is the pivotal biological mechanism associated with both protodomestication and speciation is not only a new concept, but one that challenges the accepted supremacy of genes in evolution.

My explanation for protodomestication as a natural speciation event, as I will show in the chapters that follow, is based on a scientifically sound and testable hypothesis that does not contradict the basic tenets of evolutionary biology: it simply expands on the concepts that Charles Darwin presented almost one hundred and fifty years ago. It explains, in explicit and comprehensible terms, precisely how all animals change over time—in a way that truly revolutionizes our understanding of the process.

No one else has attempted to do such a thing before. The journey through this argument will necessarily take you through uncharted territory. Questioning the dogma of domestication is just the beginning. As I've shown, exposing dogma for what it is leads to new, more productive questions. My plan is to guide you through the process of discovery as I experienced it myself, to show you what can happen when you raise questions that replace fiction with fact.

Recommended reading

Budiansky, S. 1992. *The Covenant of the Wild: Why Animals Chose Domestication.* Weidenfeld and Nicolson, London.

Clutton-Brock, J. 1981. *Domesticated Animals from Early Times.* British Museum (Natural History), Heinemann, London.

Coppinger, R. and Feinstein, M. 1991. 'Hark! Hark! The dogs do bark...' and bark and bark. *Smithsonian* Vol. 21(10):119-129.

Coppinger, R. and Coppinger, L. 2001. *Dogs: A New Understanding of Canine Origin, Behavior, and Evolution.* University of Chicago Press, Chicago.

Diamond, J. 1999. *Guns, Germs and Steel: The Fates of Human Societies.* W.W. Norton and Company, New York.

McKinney, M.L. 1998. The juvenilized ape myth – our "overdeveloped brain." *BioScience* Vol. 48(2):109-116.

McNamara, K. 1999. Embryos and evolution. *New Scientist* 16 (October): *Inside Science,* 4 pg. pullout.

Nova 2004. *Dogs, Dogs, and More Dogs.* A documentary available on tape and CD [includes interviews with Ray Coppinger on domestication and rare original footage of the Russian fox experiment]. WGBH Boston Video www.shop.wgbh.org

Mason, I.L. 1984. *Evolution of Domesticated Animals.* Longman Co., London.

Morey, D.F. 1994. The early evolution of the domestic dog. *American Scientist* (July/August):336-347.

Pringle, H. 1998. Reading the signs of ancient animal domestication. *Science* 282 (Nov. 20, issue 5393):1448.

Serpell, J. (ed.) 1995. *The Domestic Dog: Its Evolution, Behaviour and Interactions with People.* Cambridge University Press, Cambridge.

Trut, L. 1999. Early canid domestication: the fox-farm experiment. *American Scientist* Vol. 87 (March/April):160-169.

Weidensaul, S. 1999. Tracking America's first dogs. *Smithsonian Magazine* Vol. 29 (March):44-57.

Chapter 3

Thyroid Hormone: What It Is, What It Does

Thyroid hormones are unique in that they exert effects within almost every tissue of the body throughout the life of an individual.

M. E. Hadley (2000:321)

The power of thyroid hormone

After what seemed like an inordinate amount of reading, I found out that all of the physical and behavioural characteristics that change when wild animals become domestic are controlled by thyroid hormone (whether the animals are wolves, wild goats, or the foxes described in the previous chapter). In fact, the range of body functions influenced or controlled by thyroid hormone is truly staggering (Figure 3.1).

Among other things, thyroid hormone controls both brain and body growth from just after conception to adulthood (including development of the skeleton and teeth), all aspects of reproduction, and all the steps involved in basic metabolism (including the conversion of food into energy the body can use, the storage of any excess food energy as fat, *and* the mobilization of fat back into energy for immediate use). Thyroid hormone also mediates all aspects of hair growth and colour (both the initial production of hair and any seasonal changes that occur, such as annual moults), the metamorphosis of body forms (such as the transformation of tadpoles to frogs and certain body form changes in fish), and the seasonal sleep of hibernation. In addition, thyroid hormone synchronizes the body's responses to stress of all kinds, including physiological responses to external and internal stresses (such as changing temperature and viral attack) and behavioural responses to psychological stresses (such as fear of predators and conflicts with others of the same species). In other words, thyroid hormone is responsible for coordinating all of the other hormonal responses necessary for an individual to adapt to conditions of the environment that change on a daily and seasonal basis (see Box 3.1 at the end of this section for more details).

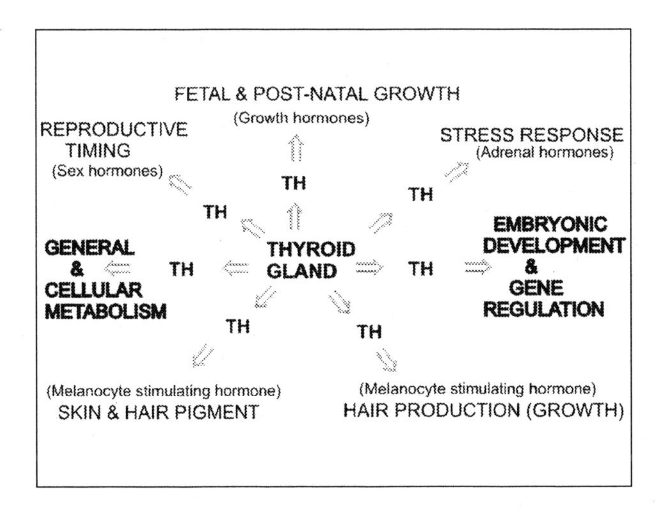

Figure 3.1 Summary of thyroid hormone (TH) effects. Those in bold involve T4 or T3 directly (primary effects, without intermediates), while the others are generated via interactions with other hormones

Thyroid hormone can do this job of coordination because it must be present at two essential stages: at the point where other hormones are released from their respective endocrine glands (their "sites of release") *and* in the tissues and cells where those released hormones have their effects (their "sites of action"). Thyroid hormone also has a direct effect on the day-to-day functions of certain cells and tissues themselves (especially in the brain). Of special importance is the fact that the function of a thyroid hormone-dependent gene can be as easily altered by changing levels of available hormone as it can by mutations in the gene itself. Such multi-faceted roles for thyroid hormone are what make it so essential—and so difficult to study.

One unique feature of thyroid hormone is that it has the same molecular structure in all organisms—it can be absorbed through the digestive tract from food, just like a vitamin. Another unique attribute is that the thyroid hormone required by an embryo for body and brain growth comes initially from its mother—either directly (through the placenta) or indirectly (from thyroid hormone deposited in egg yolk). As a consequence, the need animals have for thyroid hormone connects them to the environment *and* to the generations that came before and after in a way that's independent of specific gene actions.

For the moment, this is really all you need to know about thyroid hormone. So go ahead and move on to the next chapter if you're eager to read the entire argument: the rest of this chapter is detail you'll probably want to come back to as your mind digests the big picture. But for now, here are four points that it will be useful to remember:

1) Thyroid hormone affects virtually every bodily function in a time- and dose-dependent manner.
2) Many effects are due to thyroid hormone controlling the action of genes (much like a rheostat light switch), which means that changing hormone levels can have the same effect as a mutation to the genes themselves.
3) Because an embryo needs thyroid hormone in order to grow, the provision of thyroid hormone by mothers to offspring controls the growth of one generation to the next in a way that is independent of the genetic makeup of the embryo.
4) Thyroid hormone connects an animal to its environment (because dietary sources of thyroid hormone are provided by the habitat *and* because changing light and temperature levels affect thyroid hormone levels in the animal's body).

Thyroid hormone and iodine

Simply put, thyroid hormone is a proteinlike molecule with iodine atoms attached (Figure 3.2). While I use the term "thyroid hormone" throughout this book, in fact there are several forms of the molecule that have slightly different actions and effects because they vary in the number or placement of attached iodine atoms. I'll restrict my discussion to two of these: T_4 (with four atoms of iodine, called *thyroxine*) and T_3 (with three atoms of iodine). However, when I use the term "thyroid hormone," I really mean whichever form brings about the particular effect I'm discussing (in some cases, it isn't known which form of thyroid hormone has the most potent effect, only that one or both of them is involved—another reason for keeping the terminology general).

No real chemical effort is required to attach an iodine atom to its core molecule—unlike in many biochemical reactions of this type, the help of an enzyme isn't necessary. Because of the natural affinity that iodine has for this core protein-derived molecule, thyroid hormone is a much more common molecule than is usually appreciated.

You may have read or been told that kelp is an especially good source of iodine; what you probably didn't know (neither did I, until very recently) is that the iodine in kelp is stored as ready-made thyroid hormone. Other types of algae besides kelp also contain relatively high quantities of thyroid hormone (331, 332). In fact, it appears that the easiest way to store iodine, for all organisms (including plants), is as ready-made thyroid hormone in one form or another (some plants and invertebrate animals like jellyfish store thyroid hormone in a form that has only two iodine atoms attached) (199, 367-369). It may eventually be found that thyroid hormone is common in many types of plants, once researchers start looking for the entire thyroid hormone molecule and not just iodine. I'll discuss the implications of this later in the chapter.

Box 3.1 Basic thyroid hormone function

The thyroid gland is ubiquitous among vertebrates. It arises early in the vertebrate evolutionary sequence (i.e., it is present in chordates onward), emanating from the endoderm of the cephalic portion of the alimentary canal of the embryo. The thyroid gland (or glands, where it is has become a paired organ) is composed of numerous follicles that store the protein thyroglobulin, which act as a substrate for tyrosine iodination (thyroglobulin is derived from the essential amino acid phenylalanine, itself derived from dietary proteins; iodine is an essential mineral obtained from dietary sources) (295, 296).

Aside from a well recognized role in general metabolism (242), thyroid hormone is perhaps best known for its essential influence on metamorphosis and adaptive coloration in amphibians and fish (189, 285, 324, 621, 782, 783), and on mammalian hibernation (749, 750). However, the total range of influence of thyroid hormone on ontogenic development (from early embryonic through postnatal stages) and adult physiology in all vertebrates is astounding. Through a cascade of direct and permissive (ancillary) effects on regulatory genes and basic cell functions (which may also incorporate the influence of other hormones), thyroid hormone (in either its T4 or T3 state) is able to influence virtually all biological functions (227, 296).

Thyroid hormone is known to be essential for: 1) early embryonic cell migration, differentiation and maturation; 2) both embryonic and postnatal growth; 3) development of the entire central nervous system, including the eyes and brain; 4) hair growth; 5) adrenal gland function; 6) skin and hair pigment production; and 7) development and function of the gonads. Other "maintenance" functions include: 8) conversion of carotenes to vitamin A; 9) absorption of intestinal glucose; and 10) manufacture of cholesterol (173, 174, 210, 292, 296, 348, 463, 480, 556, 669, 743).

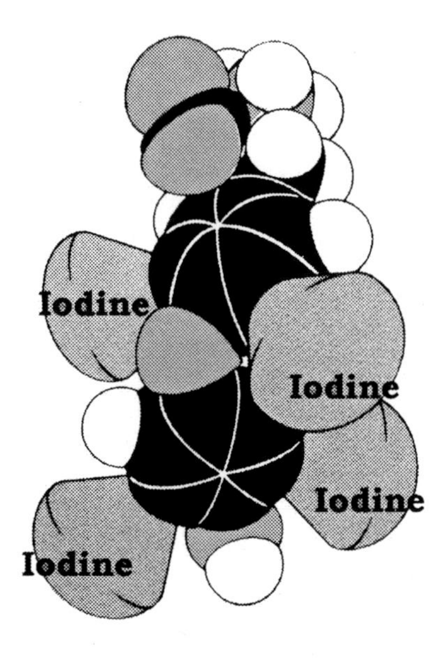

Figure 3.2 3-D model of a T_4 molecule (adapted from Stryer 1988)

In mammals, virtually all manufacture and storage of thyroid hormone takes place in the thyroid gland, although some gets made in other tissues and quite a lot is actually stored in the liver, kidneys and brain (see Boxes 3.2 and 3.3 for more details). In humans, the thyroid gland is situated in the throat, just below the larynx (Figure 3.3). All vertebrate animals have a thyroid gland: mammals, birds, reptiles, amphibians, and both primitive and advanced forms of fish. In many of these, including humans and other mammals, the thyroid gland has become a paired organ.

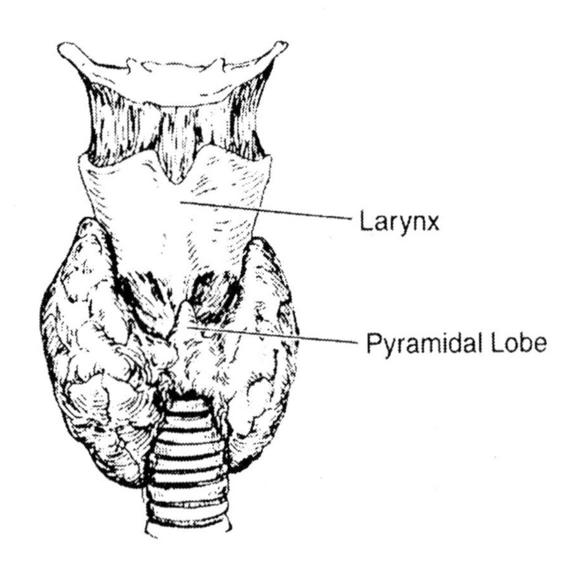

Figure 3.3 In humans, the paired thyroid glands are located just below the larynx in the throat (from Hadley 2000).

During embryonic growth, the thyroid gland forms from the same tissues that later generate the digestive tract. This origin of the thyroid gland from digestive-type tissue explains the unique feature mentioned earlier (which I'll discuss in more detail later): only thyroid hormone can be readily absorbed from the digestive tract and used as if it were self-produced (which means you can take replacement or supplemental thyroid hormone as a pill—and also absorb thyroid hormone from the food you eat).

The iodine required for thyroid hormone manufacture is an essential mineral derived from dietary sources and stored in the thyroid gland. Because iodine is essential for thyroid hormone production, it is essential for life: all vertebrate animals (and many invertebrates as well), must have a source of iodine for survival. Iodine is initially a component of the ocean but because it's volatile, it escapes readily into the atmosphere.

Once in the atmosphere, iodine can be incorporated into the precipitation that later falls on the land and into freshwater streams and lakes, making it part of the eternal cycle of water. Iodine-laden water is taken up by plants through their leaves and roots (849). Herbivorous animals incorporate iodine by eating plants that have utilized iodine-laden water or by drinking water that contains dissolved iodine; carnivores get their iodine from the ready-made thyroid hormone that's present in the blood and tissues of the prey animals they eat.

Some parts of the world appear to have low levels of naturally-occurring iodine, particularly those situated in the middle of continents far from the sea or where precipitation is low. Contrary to widely accepted belief, the incidence of iodine deficiency disorders (IDD) worldwide is not correlated with concentrations of iodine in the soil—rather, IDD appears to be dependent on localized differences in seasonal precipitation and perhaps run-off patterns of freshwater rivers and streams (244, 715). Regardless of the proximate cause, however, insufficient dietary iodine usually manifests itself as goitre, a swelling of the neck caused by an overworked and enlarged thyroid gland.

Women with goitre (who thus have insufficient thyroid hormone production, a condition called hypothyroidism) often have much reduced fertility. If they do become pregnant, however, the insufficient amounts of thyroid hormone they produce cause them to give birth to infants who not only are moderately to profoundly mentally retarded but also display markedly stunted physical growth (called congenital hypothyroidism). These birth defects occur because a fetus needs specific amounts of thyroid hormone from its mother in order for normal body and brain growth to proceed. A fascinating study that illustrates the importance of iodine to reproductive health and fetal growth was published a few years ago by a physician working for the World Health

Box 3.2 Thyroid hormone manufacture, storage & transport

Iodine-bound tyrosine derivatives are stored in the thyroid gland until *thyroid stimulating hormone* (TSH) triggers thyroglobulin to be hydrolyzed and released—with four iodine molecules attached—as *thyroxine* or *tetraiodothyronine* (T_4). Unlike many hormones, such as growth hormone and prolactin (615), the chemical form and structure of the thyroxine molecule is identical among vertebrates and even across phyla, and does not require an enzyme catalyst (199, 348). Therefore, small amounts of thyroid hormone still form in the absence of a thyroid gland.

Some thyroxine is deiodinized and released from the thyroid gland with one less iodine molecule attached, as *triiodothyronine* (T_3), although most conversion to T_3 occurs in tissues. Especially high levels of conversion are observed in the liver, brain, testis, brown fat, anterior pituitary and placenta (60, 348, 696). There is significant storage of T_4 in organs besides the thyroid glands, especially the liver but also the kidneys, brain and brown fat. Fairly large amounts of T_4 may be excreted in urine and feces (348).

Both thyroid hormones bind to several plasma proteins for circulation in the bloodstream. In humans and mice, these thyroid hormone-binding proteins are *thyroxine-binding globulin* (TBG), *serum albumin* and *transthyretin*. Transthyretin is particularly important in moving thyroid hormone within cerebrospinal fluid (246) and also transports retinoic acid, the vitamin A derivative that shares some developmental roles with thyroid hormone (204). The recognition of transthyretin as a common carrier molecule for both thyroid hormone and retinoic acid may be significant to several biochemical, physiological and developmental functions common to both. In particular, the molecular structure of thyroid hormone and retinoic acid is especially similar in the region of their nuclear receptor DNA binding sites (537). Another set of transporter molecules, only recently discovered, appears to be responsible for moving thyroid hormone from the blood into the brain—one of these transporter molecules is an "MCT" protein, the others are "Oatp" polypeptides (372).

Organization (WHO).

Dr. Glenn Geelhoed was participating in a campaign by the WHO to eliminate iodine-deficiency disorders in Africa (258). He began work in a village in Zaire where the people had an extraordinarily high incidence of goitre (upwards of 90% were affected) and congenital hypothyroidism (about 10% were affected). The birthrate for the village overall was very low, and as well as being iodine-deficient, most of the villagers were chronically undernourished. An injection of iodine was given to all villagers—one shot that supplied enough iodine to last three years.

When the doctor and his assistants returned after three years to assess the effects of their treatment, they were utterly astounded by what they saw. Most shocking was the reshaped landscape surrounding the village: acres and acres had been cleared of trees. What prompted this dramatic change? It turned out that iodine supplementation increased the energy levels of the villagers (because their bodies could finally manufacture enough thyroid hormone to boost their basal metabolic rate) and with this increased energy came increased appetites. They needed more food, which required more cleared land. In addition, there were more mouths to feed: with normal thyroid hormone production now possible, fertility in the village women increased almost immediately and the birthrate sky-rocketed.

For the first time in hundreds or even thousands of years, the villagers not only were motivated to clear significant areas of land for growing food crops but also had the energy needed to get the job done. They were happy to feel so much healthier and to have so many healthy children, but socially, they were struggling to cope with the logistics of growing so much more food. While the program was clearly a *medical* success story, no one had anticipated such a dramatic social outcome from a single dose of iodine.

Box 3.3 T_3 vs T_4 action

T_3 degrades more quickly (i.e., has a much shorter half-life) but is more metabolically active than T_4. T_3 appears to be the more physiologically relevant form (is more active in metabolism and gene regulation), in part due to its greater affinity (10-15X higher than T_4) for binding to some receptors (347). *In situ*, tissue-specific conversion of T_4 to active T_3 (controlled by the deiodinase enzyme D2) appears to be essential for thyroid hormone-controlled effects in certain tissues, including the pituitary and the brain (21, 115, 348, 372, 377, 548). The genes that produce this and other deiodinase enzymes (which all require the mineral *selenium*), are themselves dependent on thyroid hormone for expression and are an integral part of the thyroid hormone axis (348).

However, while T_3 may be the more metabolically active form, it has also been demonstrated that T_4 by itself has critical effects, particularly on early growth and development. Although the precise mechanism behind these actions is not well understood (93, 94, 437, 348), it is likely that some genes have a higher affinity for T_4 than for T_3.

The grand hormonal cascade

The production of hormones responsible for coordinating body functions begins in the brain, where electrical stimulation of receptors in the retina and *central nervous system* (CNS) relays signals to the pineal gland (a small organ in the brain) and to the specialized bit of brain tissue known as the *hypothalamus*. The pineal is well known for its ability to translate electrical signals into biochemical messages that are relayed down the hormonal cascade (298, 831), as illustrated in Figures 3.4 and 3.5 (see also Box 3.4). Although the pineal gland was once thought to be entirely responsible for sensing changing environmental conditions of all kinds and for initiating appropriate physiological responses (296, 411), it now appears this is not the case.

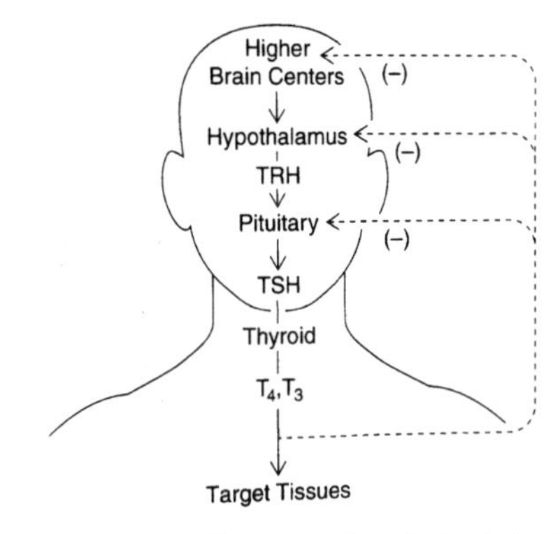

Figure 3.4 Basic cascade of thyroid hormone secretion, from control centres in the brain through to thyroid gland stimulation by TSH and release of thyroid hormone into the blood stream…where it travels via the blood stream to target genes, cells and tissues (from Hadley 2000)

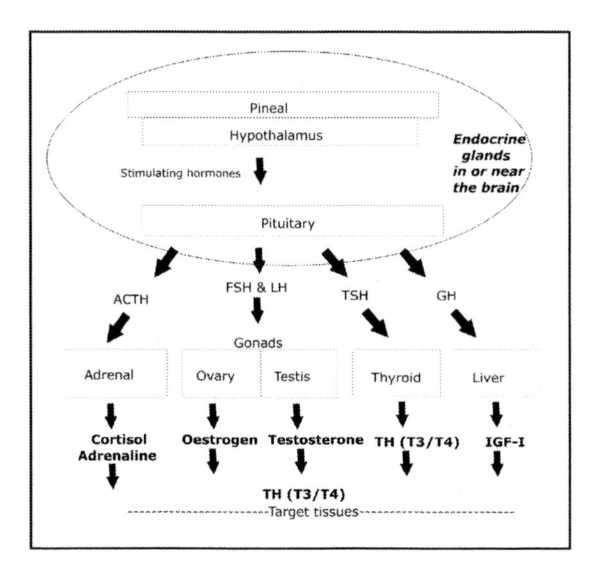

Figure 3.5 Hormonal cascade in somewhat more detail, showing hormones mentioned in the text. ACTH is adrenocorticotropin (aka adrenal cortical stimulating hormone); FSH is follicle stimulating hormone (follitropin); LH is luteinizing hormone (lutropin); TSH is thyroid stimulating hormone; GH is growth hormone (aka STH, somatotropin); TH is thyroid hormone; T3 is triiodothyronine; T4 is thyroxine; IGF-I is insulin-like growth factor-I; MSH is melanocyte stimulating hormone (a-melantropin).

We now know that in mammals, and perhaps in other animals as well, a direct neural connection exists between the hypothalamus and the retina of the eye. This direct hypothalamus–retinal connection means that stimulation of the pineal gland can be bypassed or augmented by hormones produced by direct stimulation of the nerves that connect the eye to the hypothalamus itself (626, 664).

Either way, thyroid hormone is secreted into the blood stream by the thyroid glands as the third step in a hormonal relay: 1) *thyrotropin releasing hormone* (TRH) is released from the hypothalamus; 2) TRH induces the release of *thyroid stimulating hormone* (TSH or *thyrotropin*) by the pituitary gland; 3) TSH from the pituitary stimulates thyroid hormone secretion by the thyroid glands (see Figures 3.4 and 3.5, and Box 3.5 for more details).

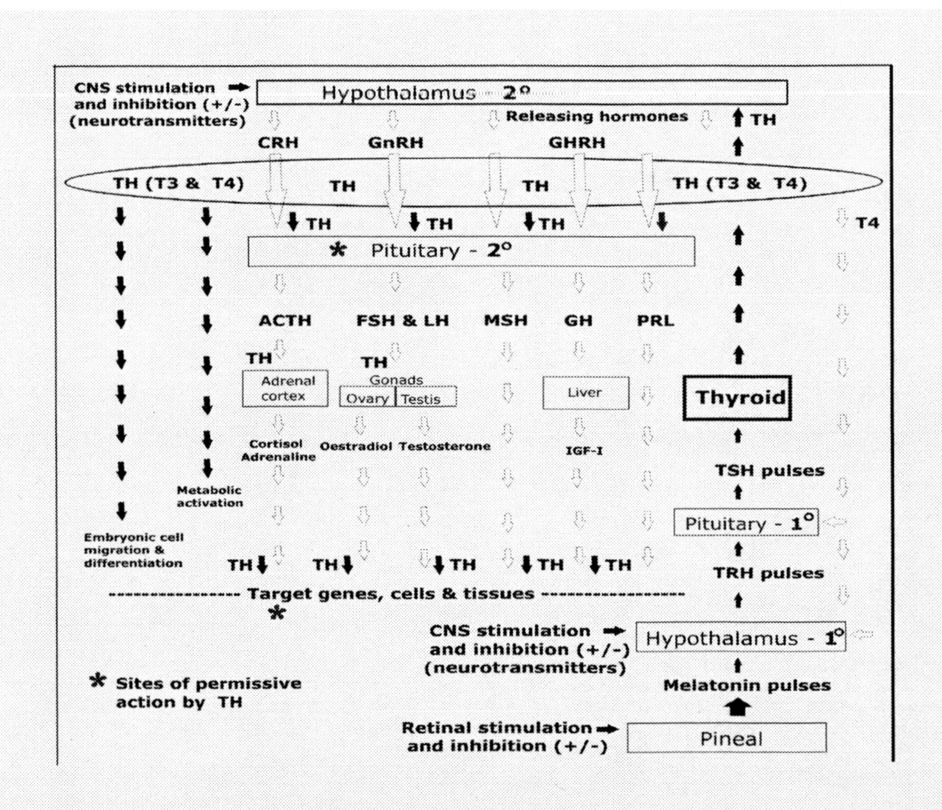

Box 3.4 Complex details of the hormonal cascade

Direct (primary) and permissive (secondary) actions of thyroid hormone (TH), which begins with neurohormone secretion from brain tissues (including the pineal and the hypothalamus) in response to electrical stimulation via the central nervous system (CNS) and the retina, lower right. TH release is actually the first response to any stimulus, because TH must be present at the pituitary for the release of many other hormones, and again at the site of target tissue, for other hormones to have appropriate effects. Acronyms are the same as in Figure 3.5, plus these: TRH is *thyrotopin-releasing hormone;* CRH is *corticotropin-releasing hormone;* GnRH is *gonadotropin-releasing hormone;* GHRH is *growth hormone-releasing hormone;* PRL is *prolactin* (296, 348).

Thyroid hormone is actually released in a rhythmic, pulsatile manner because the initial stimulation of neurons in the brain are intermittent in nature—I'll address this aspect of thyroid function in more detail in Chapter 4 as this rhythmic generation of thyroid hormone is integral to my theory.

Thyroid hormone and brain function

Thyroid hormone, specifically T_3, is essential for adult brain function. T_3 is critical to the communication that goes on in the adult brain, in part because it has a fundamental role in controlling brain-specific genes, but also because it's required for brain synapses to function properly (see refs. 326, 372 and Boxes 3.6 and 3.7). The essential fatty acid known as DHA (*docosahexaenoic acid*) is also necessary for brain development and function because it's required for the production of *transthyretin*—the hormone transport molecule that moves thyroid hormone around the brain (204, 340, 341, 343, 397). In other words, if a portion of the brain doesn't receive the thyroid hormone it needs because the transport mechanism isn't working efficiently, synapse and gene function will be compromised. This relationship between thyroid hormone and DHA means that sufficient amounts of the essential fatty acid *and* T_3 must be present for optimal brain function. Because most brain activity specifically requires T_3, a critical aspect of brain function is local tissue conversion of T_4 to T_3, which is accomplished by a specific enzyme (D2 *deiodinase*; see Box 3.3 for more details).

During fetal development, severely deficient levels of thyroid hormone supplied by the maternal system (most commonly seen in cases of acute iodine deficiency) can cause the profound form of mental retardation known as *cretinism;* even mild fetal iodine insufficiency can cause detectable brain defects, including deafness (116, 296, 258, 651). In some cases, particularly in deafness, these defects can result from localized rather than general T_3 deficiency, because the cochlea requires specific amounts of T_3 at specific times for proper formation (372, 548). More details on the role of thyroid hormone in brain development are found later in this chapter in the section on general fetal growth (also see Box 3.7 in that section).

Thyroid hormone and coat colours

Although the size, shape and reproductive changes in biological traits associated with protodomestication have been shown to be under the control of thyroid hormone (as I discussed in Chapter 2), the appearance of piebaldness (black or any other solid colour marked with white) is the hardest to explain in this context. Think spotted cows, horses with white feet, and totally white sheep. The high incidence of whitespotting or piebaldness in domestic animals has always been somewhat of an enigma and is usually assumed to be a consequence of deliberate selection (128).

But it turns out that piebald markings are quite rare in most wild mammals. Piebaldness is also rare in the ancestors (or in close living relatives) of all domestic animals, although it is the norm in a small number of groups (such as skunks and the panda). This low natural incidence of piebaldness in wild populations makes the prospect of increasing the proportion of individuals with this trait (via natural or human selection) not just improbable, but almost impossible. However, the early appearance of piebaldness in the silver foxes experimentally selected for stress tolerant behaviour (as discussed in the previous chapter) suggests that piebaldness could be an inevitable consequence of protodomestication. To explain why this is so, we need a short discussion on how pigment and patterned coats are generated.

Pigment (technically called *melanin*) is produced in special cells (*melanocytes*) that exist within skin, hair follicles and other body tissues. Many different factors can affect the activity of pigment cells during skin and hair colour production, including hormonal influences that come from the pituitary (via MSH, *melanocyte-stimulating hormone*) as well as spontaneous mutations affecting the genes and various cofactors involved (Figure 3.6). Small genetic mutations or varying amounts of hormone result in different coat and skin colours being produced (319, 635, 674, 688).

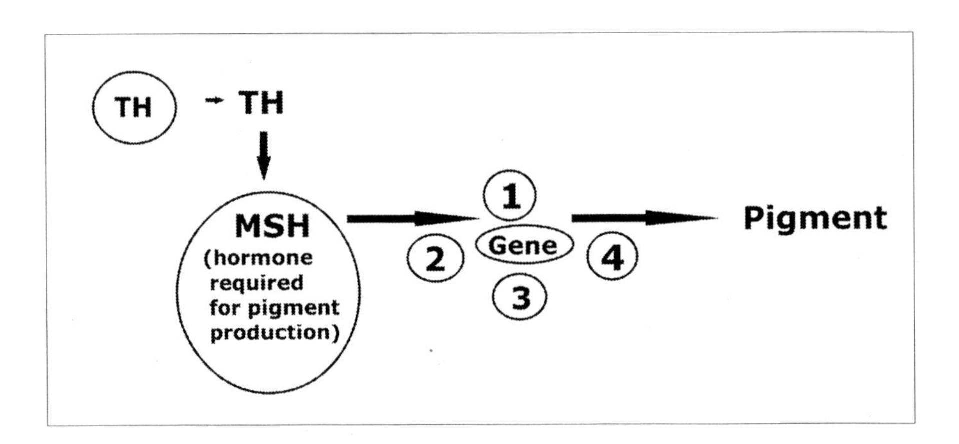

Figure 3.6 A simple model for the gene+hormone interaction required for pigment production in hair and skin cells, where points of variation are circled. Pigment production can vary due to changes in the gene (mutation) or hormone levels (of either thyroid hormone, TH, or melanocyte stimulating hormone, MSH), or any of the required "co-factors" (1–4) needed for the gene to produce pigment.

Patterning of colour, however, is determined during fetal development (689). The explanation that so far best accounts for colour patterns like whitespotting is that during early embryonic development, pigment cells start from a few locations and move throughout the embryo. Seven pairs of immature pigment cells seem to be required, and these migrate from their origins at the central mid-line of the tiny embryo (called *neural crest tissue*) to all areas of the body. Through controlled migration and multiplication of these immature pigment cells and their subsequent maturation into functional pigment cells, the entire embryo becomes completely pigmented (Figure 3.7).

Typically, colour appears to spread from two single patches of pigment-producing cells (one at the base of the tail and another between the eyes—numbers 8 and 7, respectively, in Figure 3.7) and from six pairs of symmetrical patches over the hips, lumbar region, rib cage, ears, eyes and nose—fourteen areas in all. Regions in between these centres that do not receive pigment-producing cells remain white, either because cell migration has been delayed or because multiplication of the cell patches prematurely stops (see Box 3.5 for more details). The result is a coat colour pattern that is variously spotted (Figure 3.8).

For example, the piebald coat patterns in the silver foxes from Belyav's selected population (which, as discussed in the previous chapter, are remarkably similar to the pattern seen in a typical Border collie) were ultimately determined by researchers to be caused by a one-to-two-day delay in the migration rate of pigment-producing cells in the embryo (602). Piebald coloration can vary from only a spot or two of white on a solid colour background to an all-over white that is essentially one big spot (such as seen in the Samoyed dog and polar bear).

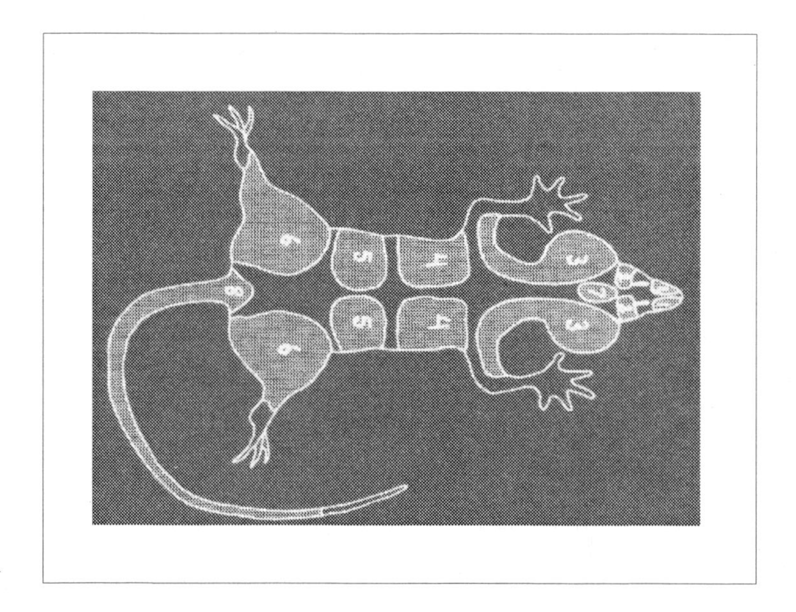

Figure 3.7 The fourteen regions that generate pigment cells during embryonic growth, showing how easily disruption of the spread of these cells can lead to an animal with white spots on certain areas of its body (from Silvers 1979).

Piebaldness has been a very useful experimental indicator of early developmental errors in the mammalian embryo. We now understand a bit more than we once did about how chance mutations in some genes affect this process.

There are at least three genes which can cause disruption of the normal migration of pigment-producing cells during early development. Specific mutations in each of these genes result in a distinct piebald pattern of pigmentation and in some particular neurological disorders.

Figure 3.8 Some examples of piebald coat patterns, including the extreme examples (bottom row) that are essentially "one big spot." (Orca photo by Robert Pitman, NOAA; potbellied pig by Don Harper; polar bear by US NOAA; Samoyed "Kirby," Kabear Kennels, USA).

Nevertheless, experimental laboratory mice with these mutations can't possibly possess the same genetic makeup as naturally piebald wild animals or most piebald domesticates. It's pretty clear, for example, that the extreme piebaldness ("one big spot") that's the normal colour for most white domestic and wild animals that have black-eyes (such as many sheep breeds, Park White cattle, the Samoyed dog, the polar bear, and the mountain goat, *Ovis Canadensis*) is not associated with the lethal digestive system defects or the deafness linked with the mutant genes in all-white experimental animals (14, 53, 54, 345, 674). Obviously, piebald coat patterns can result from several distinct abnormalities, not all of which are genetic. Unfortunately, no detailed scientific studies appear to have been done on the specific types of piebaldness possessed by wild animals, and the role of thyroid hormone in this process has not been specifically investigated (see Box 3.5 for more details).

However, because piebald coat colour patterns are now known to reflect changes in embryonic growth, their consistent appearance in domestic animals suggests that piebaldness may be an incidental consequence of the changes in growth rates known to be associated with protodomestication—and this may be true in many wild animals as well. More research definitely needs to be done on this topic and you'll see why when I discuss this issue in Chapter 6, in regards to the speciation of polar bears from brown bears.

However, piebald coat colours are not the only external hallmark of domestic animals; another is the disappearance of the alternating bands of colour in individual hairs that give wild animals a grizzled appearance (called *agouti*). Domestic animals almost always have coat colours with non-banded hairs in shades of solid red, yellow, brown and black. Why is this important? Since pigment production is governed in part by hormonal influences, it appears that this hair colour change could also be an incidental by-product of the changes in growth rates associated with protodomestication (if those indeed are controlled by changing thyroid hormone levels).

This leads us back to genes and the requirement many of them have to receive hormonal input in order to function properly. One of the colour-producing genes that operates in pigment-producing cells has been found to require a hormone (MSH) that's itself dependent on thyroid hormone (83). MSH stimulates the manufacture of an enzyme required for production of pigment in the hair follicle. High levels of this enzyme are required for the manufacture of brown and black pigments, leaving yellow or red as the "default" pigment that is produced by low enzyme levels (80).

Box 3.5 Thyroid hormone & piebald coloration

During development, embryonic neural crest tissue is the source of several specific cell lineages (in addition to epidermal and choroidal pigment cells), including neurons and glia of the peripheral nervous system, neuroendocrine cells, and pharyngeal-arch-derived tissues of the face and neck (345). Specific mutations causing failure of neural crest precursors of epidermal melanocytes and enteric ganglion neurons to migrate properly in the embryo have been shown to result in nerve supply defects of the intestine (*aganglionic megacolon*) in humans, mice and horses that are associated with various levels of piebaldness (76, 428, 820). At least three genes are known to cause disruption of the normal migration of pigment-producing cells during early development: *piebald* (S); *dominant whitespotting* (W), and *steel* (Sl). Inner ear sensory cells also originate in the neural crest (53, 56, 141, 689, 820).

Both T_4 and retinoic acid (a vitamin A derivative) have been identified as essential to the orderly movement of cells out of the neural crest during early development (42, 56, 141, 345, 581, 582). The T_4 needed for neural crest cell migration is required by the embryo very early in development and, in mammals (including humans), the maternal thyroid gland has been shown to be the source of the T_4 utilized (42, 115, 593, 598, 714). The fetal thyroid gland is not functional until fairly late in development—about 18–20 days for the rat, about 50 days for the ewe (538), and 10–12 weeks for humans (593)—at which time it begins to augment rather than replace the maternal contribution (115). Consequently, although a direct correlation between disruption of thyroxine production and piebaldness has not yet been demonstrated, the circumstantial evidence in favour of such a relationship is very strong.

As thyroid hormone is known to control the genes that produce pigment, levels of available thyroid hormone must influence hair and skin colour production (Figure 3.6). Indeed, experiments on deer mice have shown that there are significant differences in thyroid hormone levels between mice with normally banded hairs and those with solid-coloured, non-banded hairs (318, 434).

Although hormonal changes can generate non-banded hair pigmentation, solid colours can also be produced by spontaneous mutations in any of the genes that operate within the pigment-producing cell: some mutations can also produce red, yellow, brown and black coat colours (389, 635, 667, 692). In other words, defective or hyperactive genes necessary for pigment manufacture within the hair follicle can produce coat colours that are similar (or identical) to those produced by high or low levels of available hormone.

I would argue that the prevalence of these non-banded solid colours in all domestic animals (14, 674), especially when they don't occur (or are extremely rare) in wild populations of their ancestors or close extant relatives, suggests that solid colours are another inevitable consequence of the growth rate changes associated with protodomestication. I suggest that a shift in the time at which thyroid hormone was made available or in the amount of thyroid hormone delivered during embryonic growth changed first (as a consequence of protodomestication), and that over time, mutations in some genes occurred and were perpetuated by artificial human selection.

Both types of solid coat coloration thus probably exist in modern domestic animals: some have solid coats due to genetic mutations, others have solid coats because of slight differences in thyroid hormone levels (surprisingly, humans and other primates, like domesticates, also lack typical mammalian banded hairs (637, 674), likely the result of an ancient speciation event—more on that in Chapter 8). This dual "gene+hormone" control over evolutionarily significant characteristics is a topic I'll return to again and again over the course of this discussion because it calls into question the role genetic mutations really play in evolutionary change.

After all this talk about coat colours, however, you may wonder where albinos fit into the story. *Albino* coloration is a biochemical defect in the pigment-producing machinery that's usually associated with a number of rare genetic mutations. Albinism results in defective pigment production in all tissues, including the iris of the eye (hence the eyes of albinos appear either red or blue, depending on the type of animal and the extent of the defect). Albinism occurs occasionally in virtually all domestic and wild animal populations (319, 457); the colour typical of Siamese cats, for example, is a form of albinism that also occurs in some isolated populations of bears. Some of the mutations responsible for albino coloration have been perpetuated in domestic animals through artificial selection, and in wild populations of animals (and in human communities) through geographic isolation. However, albino coat colours are not consistently associated with domestic animals and are therefore not considered a hallmark of the protodomestication process.

Thyroid hormone and behaviour

In addition to controlling the development of specific physical traits, thyroid hormone can also affect behaviour. The most obvious behavioural consequences come via direct effects of thyroid hormone on day-to-day brain function (discussed earlier in the section on thyroid hormone and brain function). Thyroid hormone also influences behavioural responses to stress by stimulating the production of adrenal hormones. The genes that produce so-called "stress hormones" (adrenal *catecholamine* proteins, including *adrenalin* and *dopamine*) are all thyroid hormone dependent (296, 237, 488, 669). This mechanism works similarly to the one governing pigment production (discussed in the previous section), which is also thyroid hormone dependent (Figure 3.6). Because of the dependency of stress-hormone-producing genes on thyroid hormone, behaviour relating to an animal's stress response is fundamentally under thyroid hormone control.

Stress responses of animals are known to vary quite a bit between individuals (72). For example, in rats and wolves certain individuals have been found to produce less adrenalin or react less than others when exposed to similar stress and stimuli (240, 325, 327). Think back to the trait Dmitri Belyaev used in his experiment on silver foxes to separate potentially domesticatable animals from the rest: his selection was based entirely on the response of each animal to a particular kind of stress (the "gloved hand intruder"). Similar differences in responses to stress have been demonstrated between different breeds of dogs that have particular physical characteristics or shapes (32, 717).

Endocrinologist Mac Hadley (295, 296) defines stress as the state resulting from events (stressors) of external or internal origin, real or imagined, that tend to affect the homeostatic state; any condition tending to elevate plasma catecholamine levels in response to exogenous or endogenous stimuli (in other words, so-called "groundless" fears created by the imagination are stressful because they can raise levels of the hormones that stimulate the adrenal glands).

The thyroid response to stress is rapid and immediate: levels can skyrocket or plummet very quickly. Therefore, the ability to respond to the varied stresses that are fundamental to survival, especially those of external origin (like predators), should be governed by an individual's particular thyroid hormone physiology. And since the interpersonal social pressures that affect communally-living animals represent other kinds of stress (which also generate particular behavioural responses), individual differences in thyroid hormone physiology could also account for the individual differences in social dominance behaviour that we see in populations of animals like primates, wolves and whales (72, 239, 304, 536).

It seems then that differences between individuals in their thyroid hormone physiology could well account for both individual differences in behavioural responses to stress *and* differences in social dominance behaviours. I'll return to this issue of individual variation in the next chapter, where I discuss thyroid hormone physiology in more detail.

Thyroid hormone and fetal growth

As discussed previously in relation to the phenomenon of whitespotting or piebaldness, thyroid hormone is required for normal embryonic growth at all stages—probably from conception onward. The thyroid hormone required is supplied by the maternal system either directly (for mammals, via the placenta) or via reserves stored in egg yolk (for non-placental animals like birds, reptiles and amphibians) (Box 3.8). Many of the myriad small steps that make up embryonic growth, particularly for brain tissue, are known to be controlled by thyroid hormone (372, 839, 840). (See Boxes 3.6 and 3.7 for more details.)

For example, if thyroid hormone is not supplied in an appropriate time- and dose-dependent manner during embryonic growth, the migration and maturation of emerging brain cells is altered, resulting in permanently impaired brain function (both too much and too little can have equally devastating effects). Experiments on newborn rats suggest that thyroid hormone also controls the absolute size of the brain through diminished development of certain cell types (372). The hearing apparatus is particularly sensitive to thyroid hormone disruption, both before and after birth.

Perhaps now is an appropriate time to point out that embryonic growth in rats and mice—the animals used most often as laboratory proxies for scientific studies of human fetal development—is quite different from that in humans: rats and mice are about 12–14 days old before their body and brain growth is equivalent to that of humans at birth (21). This means that the growth of rats and mice in the two weeks after birth is equivalent to human fetal growth during the last three months of gestation. Scientists can easily vary conditions associated with growth in newborn rats and mice experimentally to mimic affects on human infants during the final stages of pregnancy. These differences are of course taken into account when designing experiments with rodents and when interpreting what the results mean in human growth terms.

Direct and gene-promoting actions of thyroid hormone are not the only way that thyroid hormone effects growth and development: remember that thyroid hormone is also required for the production and actions of other hormones. Growth and sex hormones, for example, are required for growth of the skeleton (43): both are dependent on thyroid hormone for their release from the pituitary as well as for their actions on bone.

For animals that develop internally, all required hormones (as well as vitamins, essential fatty acids and growth factors) are supplied by the mother, via the placenta. The placenta itself may be dependent on T_3 for its formation during early pregnancy, and later is capable of producing several hormones required by the fetus that can't pass through from the maternal system.

For animal embryos that develop externally (inside an egg), all required hormones and other essential nutrients are deposited in the egg yolk by the mother (184, 203, 364, 458, 569, 601). The exact amounts of hormone incorporated are controlled by the maternal system, as shown by colleagues Morgan Wilson and Anne McNabb (814), who found that the T4 concentration of egg yolk varied with the thyroid hormone status of the hen in Japanese quail. This process puts non-mammalian vertebrates—such as birds, reptiles, amphibians and fish—in a similar position as placental mammals in regards to the mother maintaining control over early embryonic growth of her offspring.

Marsupials such as kangaroos, which spend part of their embryonic growth period outside their mother's womb but inside her pouch firmly attached to a teat, get the thyroid hormone they require for late fetal development from her milk (630).

The demonstrated role of thyroid hormone in embryonic growth provides a really significant insight into how control over species-specific growth can be achieved. Because of the critical role played by maternal thyroid hormone in early embryonic development (both directly and indirectly), it's clear that the precise endocrine physiology possessed by a mother (or passed along by her into her eggs or milk) must control the early development of her offspring—and continues to influence growth until they are born (86, 115, 595, 814).

Box 3.6 Direct vs. indirect effects of T_3

As a gene regulator, T_3 has been found to bind to both nuclear and mitochondrial thyroid hormone receptors to form a *ligand-receptor complex*. This T_3-receptor complex then binds to a specific DNA sequence located within the promoter region (the *thyroid responsive element*, or TRE) of a number of genes, triggering the transcription of gene products (enzymes and proteins) within cell nuclei and mitochondria (21, 115, 407, 682, 834). Thus thyroid hormone, in its T_3 state, has been found to influence the transcription of a wide variety of genes, including those involved in the synthesis of lung surfactants, nerve and epidermal growth factors, and a large number of critical brain function proteins (21, 252, 372, 406, 566, 614).

Direct (non-genomic) effects of T_3 have also been demonstrated; that is, some interactions do not involve the binding of the hormone to a receptor or TRE. In particular, these non-genomic effects have been identified in cell and mitochondrial membranes and have been shown to control several essential functions. For example, T_3 has been found to stimulate Ca^{2+}-ATPase production in cell membranes, to increase oxidative phosphorylation in mitochondria and to induce the synthesis of Na^+/K^+-ATPase needed to activate the so-called "sodium pump" that produces heat in mitochondria (169, 296, 348, 682, 834).

Of particular importance to this discussion is an essential role for T_3 in *steroidogenesis* (the synthesis of steroid hormones from a cholesterol substrate, which takes place in mitochondrial inner membranes and is required for manufacture of all glucocorticoids, catecholamines, testosterone, and estrogen). T_3 has been shown to stimulate *steroidogenic acute regulatory protein (StAR)* gene expression in a time- and dose-dependent manner (increasing production of the enzyme needed to convert cholesterol into pregenolone, the first step in steroid hormone manufacture) in Leydig cells of testes of male rats, thus significantly increasing testosterone production (357, 488, 358). Although this has yet to be proven a general phenomenon affecting all steroidogenic cells, T_3 has been shown to regulate StAR protein-mediated steroidogenesis in adrenal cortex tissue and to induce gonadal growth in Japanese quail (361, 842).

Only from that point during development when the fetal thyroid gland becomes functional (which varies from species to species) do genes controlling thyroid hormone function that were contributed by the father have an opportunity to be expressed in their offspring.

The distinct stages at which offspring growth is affected by each set of parental genes controlling thyroid hormone function may provide a partial explanation for why the physical form of offspring from hybrid crosses differs depending on which species is the dam. For example, consider the hybrid offspring from horses and donkeys: given the same breeds or types of parents in both cases, when the female is a horse and the male a donkey, the offspring is called a *mule* because it has certain consistent physical characteristics (including larger size) that distinguish it from a *hinny*, the offspring that results when the female is a donkey and the male a horse (33, 205). A number of similar examples are known among wild animals of all kinds, including mammals and fish.

It's become patently clear that we know precious little about developmental timing and growth rates in different species of animals. Given the importance of changing rates of growth and timing of sexual maturation to evolution, we are in rather desperate need of more research in this area (509). But with that said, we still have to remember two other factors that are essential for evolution to occur: individual variation and changes to the environment. Changing environments often mean changing food supplies, even if the environmental change is voluntary, and I discuss this connection between thyroid hormone and food supply in the next section.

Box 3.7 Thyroid hormone & fetal growth

Embryonic neural crest tissue is the source of several essential cell lineages that are dependent on retinoid acid and T_4 for properly timed migration, proliferation and maturation, including epidermal and choroidal pigment cells, neurons and glia of the peripheral nervous system, neuroendocrine and inner ear sensory cells, and pharyngeal arch-derived tissues of the face and neck (42, 141, 345, 581). In the developing digestive system, thyroid hormones are known to be responsible for the differentiation of the epithelial lining of the small intestine (where nutrient absorption occurs) and to affect the timing of tooth eruption and tooth enamel formation (551, 596).

T_3 has been shown to be critical for the expression of various myosin isoforms during embryonic, neonatal and adult muscle fibre formation (253, 712). Both T_4 and T_3 have been identified as essential, in animal models such as the rat, for oligodendrocyte differentiation, axonal myelination, dendritic and axonal growth, neurotransmitter regulation and synaptogenesis in the central nervous system (20, 21, 115, 194, 252, 372, 437, 578, 696).

Maternal T_3 stimulates the production of *epidermal growth factor* (EGF) as well as *oestradiol* (via affects on steroidogenesis) in the placenta. In fact, maternal T_3 may play a key role in the development of the placenta itself (392). In addition, all known thyroid hormone receptors (TRs) are expressed by the placenta, which manufactures a number of other substances critical to development, including steroid hormones (such as *progesterone, estrogen, testosterone*) and *placental lactogen* (a growth factor related to GH required by the fetus) (115).

Thyroid hormone and nutrition

One aspect of this model that merits further discussion is the relationship between thyroid hormone production and nutrition. There are several ways in which nutrition can impact thyroid hormone physiology in all animals, including humans. One is through variations in dietary sources of the iodine required for thyroid hormone production (discussed at the beginning of this chapter); another is through consumption of dietary *flavinoids* that mimic thyroid hormone. Flavinoids are compounds found naturally in many foods, such as cassava (*Manihot esculanta*), maize (*Zea mays*), sorghum (*Sorghum bicolor*), sweet potato (*Ipomoea batatas*), cabbage (*Brassica oleracea*) and other members of the genus *Brassica*, as well as fruit of the genus *Prunus* (especially apricots, cherries and almonds). Some flavinoids (but not all) are converted by the gut into a substance that interferes with normal thyroid hormone function via *competitive exclusion* (in other words, these compounds are able to act in place of thyroid hormone in many essential functions because they have a similar molecular structure in the particular region of the molecule that reacts with other molecules and hormone receptors (584).

Box. 3.8 Summary of maternal supply of hormones to embryo

Maternal Circulation				**Placenta [Yolk]**				Fetal Circulation
TRH	→	→	→	**TRH** [?]	→	→	→	TRH
TSH	→	I						TSH
T_4	→	→	→	**T_4** [T_4]	→	→	→	T_4
T_3	→	→	→	**T_3** [T_3]	→	→	→	T_3
				SH [SH]	→	→	→	SH
GH	→	I						
				IGF-I [IGF-I]	→	→	→	IGF-I
				TRs [?]	→	→	→	TRs

Substances known to pass through the placenta or manufactured by it are indicated in bold, substances known to be stored in egg yolk in square brackets [ ] and substances which which do not cross the placenta are indicated by a solid thick bar.

TRH, thyrotropin-releasing hormone; TSH, thyroid stimulating hormone; T3, triiodothyronine; T4, thyroxine; GH, growth hormone; TRs, thyroid hormone receptors; SH, steroid hormones (such as estrogen & progesterone); IGF-I, insulin-like growth factor-I. Adapted from Chan and Kilby 2000.

Occasionally, both effects (iodine availability and flavinoid content of foods) may come into play, as in Zaire, Africa—including the village where Dr. Geelhoed undertook his WHO iodine deficiency work, mentioned earlier in this chapter. In many of these villages, the extremely high incidence of goitre (enlarged thyroid glands associated with low thyroid hormone output, or *hypothyroidism*, caused by insufficient dietary iodine), is made worse by a reliance on the *cassava* tuber as the primary source of calories (258).

Last, and perhaps most drastically, the thyroid function of individuals may be affected by direct consumption of thyroid hormone from prey animals and their eggs (199, 348). Thyroid hormone is present in the flesh, blood, milk, organ tissue, egg yolks and thyroid glands of all animals *and also in some plants*. Kelp and other algae, which have long been known to be a rich source of iodine, have been found to store much of this essential mineral as T_3 and T_4. This means that kelp and algae provide ready-made thyroid hormone to all animals that consume them—whether they have backbones or not. In fact, thyroid hormones are used by a wide range of marine invertebrates (such as starfish, jellyfish, sea urchins and coral) to control or modify their growth and development (199, 331, 332, 367-369, 395). Although many of these organisms use thyroid hormone from the algae they eat to control their development, some actually produce it themselves.

Vertebrates can also utilize thyroid hormone consumed in food. One of the factors that makes thyroid hormone unique among hormones is that it can easily be absorbed directly, without modification, through the digestive tract (296). The consumption of thyroid hormone-rich foods must add considerably to an animal's daily thyroid hormone load, especially in the case of carnivores and marine animals that eat kelp or other algae. Indeed, consumption of dietary thyroid hormone may explain a phenomenon I mentioned previously: the large disparity in turn-over rates for thyroid hormone between carnivores and humans (13.0 to 16.6 hours in dogs and cats vs. 6.8 days in humans). Carnivores obviously require an active metabolism capable of rapidly clearing and excreting the massive input of exogenous thyroid hormone, perhaps coupled with a relative insensitivity to temporarily high levels and/or an ability to produce similarly high levels in the absence of thyroid hormone from food.

The presence of thyroid hormone in food sources, coupled with the unique ability of animals to use it as if it were self-produced, must be exceedingly important to evolutionary change. In all animals, the changes in diet that commonly accompany habitat shifts during colonization events may contribute to the shifts in growth rates associated with speciation events. The wide availability of thyroid hormone in dietary sources is yet another factor that strongly implicates thyroid hormone as the biological mechanism driving evolutionary change, because it provides an especially strong link with environmental conditions that change over time.

In the next chapter, I discuss the essence of my theory—how individual variation in *thyroid hormone rhythms* can drive evolution.

Recommended reading

Arem, R. 1999. *The Thyroid Solution*. Ballantine Books, New York. [written in clear lay terms, by a physician specializing in thyroid hormone]

Bassett, J.H.D. and Williams, G.R. 2003. The molecular actions of thyroid hormone in bone. *TRENDS in Endocrinology and Metabolism* 14(8):356-364. [still a bit complicated for some readers but good diagrams]

Hadley, M.E. 2000. *Endocrinology, Fifth Edition*. Prentice-Hall, Inc., Englewood Cliffs. [a textbook with good coverage of all hormones]

Riddle R.D. and Tabin, C.J. 1999. How limbs develop. *Scientific American* (February):74-70. [no hormones, just the basics of embryonic growth]

Box 3.9 References for effects & actions of thyroid hormone

Effects & actions of thyroid hormone	References
general metabolism/nutrition (genomic & non-genomic)	Franklyn 2000; Hadley 2000; Hulbert 2000; Yen 2001; Yoshida et al. 1997; Ahima & Flier 2000; Kershaw & Flier 2004; Douyon & Schteingart 2002; Armario et al. 1987; Knudsen et al. 2005; Mantzoros et al. 2001; Kok et al. 2005; McNabb & King 1993; Michalopoulou et al. 1998; Segal & Ingbar 1989; Silva 1995; Staub 1998; Doulabi et al. 2004; Young et al. 2005
metamorphosis & adaptive coloration in fish	Dickhoff 1993; Gavlik et al. 2002; Hadley 2000; Hulbert 2000; Ebbesson et al. 2000; Grau et al. 1985; Liu & Chan 2002; Specker et al. 2000; Power et al. 2001
metamorphosis & adaptive coloration in amphibians	Denver 1999; Huang et al. 2001; Hadley 2000; Rose 2005; Hulbert 2000; Eliceiri & Brown 1994; Elinson 1987; Hayes 1997; Ishizuya-Oka et al. 2001; Schreiber et al. 2001; Shepherdley et al. 2002; Stolow & Shi 1995; Yaoita et al. 1990; Voss 1995; Voss & Shaffer 1997; Wright 2002; Wright et al. 2003 (X2)
mammalian hibernation	Tomasi & Mitchell 1994; Hulbert 2000; Hadley 2000; Tomasi et al. 1998
embryonic differentiation & maturation	Satoh & Sairenji 1997; Pavan & Tilghman 1994; Zoeller 2003; Jones et al. 2005; Martinez & Gomes 2005; Wilson & McNabb 1997
embryonic and postnatal growth	Bassett & Williams 2003; Piosik et al. 1997; Kilby et al. 1998; Hulbert 2000; Amma et al. 2001; Buyse et al. 1990; Blachet et al. 2001; Cudd et al. 2002; Dawson et al. 1994, 1996; Gardahaut et al. 1992; McNabb & King 1993; O'Steen & Janzen 1999; Morreale de Escobar et al. 1985; Pickard et al. 1993; Piosik et al. 1997; Smallridge & Ladenson 2001; Schew et al. 1996; Weiss & Refetoff 1996

development & function of the central nervous system (including eyes, ears & brain)	Chan & Kilby 2000; Bernal 2002; Yen 2001; K⊠hrle 2000; Chan & Rovet 2003; Brent 2000; Barres et al. 1994; Brucker-Davis et al. 1996; Eravci et al. 2000; Farwell & Leonard 2005; Brosvic et al. 2002; Anderson et al. 2003; Jones et al. 2005; Ng et al. 2005; Garcia-Segura & McCarthy 2004; Kalsbeek et al. 2000, 2005; Koibuchi et al. 1996; Lavado-Autric et al. 2003; Manzano et al. 2003; Henley & Koehnle 1997; Nunez 1985; Oppenheimer & Schwartz 1997; Vermiglio et al. 2004; Zoeller 2003; Zoeller et al. 2002, 2005
tooth eruption & tooth enamel formation; bone formation & growth	Pirinen 1995; Noren & Alm 1983; Barnard et al. 2003; Bassett & Williams 2003; Lewinson et al. 1989
skin & hair pigment production; coat colour & hair growth	Hadley 2000; Burchill et al. 1993; Gunaratnam 1986; Bultman et al. 1992; Bennett 1991; Lapseritis & Hayssen 2001; Hayssen 1998; Pavan et al. 1995; Pavan & Tilghman 1994
behaviour correlations	Lapseritis & Hayssen 2001; Morgan et al. 2000; Hadley 2000); Pfaff et al. 2000; Dellovad et al. 1996; Hayssen 1998; Hauser et al. 1997, 1998; Henley & Koehnle 1997; Knap & Moore 1996; Kalsbeek et al. 2000; Lanier et al. 2001; Vermiglio et al. 2004
stress response	O'Connor et al. 2000; Li et al. 2005; Hadley 2000; Jefcoate 2002 ; Baumgartner et al. 1998; Li et al. 2005; Fowden et al. 2001; Manna et al. 1999; Schriebman et al. 1993; O'Malley et al. 1984; Silva & Larsen 1983, 1985; Silva 2000
development & function of the gonads	Yoshimura et al. 2003; Bandyopadhyay et al. 1996; Shi and Barrel 1992; Chastel et al. 2003; Chan et al. 2001; Hayssen 1998; Bhattachacharya et al. 1996; Anderson et al. 2002; Longcope 2000; Manna et al. 1999; Jana & Bhattachacharya 1994; Skibola et al. 2005; Stephanou & Handweiger 1995; Southren et al. 1974; Yoshimura et al. 2003; Zhao et al. 2005
transcription of gene products (enzymes and proteins) within cell nuclei and mitochondria	Chan & Kilby 2000; Shin & Osborne 2003; K⊠hrle 2000; Yen 2001; Zhao et al. 2005; Kalsbeek et al. 2005; Ng et al. 2005; Evans 1988; Brent 2000; Brent et al. 1991; Davis et al. 1989; Huang et al. 2001; Raja et al. 1991; Satoh & Sairenj 1997; Schilthuis et al. 1995; Song et al. 1995; Timmer et al. 2003

daily & seasonal thyroid hormone variation; thyroid hormone pulsatility	Gancedo et al. 1995, 1996, 1997; Shi & Barrel 1992; Tomasi & Mitchell 1994; Wright et al. 2003a; Duckett et al. 1989; Lien & Siopes 1996; Kemppainen & Peterson 1996; Campos-Barros et al. 1997; Kalsbeek et al. 2005; Cogburn & Freeman 1987; Bitman et al. 1994; Custro et al 1994; Baumgartner et al. 1993; Meddle & Follet 1997; Kuhn et al. 1983; Gupta & Premabati 2002; Hansen et al. 2004; Kaptein et al. 1994; Klandorf & Sharp 1985; Lapseritis & Hayssen 2001; Hayssen 1998; Lucke et al. 1977; Goichot et al. 1994; Nunez et al. 1970; Gomez et al. 1997; Milne et al. 1990; Richardson et al. 2005; St. Aubin et al. 1996; Doulabi et al. 2004
pulsatility in other hormones	Windle et al. 1998 (X2); Aarseth et al. 2003; Adcock et al. 1998; Baumgartner et al. 1993; Brabant & Park 2000; Brabant et al. 1990; Butler 2000; Copinschi et al. 2000; Gillespie et al. 2003; Goichet et al. 1998; Gomez et al. 1997; Greenspan et al. 1986, 1991; Haisenleder et al. 1992; Keenan et al. 2003; Kok et al. 2005; Lightman et al. 1998; Licinio et al. 1998; Mantzoros et al. 2001; Mathew et al. 2003; Pincus 2000; Robinson 2000; Russell-Aulet 2001; Schaefer 2000; Terasawa 2001; Thompson et al. 1992; Velduis 2000; Velduis et al. 2001, 2005; Velduis & Bower 2003; van Coevorden et al. 1991; van den Berghe et al. 1999; Watts et al. 2004; Wright M. 2002; Wright et al. 2003b

Chapter 4

Orchestrating Life: Thyroid Rhythms

Rhythms of life—an overview

The broad range of functions that thyroid hormone influences make it critical to life itself, but the feature that gets lost in describing its actions and effects is the way it's secreted. A little-known but essential characteristic of thyroid hormone is that it's released in pulses every few minutes—bursts of hormone that can vary widely in intensity. In this chapter, I discuss the nature of this rhythmic secretion and the significance of the phenomenon to evolutionary change. As in the previous chapter, these first few pages are intended to give you enough basic information to proceed to the next chapter if you're eager to hear the rest of the story. Do come back for the details later, because you're sure to find some of them intriguing.

The pulsating nature of thyroid hormone release is absolutely critical to this story. Virtually all genes, cells and tissues that require thyroid hormone (as described in the previous chapter) respond to it in a dose- and time-dependent manner. This suggests that precision in timing and absolute amounts of thyroid hormone delivered must be critical to species-specific growth and body functions, and therefore, that thyroid hormone secretion patterns must be species specific. Indeed, evidence to date suggests that the rhythmic pattern of secretion, for many hormones, is specific for each species, and each breed of domestic animal, as illustrated in Figure 4.1.

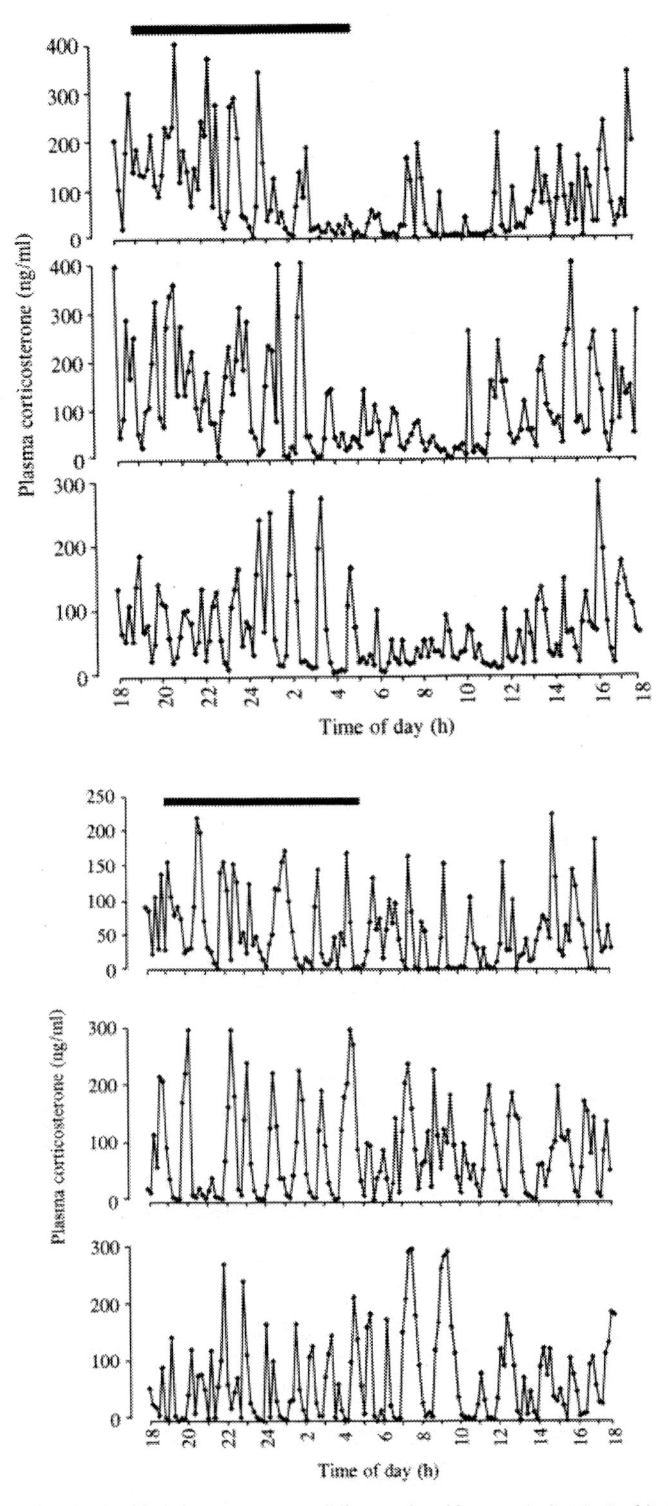

Figure 4.1 An example of the kinds of daily hormone pattern differences found between individuals of distinct animal breeds (a proxy for closely-related species). These patterns are profiles of the hormone corticosterone from three each of two physically and behaviourally distinct breeds of rat. (From Windle, R.J., Wood, S.A., Lightman, S.L. & Ingram, C.D. 1998. The pulsatile characteristics of hypothalamo-pituitary-adrenal activity in female Lewis and Fischer 344 rats and its relationship to differential stress responses. Endocrinology 139(10):4044-4052, figures 1 & 2. Copyright 1998, The Endocrine Society.)

I contend that all animals within a population share the same basic pattern of thyroid hormone secretion—the same particular *thyroid rhythm*—in the same way as species of song birds share a particular pattern of vocalization we call a song. And because thyroid hormone is critical to fetal development, when thyroid hormone from a mother passes through the placenta to her growing fetus in this species-specific rhythm, it results in the fetus growing at a precise species-specific rate. After birth, the species-specific pattern of thyroid hormone that's produced by the thyroid gland of the newborn itself (its *own* thyroid rhythm) results in continued species-specific growth until maturity.

In other words, I'm suggesting that the species-specific flow of thyroid hormone from the mother is largely what causes a chimpanzee fetus to grow into a chimpanzee rather than a human, despite the fact that 99% of its genes are identical to ours. The newborn chimp produces thyroid hormone in its own chimp-specific way and so continues to develop into a chimp and function like a chimp—perpetuating chimps that look and act virtually the same generation after generation.

The interaction of this species-specific thyroid rhythm with reproductive hormones of both sexes provides the first really plausible explanation for how and why there are often pronounced species-specific differences in size and shape between males and females in many animals (*sexual dimorphism*). It also explains how seasonal readiness of reproductive organs, and thus timing of mating and birth, can be coordinated between the sexes and among individuals within a population.

It's clear that rhythms of hormone secretion must have a genetic basis, but precisely how this works is not well understood. We do know that many genes responsible for pulsating hormone release are located in brain cellls. Hormonal rhythms seem to be initiated by complex electrical interactions between particular cells in the hypothalamus (so-called "clock genes") and neurohormones released by the pineal gland. The interaction of these clock genes with one another appears to control the basic circadian rhythm—the amazing mechanism that causes so many biological functions to run on a 24-hour day. However, it's clear that while circadian rhythms affect hormonal rhythms, the two are not identical. As a consequence, it remains to be seen precisely which genes are responsible for generating thyroid rhythms, especially since it's now known that the thyroid gland has a direct nerve connection to the the brain and can therefore trigger thyroid hormone release (or turn it off) independent of the hormonal cascade. However, for now I'll call the genes may be involved in generating thyroid rhythms, clock and rhythm genes, and the brain cells that contain these genes, clock and rhythm cells.

The DNA composition of clock and rhythm genes is not the only factor affecting their function: because their output is largely electrical in nature, the precise physical relationship of clock and rhythm *cells* to each other appears to be important as well (just as twelve people standing shoulder to shoulder can communicate by whispering, but put all or some of them even three feet apart and voices have to be raised to be heard). Such changes in the physical relationship of clock and rhythms cells to each other can arise only from slight differences in development of the brain during embryonic growth (which we now know is governed largely by thyroid hormone produced by the mother).

As a consequence, slight individual variations in species-specific thyroid rhythms are bound to occur, due to minor mutations in the genes of the clock and rhythm cells or as a result of minute changes in the relationship of these cells to each other—or both. For example, slight individual differences in hormone rhythms are apparent between the three rhythm profiles of each breed depicted in Figure 4.1. I contend that such individual variations in thyroid rhythms are ultimately converted into much of the noticeable differences in physical appearance, skeletal conformation, behaviour and reproductive characteristics within a population of animals that we call "individual variation." In other words, the distinct traits that make each individual unique do not always vary because of DNA mutations in specific genes but because the function of those genes is controlled by individually unique thyroid hormone rhythms, as shown in Figure 4.2.

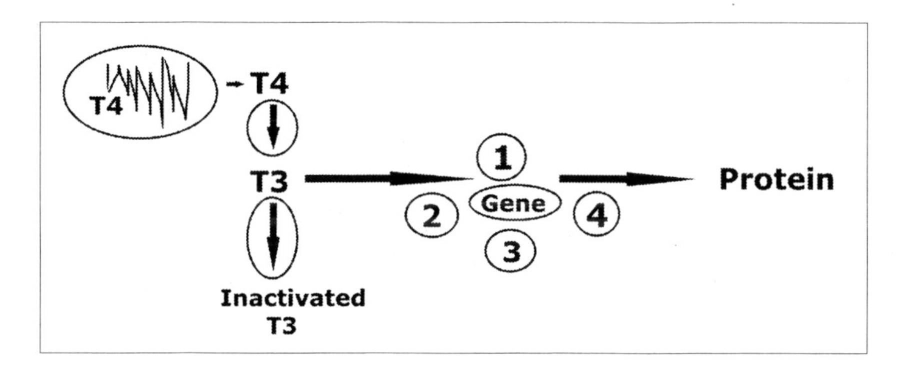

Figure 4.2 A model for the eight or so points (circled) at which the function of a gene that requires thyroid hormone can be regulated, causing variation in the amount of protein produced (either more or less, sort of like a rheostat on a light switch). These points include the initial rhythmic release of T_4 from the thyroid gland (far left) and the DNA composition of the gene in question. Circled arrows are conversion steps controlled by additional enzymes, which themselves are regulated by thyroid hormone. Other circled points (1-4) are gene-moderating factors often required for the gene to function properly—the names of these are less important than the fact that at least four of them may be involved—and if any one of these vary, the amount of protein produced by the gene (its overall function) will also change. (These gene-moderating factors include: 1 various corepressors or coactivators; 2 retinoid X receptor; 3 thyroid hormone receptor; 4 an upstream promoter transcription factor.)

Understanding that the species-specific physical relationship of clock and rhythm cells to each other can be modified in an individual by its mother's thyroid rhythm (which may be reacting to some dietary change or stress experienced during pregnancy) makes it easier to comprehend *how* environmental change can contribute to evolutionarily significant changes that are not (technically speaking) genetic in nature. These minor variations in species-specific rhythms (due to small mutations in clock and rhythm genes *or* slight differences in brain development) are critical to evolution because they supply the raw material for natural selection.

Because individual thyroid rhythms are unique, all animals are slightly different from each other in many ways, such as size, behaviour, stress-response and reproductive physiology. Such slight differences may result in some animals being better equipped for survival than others, depending on how they react to, or deal with, new environmental conditions. For example, think about what happens when part of an animal population moves into a new territory. Although the new habitat may offer many advantages to the colonizers, it also presents new stresses that have to be dealt with by each colonizing individual. I propose that new habitats are more likely to attract physiologically stress tolerant individuals than less stress tolerant ones. As a consequence, the colonizing group consists of physiologically similar animals only (specifically, those with *particular* thyroid rhythms). A new thyroid rhythm pattern becomes established for the colonizing group and as a result, its descendants undergo rapid changes in morphology, behaviour and life history. This is how a new species is created.

For those who are skipping the rest of the chapter for now, here's a summary of what you need to remember:

- Thyroid hormone rhythms appear to be species-specific, but individuals within each species have a slightly different (unique) rhythm.
- These individually-unique thyroid rhythms appear to be responsible for generating the slight differences in physical, physiological and behavioural characteristics that constitute "individual variation"—without this individual variation, evolution could not occur.
- Clock and rhythm cells in the brain control rhythmic secretion patterns of thyroid hormone: the genetic makeup *and* physical relationship of these cells to one another together are responsible for the precise profile of individually-unique and species-specific rhythms.
- Colonization of new habitats is the most common scenario leading to speciation events—including animal domestication.
- New habitats attract physiologically stress tolerant individuals (animals with *particular* thyroid rhythms) preferentially over less stress tolerant ones—as a consequence, founder populations are composed of hormonally similar animals, not a random assortment.
- During colonization of new habitats, rapid changes in morphology, behaviour and life history occur in descendant populations as a new thyroid rhythm pattern for the group is established, usually within twenty generations—this new rhythm defines the new species.
- Thyroid rhythm theory provides an explanation for the speciation of all mammals, birds, reptiles and amphibians.
- One of the most important aspects of this revolutionary new theory is that it provides a way for variation to accumulate within a population other than through genetic mutation; in addition, it's testable by experimental methods.

What sets the rhythm?

Thyroid hormone is released in a rhythmic manner in all vertebrates because of pulsating electrical stimulation of glandular portions of the brain. The *pineal* gland receives intermittent electrical signals from the retina and other neural receptors, while the *hypothalamus* gland can be directly stimulated by electrical signals from other parts of the brain. Both of these glands translate electrical input messages into chemical output messages, which is what the rest of the body is waiting to hear. The pituitary gland is the grand intermediary in this chemical messaging scheme: the pituitary receives the pulsating hormonal signals from the hypothalamus and responds by producing a number of pituitary hormones. These pituitary hormones interact with other hormone-producing organs, including the thyroid gland (as described in the previous chapter and depicted in Figures 3.4, 3.5 and Box 3.4). In this way, pulsating electrical signals received by the two neurohormone-producing organs in the brain cause thyroid hormone to be secreted from the thyroid gland into the bloodstream in a distinctly rhythmic manner (288, 289, 298, 662). Although another method of thyroid hormone release can pass this hormonal cascade (which I describe in more detail later in this chapter, Box 4.2), this brain-to-pituitary hormone relay appears to be the dominant mode of generating thyroid hormone rhythms.

Box 4.1 Rhythmic thyroid hormone secretion & interactions

Hormone pulsatility originates high in the hormonal cascade, where electrical stimulation of receptors in the retina and the *central nervous system* (CNS) relays signals to the pineal gland (a small organ in the brain) and the hypothalamic region of the brain (which forms the walls and lower part of the third ventricle). The pineal is well known for its ability to translate electrical signals into biochemical messages (*neurohormones*, including *melatonin*, *serotonin*, and *noradrenalin*): pulses of neurohormones produced by the pineal gland in response to electrical stimulation from the retina and the CNS prompt intermittent stimulation of the hypothalamus. The hypothalamus secretes hormone-releasing hormones, including *thyrotropin releasing hormone* (TRH), that are relayed down the hormonal cascade (298, 831).

Pulsatile release of pineal melatonin (*or* direct stimulation of the hypothalamus by the brain, see Box 4.2) stimulates pulsatile secretion of TRH. TRH pulses kindle bursts of *thyroid stimulating hormone* (TSH or *thyrotropin*) from the pituitary (308). TSH pulses stimulate pulsatile release of thyroid hormone from the thyroid. Pulsatility is a feature of all hormones and may be essential to their physiological activity (296, 298).

In non-mammalian vertebrates, *corticotropin-releasing hormone* (CRH, or *corticotropin*), which in mammals appears to act exclusively on adrenal tissue, also elicits release of TSH, and subsequently, thyroid hormone secretion (178, 183): this secondary control mechanism over thyroid hormone release has been shown to give amphibians in particular an important additional set of hormonal cues for regulating metamorphic processes that can be exquisitely timed to rapidly changing environmental conditions (such as the sudden drying or flooding of a pond).

The pulsating nature of hormone release is probably not accidental: indeed, it appears that rhythmic production may be essential to the proper activity of all hormones. It's been found that continued exposure of tissues to the neurohormone *melatonin* soon makes them unresponsive, so that increasing levels of it eventually have no effect at all (296, 298). You may have experienced such a phenomenon yourself, but in olfactory terms. Say you're bombarded by an overpowering smell of some kind in a small enclosed space. If you stay around long enough, you'll hardly notice it at all. However, if you duck outside for a few minutes and return, the smell will be as strong as ever. Hormone pulsatility gives cells and tissues this same refreshing break and prevents overexposure of tissues to the important message the hormones must relay.

In all animals, the precise pattern of thyroid rhythms are known to change both seasonally and daily according to other physiological demands (131, 220, 250, 293, 416, 453, 480, 526, 665, 680, 750). Fluctuations in thyroid rhythms also occur with age, reproductive stage, psychological state and general health (718, 768). As amounts of thyroid hormone released are known to vary relative to the many factors described above, static measurements (single samples) of blood thyroid hormone concentrations (or of its pituitary harbinger, *thyroid stimulating hormone*, TSH), which are often used to characterize thyroid function, can't actually be reliably compared between species or even between individuals (238, 246). In part, this is because any two samples drawn are unlikely to have been taken under identical conditions.

However, a few tests exist that may serve as proxies for the existence of thyroid rhythms because they document differences in thyroid hormone metabolism between species. One such test measures the time taken for a measured amount of thyroid hormone (in its T_4 form) to become reduced to half its original value (called its *half-life*), a test that reflects differences in thyroid hormone metabolism *after* thyroid hormone is released. Here we see that the average half-life of T_4 is 13 hours in dogs, 16.6 hours in cats, and 6.8 *days* in humans (379)—in other words, *carnivores* like dogs and cats can (or must) replenish T_4 levels much more often than omnivorous humans (I'll discuss such differences in thyroid hormone metabolism between carnivores, herbivores and omnivores in more detail later). Some differences between animal breeds have also been demonstrated. For example, the half-life of T_4 in the beagle (a dog that is typically *diestrous*—breeds twice a year) is twice as high as that found in the basenji (a dog that is typically *monoestrous*—breeds only once a year) (556, 672). So although comparisons within and between species for measured values of thyroid hormone are problematic, it's nevertheless apparent that significant species-specific differences in thyroid hormone metabolism exist. However, it's the nature of these species-specific differences in metabolism that I'm concerned with, particularly differences in thyroid hormone secretion patterns.

Only a few studies have actually sampled thyroid hormone levels frequently enough to document the normal daily fluctuating pattern of thyroid hormone secretion in several species. However, these studies suggest not only that marked differences do exist between species for daily rhythms of thyroid hormone secretion but that differences in daily rhythms *within* species also exist. Most studies have dealt with only one species at a time, and none were done at the recommended frequency of 5–7 minute intervals (domestic cattle, 15-min. intervals; domestic cat, 20-min. intervals; human, 20-min. intervals; a fish, 1-hr. intervals; domestic horse, 4-hr. intervals; domestic rat, 4-hr. intervals; domestic chicken, 4-hr. intervals). Nevertheless, individual variations are decidedly noticeable, especially for those measured several times per hour (66, 104, 131, 195, 272, 388, 466).

The few truly "comparative" studies examined a number of amphibian species, some closely related (several species within a genus), and others somewhat more distantly related (different species within separate genera), demonstrating that daily profiles of thyroid hormone differ significantly among species (250, 251, 416, 832).

Other research has picked up a different kind of thyroid hormone variation between species, particularly for birds. What these studies show is that the spike of thyroid hormone that occurs in all birds just after hatching differs significantly between newborns of species that produce rapidly-maturing (*precocious*) young, such as the Japanese quail, *Coturnix japonica*, the bluenecked ostrich, *Struthio camelus*, and the emu, *Dromaius novaehollandiae*, and those with slowly-maturing (*altricial*) young, such as the European starling, *Sturnus vulgaris* (67, 306, 665). All newborn animals have much higher circulating thyroid hormone levels than adults of the same species because of the rapid growth that occurs just after birth or hatching (348).

Thyroid rhythms as conductors: hormone pacemakers

Virtually all pituitary and steroid hormones that emanate from hormone-producing organs such as the gonads, the adrenal gland and the liver (including testosterone, estrogen, growth hormone, insulin and adrenalin), are known to be secreted in rhythmic fashion (e.g., 818)—and all have been shown to be dependent on thyroid hormone for release. For example, there is compelling evidence (based on experimentation aimed at understanding how such interactions work at the molecular level) that thyroid hormone has a critical role in the secretion and actions of GH/IGF-I (*growth hormone* and its active analogue, *insulin-like growth factor-I*)—that thyroid hormone is necessary for the rhythmic release of growth hormone-releasing hormone from the hypothalamus all the way through to tissue actions on developing bone and muscle; the same relationship holds for thyroid hormone and adrenal hormones (glucocorticoids and catecholamines), as well as for prolactin (a reproductive hormone produced by the pituitary gland).

I suggest that, as a consequence of this dependent relationship, rhythms of thyroid hormone may be the primary pacemaker that drives or augments rhythmic production in all the others. Some of this pacemaker control may come from the direct nervous connection now known to exist between the SCN of the brain and the thyroid gland itself. While a pacemake control function for thyroid hormone over others in the hormonal cascade remains to be conclusively demonstrated, there is very strong circumstantial evidence that such a central control mechanism not only exists but has substantial biological and evolutionary significance.

It has long been suggested that thyroid hormone is the biochemical agent responsible for coordinating the body's total adaptive response to both short-term (daily) and long-term (seasonal) changes in environmental conditions (296), allowing fully coordinated (and often rapid) adjustment of such actions as timing of reproductive function and increased metabolism in response to changes in light and temperature. The combined evidence that thyroid hormone release is distinctly rhythmic in nature and that thyroid rhythms vary according to both external (environmental) and intrinsic (physiological) factors suggests to me the distinct possibility that shifts in timing and intensity of thyroid hormone pulses (i.e., changing thyroid rhythms) are the biological mechanism through which individual short-term adaptation is achieved. This conclusion is supported by new evidence that thyroid hormone controls growth and development at the molecular level in a time- and dose-dependent manner (252, 437, 850). Thyroid rhythms appear to be the biological mechanism that's responsible for allowing vertebrates to change permanently in response to changing environmental conditions over evolutionary time.

The rhythmic secretion of thyroid hormone exerts a strong influence throughout the entire endocrine system. While all hormones (including growth hormone, testosterone, estrogen, insulin) are secreted in a pulsatile fashion (88, 89, 100, 135, 454, 658, 663, 594, 662, 718, 772, 818), it's apparent from detailed study that at least some of these rhythms are not independently generated or maintained. For example, in rats, birds and humans, thyroid hormone has been demonstrated to be necessary for generating the pulsatile production of GHRH (growth hormone releasing hormone) by the hypothalamus (20, 29, 309, 839) and thus required for pulsatile GH release from the pituitary (638, 772); in rats, thyroid hormone has been demonstrated to be required for the pulsatile release of ACTH (adrenocorticotropic hormone or corticotropin) from the pituitary (546) and thus essential for initiating the pulsatile release of catecholamines and glucocorticoids from the adrenal gland (see previous chapter, Box 3.9).

In summary, I contend that thyroid rhythms are strongly implicated as the mechanism that drives evolutionary change in vertebrates, not only because of the crucial role of thyroid hormone in embryonic, fetal and postnatal growth and development (via effects on regulatory genes and non-genomic cellular processes), but also because thyroid hormone is the only known factor demonstrated so far to link (through hormonal interactions and interdependence) the morphological, reproductive and behavioural traits known to change in coordinated fashion over evolutionary time.

Thyroid rhythms and individual variation

It's my contention that species-specific rhythms of thyroid hormone secretion may be the crucial underlying factor in protodomestication and speciation involving changes in growth rates and timing of sexual maturation. Virtually all target genes, cells and tissues that require thyroid hormone during the embryonic and postnatal growth periods respond to it in a dose- and time-dependent manner (252, 437, 850), suggesting that species-specific precision in timing (the frequency of thyroid hormone pulses) and absolute amounts of thyroid hormone secreted (amplitude of thyroid hormone pulses) must be critical for species-specific development (a similar dose- and time-dependent effect has been demonstrated for *juvenile hormone,* the substance which controls growth in insects) (550). I suggest that, as a consequence, very slight individual variations in thyroid rhythms between individuals in wild populations could produce small physiological differences that are ultimately manifested as noticeable differences in physical, skeletal, behavioural and reproductive traits between individuals (biologists calls these *phenotypes*).

Manifestations of such individual rhythm differences may be most readily apparent as variations in coat colour and size. This is due to the strong influence of thyroid hormone on growth and on secretion of *melanocyte stimulating hormone* (MSH), which subsequently affects hair colour, length and thickness. In addition, although perhaps less immediately noticeable, individual differences in thyroid rhythms almost certainly underlie the slight individual variations in timing of reproductive events (including ovulation and testes maturation) and annual moult that are recognized in virtually all mammalian taxa—most of these events vary in onset among individuals by several weeks. In addition, individual variations in behavioural responses to stress (including levels of social dominance, territoriality and fearfulness) have also been documented among individuals (40, 239, 316, 580).

Provisional support for the hypothesis that individual thyroid rhythms underlie a suite of individual physical and physiological differences comes from a study on a population of young marine flounder (also known as flatfish or sole, members of the family Pleuronectiformes) (256). As mentioned previously, thyroid hormone is known to control the timing and speed of metamorphosis in many animals; in flounders, the transformation from "normal fish" shape to "flounder" shape involves thyroid-controlled metamorphosis. At a certain stage of growth, thyroid hormone surges drive the eyes of young flatfish to morph from a typical "eyes-on-either-side" position to a peculiar "both-eyes-on-the-same-side" position that's useful for an adult life spent living on the sea bottom. In a group of experimental flounders, it was found that individual differences in the timing and intensity of the thyroid hormone surge that drives this "eye morph" change correlated both with differences in timing of what's called "settling" behaviour (the time when morphed individuals take up residence on the sea bottom) and with the growth spurt that followed. As a consequence, both settling behaviour and growth of the group could be synchronized by manipulating the thyroid status of each individual within the population so that all experienced their thyroid surge together (an experiment aimed at making flounder farming easier and more efficient).

Individual variations in thyroid rhythms are also expected to generate particular intraspecific differences within populations because of the intimate relationship shown to exist between thyroid hormone, reproductive hormones and growth hormone. Reproductive and growth hormone have both been shown to be generated in individually-unique rhythmic fashion (87, 560, 638, 744, 772). Although the researchers responsible for these studies generally attribute little biological significance to such slight profile differences, I suggest that the interactions between these hormones and individual thyroid rhythms probably contribute much more to individual variation than previously thought.

For example, there has been shown to be an interdependence between estrogen, IGF-I (*insulin-like growth factor-I,* a liver protein stimulated by GH and actually responsible for most of the essential functions usually associated with growth) and thyroid hormone in the way they impact on bone growth and remodelling (43, 125, 286). In addition, since estrogen seems to have a "booster" effect on the rhythmic release of growth hormone and IGF-I (671, 774, 821), an interaction of these hormones with thyroid rhythms may explain the species-specific nature of sexual dimorphism mentioned earlier. Studies on mink and martins (176), rats (294), primates (264, 497), sea lions and fur seals (522) and Arctic foxes (603)—to name just a few—all show sex-specific differences in rates of growth and timing of sexual maturation that are consistent within each species.

However, it's also possible that in addition to (or perhaps instead of) the hormonal interactions described above, thyroid hormone may have a more direct impact on the development of species-specific sexual dimorphism. The standard explanation for why males and females develop differently presumes that sex-specific genes on chromosomes (so-called "genetic sex"—in humans, XX for females/XY for males) determine an animal's gonad type midway through fetal development—and that gonadal hormones subsequently generate all other sex-specific traits in other tissues. However, this interpretation is now being challenged. Recently, researchers have suggested that the genetic sex of an individual may in fact generate sex-specific differences in critical brain structures during very early embryonic development (well before gonads are functional) and that these sexually-distinct brain structures then control sex-appropriate development of gonads and other tissues (31, 186).

If this new interpretation turns out to be correct, given the critical importance of thyroid hormone to embryonic neural tissue growth and differentiation that has already been demonstrated, it's possible that species-specific sexual dimorphism is a result of species-specific thyroid rhythms interacting first with the genetic sex of an individual to generate sex-appropriate brain structures and subsequently with other hormones to guide development of sex-specific tissues in a species-specific manner. The involvement of thyroid hormone with growth and sex hormones (or with genetic sex determiners during embryonic development) does not explain why males are the larger sex in some animals (such as bears, lions and sea lions) while females are larger in others (such as eagles, hawks and some whales). However, if thyroid rhythms are species-specific (as I maintain must be the case), it would explain both the consistent differences in growth rates and timing of sexual maturation and their continued species-specific expression, generation after generation (497).

Indeed, the importance of individual differences in physiology to evolution is becoming more apparent as researchers attempt to understand what factors influence the success of populations over time. For example, one study found that individual physiological differences among fish larvae in their sample were extremely critical to recruitment success under certain environmental conditions: the larvae that survived to adulthood were not "average individuals" but represented a restricted fraction of the population that possessed a particular physiological makeup (154). In other words, survivors were not a random subset of the original population but a group of physiologically similar animals particularly suited to the specific set of environmental conditions that existed at the time of their early growth. Such phenomena are particularly germane to this argument, for if we substitute the phrase "thyroid rhythm" for "particular physiological makeup," we have the beginnings of a model for how individual variations in thyroid rhythms allow populations of animals to adapt to environmental conditions that change from year to year, or decade to decade.

It's apparent to me that individual variation in hormone physiology needs to be studied more carefully within an evolutionary context. Individual variations in physical and physiological traits are almost always attributed to slight mutational changes that occur randomly and continuously in genes (often called *variable alleles*), although recently a process that changes the activity of a gene without changing its DNA sequence (23) has also be attributed with this power (called *methylation*, this phenomenon is studied within an emerging field of genetics that has coopted an existing term, *epigenetics*, which formerly meant something quite different, as explained in the glossary). And while such genetic variation may well exist and appear correlated with specific physical, behavioural or reproductive traits (see also Box 5.2), these genes are not necessarily what is being shaped by natural selection. Have a look again at Figure 4.2: it illustrates that the DNA composition of any particular gene that depends on thyroid hormone for its production (which includes a huge proportion of all available genes) is only one component out of at least eight factors that ultimately effect the amount of protein that a gene produces (most genes produce a particular protein—it's what the protein does that produces the "effect" of the gene). A mutation in a gene must be exposed, through its effects on the biochemistry, physiology, behaviour or growth of the individual, before it can be subjected to natural selection (299, 801, 802).

A mutation in a single thyroid hormone dependent gene can affect only the amount of the specific protein that gene is responsible for; a shift in thyroid rhythms can alter the amount of that specific protein as well as scores of others—including those involved in reproductive timing, growth and behaviour—all at the same time. This simultaneous and multi-tiered aspect of thyroid hormone effects on gene expression means that a particular thyroid rhythm governs an enormous suite of traits that are not obviously related. As a consequence, selection for a single trait controlled by a particular thyroid rhythm (such as size, or heat tolerance) can trigger a cascade of changes in associated traits (such as reproductive timing, behavioural responses to stress and coat colour differences) that appear unrelated but are biochemically, physiologically and/or developmentally linked (11, 13). Alternatively, one could select for a particular behavioural response to stress and get reproductive timing, size and coat colour changing simultaneously.

The potential existence of distinct species-specific thyroid rhythms (with associated individual rhythm variants responsible for variations in behaviour, size, skeletal conformation and reproductive physiology) might explain the fact that some species within the genus *Canis*, such as the wolf, *C. lupus*, display a great deal of variation over the whole of their range (370, 371, 512, 845), while the coyote, *C. latrans,* does not (518, 844). If such variation indeed reflects the natural range of differences in thyroid rhythms within each species, wolves should have a larger range of hormonal variation than coyotes. I suggest this has given more evolutionary options to wolves than to coyotes. Stephen Jay Gould (2002:1271) would call this differential "flexibility for future change" (or *evolvability*), explaining why the wolf has generated dog descendants two or more times in the last 15,000 years as an adaptive response to human-dominated habitats (see discussion in Chapter 3), while the coyote has merely expanded its range (555).

I've concluded, based on the evidence presented thus far, that thyroid rhythms are a prime candidate for the biological mechanism that coordinates species-specific growth and development. This mechanism might also explain the differential ability of some populations to change and adapt over time to changing conditions while others become extinct ("evolutionary evolvability"). It's apparent from the literature that we already know some such mechanism must exist (299). Some researchers have pleaded for more attention to be given to the function of hormones in developmental processes, and one even presented a model describing the role that a critical multi-functional hormone (such as thyroid hormone) might play over the entire lifetime of an organism (184) and showing how such a hormone might "evolve" with the growing organism as it ages, reminiscent of Parker and McKinney's suggestion (580) that growth and development may be individually distinct for each animal ("individual ontogenies"). I suggest that because of both the multitude of physiological and developmental roles thyroid hormone serves and the changes in thyroid hormone metabolism that occur over an individual's lifetime, thyroid hormone is a good candidate for the control mechanism we are seeking.

Ever since the role of steroid hormones in development and gene regulation was demonstrated almost twenty years ago (210), the critical importance of hormonal interactions to evolution has been more fully appreciated (223). Real progress has been severely hampered, however, by the lack of an appropriate theoretical framework that ties critical endocrine effects and interactions into an evolutionary context, a problem that my thyroid rhythm theory addresses.

Genetic control of thyroid function

A molecule as essential across all vertebrates as thyroid hormone could not vary without disastrous repercussions, and as I've said, it's not the thyroid hormone molecule itself, but its pattern of production and utilization, that varies within and between species. Although it's clear that rhythmic secretion must have a genetic basis, not enough is known about which genes are involved or exactly how they impact thyroid hormone metabolism. While the genes that govern the incorporation of iodine into thyroid hormone operate within the thyroid glands themselves, the genes that control release of thyroid hormone are probably located in the brain.

Ultimate control over thyroid hormone secretion appears to come from complex interactions between the cells of the *suprachiasmatic nucleus* (SCN) of the anterior hypothalamus in the brain and the neurohormone *melatonin,* produced by the pineal gland. Electrical stimulation of receptors in the retina relays signals about light to the pineal gland or the SCN; stimulation of receptors in the brain and peripheral nerves (CNS, *central nervous system*) do the same regarding temperature and emotional states. The pineal gland responds to these combined signals with bursts of melatonin, which can be augmented or overridden by electrical and/or chemical output from the SCN (298, 830, 831); thus, neurohormonal stimulation from the SCN or the pineal stimulates pulsatile secretion of *thyrotropin-releasing hormone* (TRH) from the hypothalamus. Pulsatile release of TRH stimulates bursts of TSH from the pituitary, which stimulates pulsatile release of thyroid hormone from the thyroid glands.

Box 4.2 Direct stimulation of the SCN & the thyroid gland

Recent experimental evidence has documented a direct neural connection between the suprachiasmatic nucleus (SCN) of the hypothalamus and the retina, at least in mammals. The direct SCN–retinal connection provides an alternate control mechanism in these vertebrates (and perhaps others) for rhythmic secretion of both neurohormones and hormone-releasing hormones, including TRH: it means that the SCN and its hormones can be controlled by melatonin rhythms released by the pineal gland or stimulated directly by retinal nerves (626, 664).

Even more significantly, a direct connection between the SCN and the thyroid gland itself has now been demonstrated to exist, which means that thyroid hormone can be stimulated or dampened *independently of changing TSH levels produced by the pituitary* (377, 843). As electrical signals from the retina and SCN are produced intermittently, it's probable that thyroid hormone released as a result of direct nerve stimulation is also secreted in rhythmic fashion.

As already mentioned, genetic control over this pulsatile hormonal cascade seems to come from eight or more "clock genes" that reside in *circadian oscillator cells* of the mammalian SCN. The interaction of these genes with each other to generate a neurohormonal and/or electrical output appears to control basic *circadian timing,* which generates the body's 24-hour clock. Because as yet unknown "clock-modulating" genes are thought to exist as well (625, 626), the mechanism is far from being well understood. The SCN appears to be developed and functional in a number of laboratory animals by the late fetal stage (624, 681) or, at the very latest, by the early postnatal period (634). However, it cannot be emphasized too strongly that although circadian rhythms do modulate hormonal rhythms, *the two are not identical*; thyroid rhythms are known to shift during development in amphibians and fish even when light/dark cycles are constant (293, 831). Thus, it remains to be seen precisely which genes affect thyroid rhythms and which affect circadian rhythms specifically (22, 92, 162, 413, 266, 514, 628, 659, 759).

In addition, it is also apparent that the genetic basis for differences in overall thyroid hormone metabolism doesn't lie exclusively in the genes that govern the secretion of thyroid hormone. Variation is known or suspected to exist in the conformation or concentration of independent genes that produce proteins controlling thyroid hormone distribution in the blood and cerebral fluid (e.g., *transthyretin*), or that encode thyroid hormone receptors, receptor ligands and/or cofactors in target tissues (20, 347, 834, 837).

Even more significantly, it has recently been demonstrated that there are direct nerve connections between the SCN of the hypothalamus and the thyroid gland itself (see Box 4.2). This discovery provides proof for a phenomenon which has long been suspected: that control over thyroid hormone release is not governed exclusively by pituitary TSH. A direct SCN–thyroid connection explains the immediate changes in thyroid hormone release known to occur (within minutes) under conditions such as stress and cold. Importantly, *a direct SCN-thyroid gland connection provides a mechanism that explains how thyroid rhythms could exert pacemaker-like control over all other hormones.*

There are indeed many steps involved in getting thyroid hormone molecules to target genes that reside in various tissues. Discerning the variation that exists in the interdependent genes in the brain that govern thyroid secretion is much more difficult than identifying independent genes associated with thyroid hormone transport (either through blood or cerebral fluid, ref. 246), genes for thyroid hormone receptors within tissues or cells themselves (227) or genes that require thyroid hormone to regulate their function (123, 712, 761). All may be important to evolution, but I maintain that pulsatile thyroid hormone secretion is the most significant (and thus deserves full investigation first) because of the ubiquitous role that thyroid hormone availability, in a time- and dose-dependent manner, plays in developmental regulation across taxonomic groups.

Thyroid rhythm theory

Using my theory, the role of thyroid hormone in the adaptation of individuals to environmental conditions that change on a seasonal and daily basis can be expanded to explain the adaptation of populations of animals (species) to environmental change over geological time. Because thyroid hormone production is ultimately controlled by a group of genes (precisely how many is unclear, but perhaps as many as a dozen), slight individual variations in species-specific thyroid rhythms are bound to occur as a result of minor genetic mutations in this gene complex. This variation is what supplies the raw material for natural selection.

Individual variation in production patterns of thyroid hormone means that all animals are physiologically slightly different from each other; each is hormonally unique because its distinct thyroid rhythm is a variant of the species-specific pattern.

Under certain conditions, these slight physiological differences make some animals better equipped for survival than others because they react to, or deal with, certain environmental conditions slightly differently. Often, a population expands in size until eventually competition for resources (food or breeding sites) drives some animals to move into an adjacent territory. In many instances, that adjacent territory will be different in some or many respects from the original habitat. Explaining precisely how and why new species arise when colonization of new habitats occurs has always been difficult, but my theory offers a fully comprehensible and testable explanation.

The intimate role that thyroid hormone plays in the response of animals to stress is pivotal to this model. Although new habitats may offer attractive benefits to colonizers, they nevertheless present stressful conditions that must be dealt with by each colonizing individual. I propose that new habitats attract physiologically stress tolerant individuals (animals with *particular* thyroid rhythms) preferentially over less stress tolerant ones. I suggest that because only individuals with stress tolerant thyroid rhythms form founder populations, the group is composed of physiologically similar animals rather than a random assortment.

Due to the essential developmental role of thyroid hormone, rapid changes in morphology, behaviour and life history (like timing of reproduction) are predicted to occur in descendant populations as a new thyroid rhythm pattern for the group is established. The experimental domestication studies on foxes suggest that such changes may occur within twenty generations.

This new concept provides an explanation for the speciation of all mammals, including ourselves. It works for birds, reptiles and amphibians as well, because the thyroid hormone needed for fetal growth comes pre-packaged in maternally-produced egg yolk (in mammals born at a very immature stage, like kangaroos, the still-developing joey attached to a teat in its mother's pouch gets the thyroid hormone it needs from her milk). It also offers a plausible explanation for the phenomenal successes we have had in developing animal breeds over the last few centuries, which I explain in detail in the next chapter.

In a very general sense, the concept probably also works as an explanation for evolutionary change in plants, insects and invertebrate animals (like corals, starfish and clams). Although these organisms do not have thyroid glands that store and secret thyroid hormone in rhythmic fashion, they do have rhythmically generated hormones that act in a similar manner, such as *juvenile hormone* in insects (4, 46, 216, 228, 331, 332, 336, 399, 430, 550, 728). Evolutionarily significant shifts in growth and development parameters are as important to lineages of these organisms as they are in vertebrate animals

(291, 351, 464, 806). In addition, as discussed in Chapter 3, many plants and invertebrates are capable of manufacturing, storing and/or utilizing thyroid hormone in one form or another (47, 57, 119, 199, 367-369, 395), which suggests that thyroid hormone may have an exceptionally broad evolutionary role in growth and development that lies well beyond the model proposed here for vertebrates.

Testing the hypothesis

Thyroid rhythm theory assumes that individually-unique thyroid rhythm variants exist within species-specific patterns for animal populations and that these thyroid rhythm variants are the actual characteristics targeted by natural selection in instances of adaptation and colonization. The model predicts that non-random subdivision of populations often occurs during speciation, isolating particular subsets of individuals with similar thyroid rhythms within founder populations. Developmental repercussion of reduced thyroid rhythm diversity in the founding group is assumed to be responsible for generating the particular growth and development changes seen in descendant taxa.

The basic hypothesis to be tested, therefore, is that daily rhythmic thyroid hormone secretion profiles in any vertebrate species are individually variable and that these variations between individuals can be correlated with discernable physical, reproductive and/or behavioural differences (thyroid rhythm phenotypes). As a group, individual thyroid rhythm phenotypes together should generate a distinctive pattern for the population that is species-specific (or in the case of domesticates, breed-specific); that is, the average pattern for the group should be distinguishable from that of a closely-related species.

Devising experiments that can reliably test this premise will undoubtedly be difficult due to the dynamic nature of the endocrine system and its inherent sensitivity to stress of any kind. Such sensitivity presents a unique challenge to the determination of normal thyroid rhythms within and between taxa, since thyroid hormone levels must be measured frequently (at least every 5–7 minutes), under controlled conditions and for many individuals.

Experimental tests have demonstrated that automated sampling may circumvent many of the difficulties of testing the thyroid rhythm hypothesis. For one study, which measured levels of the adrenal hormone *corticosterone* in rats, a surgically implanted cannula connected to an automated blood-sampling apparatus allowed minute quantities of blood (10–20 µl) to be collected every ten minutes over a twenty-four-hour period without disturbing the animals by repeated handling (454, 815, 816). The two breeds of rats used in this study had significant differences in mean profiles of hormone production (as well as slight individual variations within breeds) in addition to significant differences in behavioural responses to a controllable stress (so-called *white noise*). The success of these experiments in demonstrating the existence of fine-scale patterns of corticosterone production and in correlating these profiles to stress responses suggests that a similar method might be suitable for testing the thyroid rhythm hypothesis (similar automated sampling methods have also been applied to studies on pulsatile hormone secretion in humans, and a range of other species—including cats, rats, horses, cattle, chickens, frogs and fish; see Box 3.9 for references).

However, using automated blood sampling devices on a broad range of taxa will not circumvent all potential testing problems. For example, current laboratory assay methods for measuring thyroid hormones in minute quantities may place limitations on the smallest samples that can be analyzed—new assay methods may need to be developed (5). The sampling apparatus itself may need modification to allow testing of a full range of animals: laboratory-housed fish species and free-ranging bears, for example, pose very different logistical problems for an automated thyroid hormone sampling and storage device.

If individual variation within species-specific profiles of thyroid rhythms can be confirmed, controlled breeding experiments (similar to those described for silver foxes in Chapter 2) will be necessary to confirm that small interbreeding groups of animals with similar thyroid rhythm profiles produce descendants of a different type within twenty generations or less. It would be most convincing if descendants of such breeding programs could have their thyroid rhythm profiles monitored as well, for these should differ from the original source population.

If it can be demonstrated that thyroid rhythms are indeed variable within certain limits for different populations (or breeds, or for certain morphs within species) *and* that heterochronic changes can be generated by interbreeding small groups of physiologically similar individuals, the final step will be to find the genetic sources of those pattern differences. Although species-specific thyroid rhythms are probably controlled by genes in the SCN that are directly associated with generating hormone pulsatility, other factors may effect thyroid hormone utilization in ways that are also species-specific, such as the different concentrations of the thyroid hormone transporting molecule *transthyretin* found in chimpanzees and humans by one research group (246). Mutations in genes controlling other aspects of thyroid-mediated actions in target tissues (such as receptors, receptor ligands and/or cofactors) are also potential causes of variation.

However, it is expected that in most cases, differences in such genes will be found to supplement, compound or confound thyroid rhythm effects rather than contribute to their initial rhythmic generation (and as a consequence, genes involved in such thyroid hormone-mediated "end-factor" processes could be selected for independently from thyroid rhythms or simultaneously, suggesting that mutations in such genes may explain the origins of some evolutionary novelties that are *not* developmental in nature). I suggest that the first step in genetic characterization of individual and species-specific thyroid rhythm profiles should be that of documenting variation in SCN output.

In the meantime, ongoing research into the regulatory mechanisms of embryonic development should unravel some of the essential molecular interactions that involve thyroid hormones and thyroid rhythms. Research on *Hox* genes that act during embryonic development has thus far revealed they respond to retinoic acid as well as to other molecules (438). In light of the known developmental regulation functions that retinoic acid shares with thyroid hormone, or in which their roles cannot be distinguished (42, 168, 210, 666, 704, 714), it would be prudent for researchers to look at the response of *Hox* genes to thyroid hormones, and/or thyroid rhythm pulses, in combination with retinoic acid. Given the critical roles recently demonstrated for thyroid hormones themselves in embryonic development, we also need to know what effects different thyroid rhythm profiles might have on any given developmental program.

Lastly, research into the physiological and genetic basis of natural piebaldness (rather than aberrant whitespotting mutants) may also be illuminating. Piebaldness, if we can come to understand exactly what it signifies, could serve as an especially useful diagnostic marker for developmental change.

Recommended reading

Antle, M.C. and Silver, R. 2005. Orchestrating time: arrangements of the brain circadian clock. *TRENDS in Neuroscience* 28(3):145-151. [still a bit complicated for many readers but good diagrams]

Short, R.V. and Balaban, E. (eds.). 1994. *The Differences Between the Sexes.* Cambridge University Press, Cambridge.

Wright, K. 2002. Times of our lives. *Scientific American* (September):58-65. [circadian rhythms]

Wise, P.M. 1998. Menopause and the brain. *Scientific American* Women's Health: A Lifelong Guide (Special Issue):78-81. [discussion of hormone rhythms]

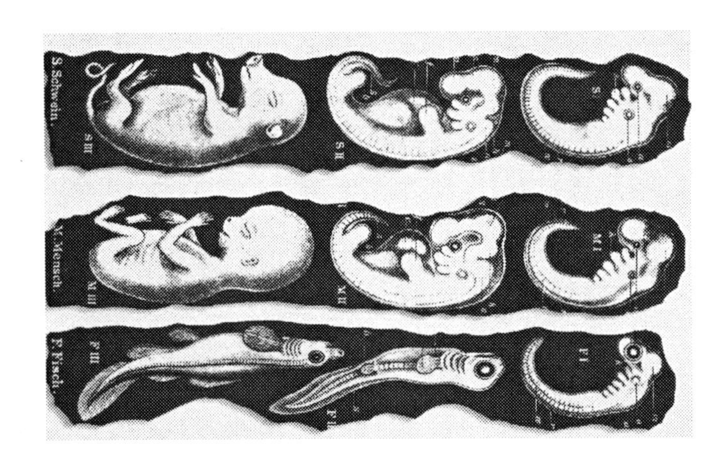

Chapter 5

Thyroid Rhythms in Protodomestication & Breed Development

Wolf to Woof

Dogs were the first and most widely distributed domesticate (see the Appendix, which summarizes major events significant to human history). We now have solid evidence from a number of studies (including genetics and behaviour) that the wolf (*Canis lupus*) is the direct ancestral species of all forms of dog (*C. familiaris*) (see ref. 262 for a discussion on the taxonomy of dogs and wolves). Despite the very close genetic relationship between them, distinctive reproductive physiology and behaviour encourage each of these species to mate only with their own kind (32, 33, 73, 267, 513, 713). When hybridization does occur, it almost always involves wolf males mating with dog females, although the reverse occasionally occurs when large breeds of dogs are involved (122, 147, 325, 350, 518, 652, 793). In addition, once wolves transformed into dogs, they couldn't go back: even after many generations in the wild, feral dogs and other domesticates retain the traits distinctive of protodomestication (325, 415).

The earliest known undisputed evidence for the dog as a distinct species is a mandible from a German grave site dated to about 14,000 BP (years before present) and reported more than twenty years ago (55, 130); there are now a few additional specimens of this age from the same region (the Czech Republic, ref. 547; western Russia, ref. 657). Several specimens from what archaeologists call the Pre-Pottery Neolithic are considered the oldest dog specimens from Israel, all of which date to ca. 11,000–12,000 BP (170, 175, 736).

During the period 10,000–7,000 BP, there are a large number of undisputed dog finds from many places around the world (534, 130), including:

- Iraq at ca. 9,250–7,750 BP
- China at ca. 9,000–7,8000 BP
- Chile at ca. 8,500–6,500 BP
- England at ca. 9,900 BP
- Germany and Denmark at ca. 9,000 BP
- French Alps at ca. 10,000 BP
- Swiss Alps at ca. 7,000 BP
- Japan at ca. 8,000 BP
- United States at ca. 10,000–8,000 BP

For a summary of dates and locations of early dog remains worldwide, see Table 5.1. For global distribution of dog remains found in two distinct time periods, see Figures 5.1 (ca. 3,500–7,000 BP) and Figure 5.2 (ca. 14,000–10,000 BP).

Early dogs show some size variation between regions, but in general appear to be similarly dingolike, unspecialized animals (130). Northern European and Chinese specimens tend to be the largest (dingo-sized), Japanese (and some North American) specimens the smallest (terrier-sized), and western Asian samples somewhere in between (145, 147). Distinct morphotypes of dogs (such as giant mastiffs, gracile sighthounds and toy-sized lap dogs) are not apparent until much later, in general not emerging until about 3,000–4,000 years ago in most areas.

Table 5.1 Summary of exact dates and locations of early dog remains worldwide

Continent	Country	Locale	Material	Dates (uncorrected)
Europe	Germany	Kniegrotte Cave	skull frag.	12,280 ± 90 BP, 13,585 ± 165 BP, on strata (ref. 547)
	Germany	Oberkassel	mandible	ca. 14,000 BP, on strata (ref. 130)
	Germany	Bedburg-Köningshoven	skull (adult)	ca. 10,000 BP, on strata (ref. 130)
	England	Star Carr Seamer Carr	skull frag. (juvenile) verts. (juv.)	9,559 ± 210, on strata, 9,490 ± 350, on strata, 9,940 ± 110 BP, on bone (ref. 130)
	France	St. Thibaud Rockshelter (Alps)	skull + femur, vert.	10,050 ± 100 BP, on bone (ref. 113)
Western Asia	Israel	Ein Mallaha	full skeleton (burial; juv.)	ca. 12,000 BP, on strata (ref. 171)
	Israel	Hayonim	full skeletons (2; burials)	ca. 12,500 BP, on strata (ref. 736)
	Russia (West, near Kiev)	Eliseevichi I (53° N, 33°E)	skulls (adult) (2)	13,900 ± 55 BP, on bone (refs.152, 657)
	Russia (West, Altai region)	Razboinichiya Cave (51° N, 85°E)	skull	14,850 ± 70 BP, on strata (ref. 152)
Eastern Asia	China (North)	Jiahu (Wuyang, Henan)	full skeletons (11; burials)	ca. 9,000–7,800 BP, on strata (ref. 152)
	China (North) (preliminary)	Nanzhunangtou, (Xushui, Hebei)	mandible	ca. 10,000 BP, on strata (ref. 152)
	Japan	Natsushima Shell Mound (Kanagawa)	not specified	ca. 9,300 BP, on strata (ref. 152)
	Japan	Kamikuroiwa Rockshelter (Ehima)	full skeleton (burial)	ca. 8,500–8,000 BP, on strata (ref. 152)
	Russia (East, on Kamchatka)	Ushki-I	full skeleton (burial)	10,360 ± 45, 10,860 ± 40 BP, on strata (ref. 152)
North America	USA (Utah)	Danger Cave	mandibles & skull frags	ca. 9,000–10,000 BP, on strata (ref. 130)
	USA (Illinois)	Koster	full skeletons (3; burials)	8,130 ± 90, 8,480 ± 110 BP, on strata (ref. 534)

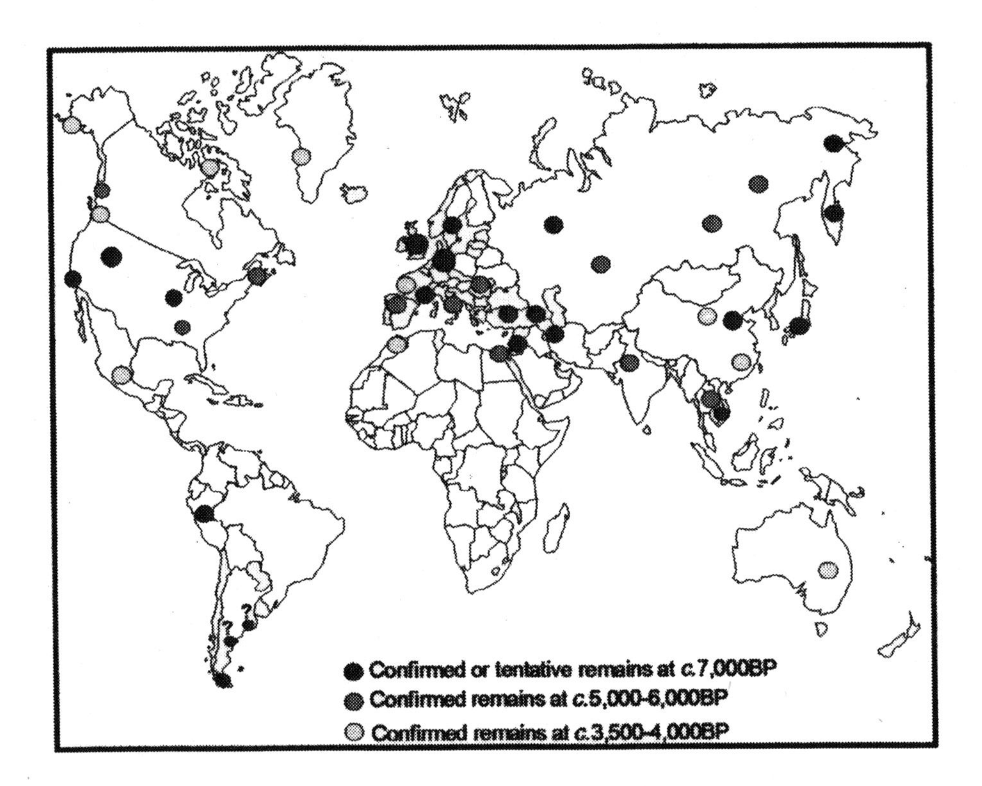

Figure 5.1 Global distribution of dog remains between ca. 3,500 and 7,000 years ago

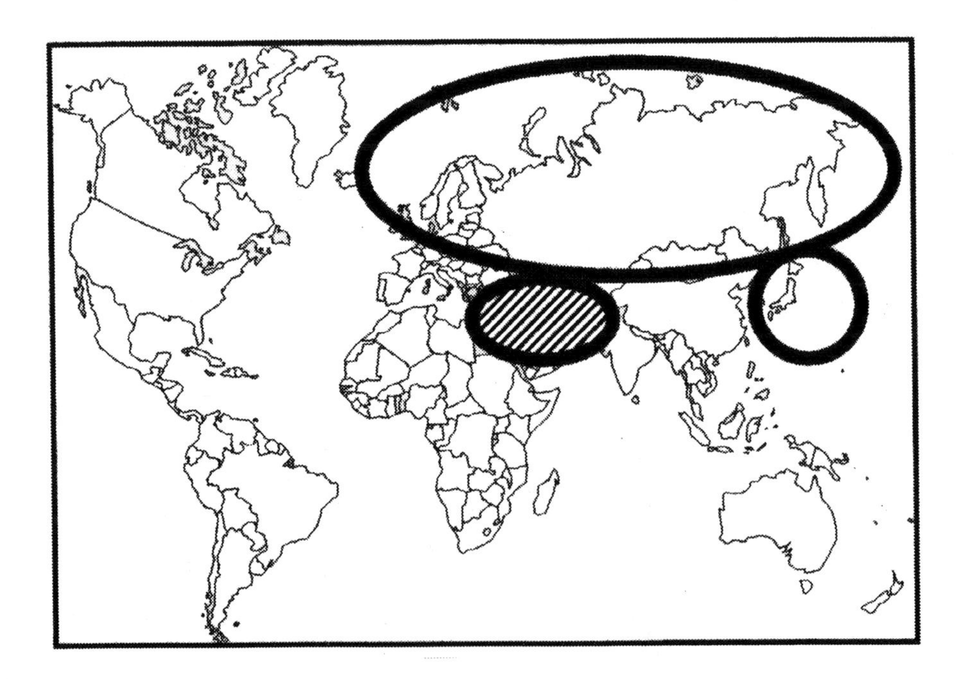

Figure 5.2 Distribution of early dogs (ca. 14,000–10,000 years ago) by relative size, with large dogs in northern Europe and Asia (large oval), medium dogs in the Middle East (small, filled oval) and small dogs in Japan (small circle)

Dogs have been shown to be juvenilized (*paedomorphic*) in relation to wolves, based on several studies of behaviour, growth and skeletal anatomy (136, 137, 531-533); in other words, the most obvious differences between the wolf and dog are the result of the dog maturing at a stage of development equivalent to a juvenile wolf in physical form, size and behaviour. Robert K. Wayne showed, in a pivotal study conducted more than twenty years ago (see also Chapter 2), that these differences have been caused by an inherited reduction in growth rate implemented during late fetal development and early postnatal growth (790-792).

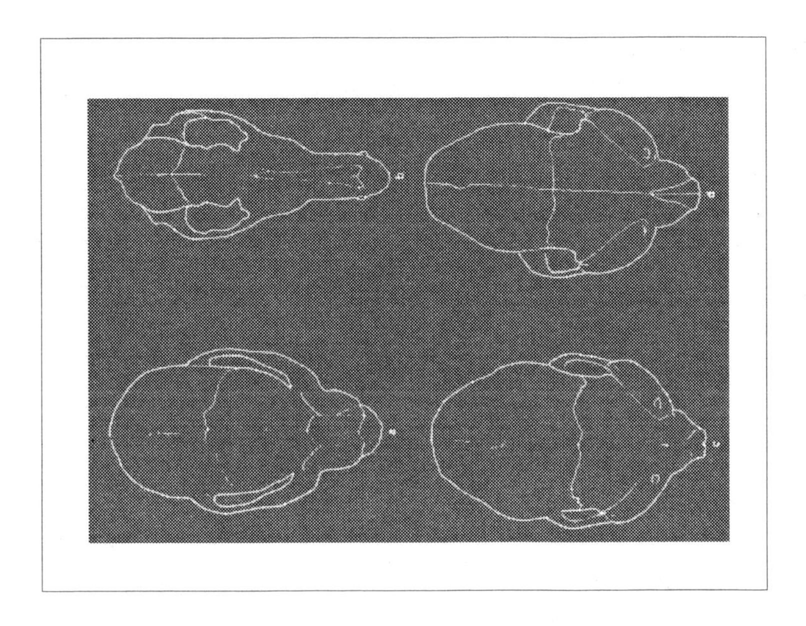

Figure 5.3 Differences in form between adult (top row) and newborn, in the wolf (left) and wildcat (right); same as Figure 2.2

Wayne undertook a series of comparative studies using several wild canid species and a few dog breeds and showed, for example, that small toy breeds (such as the Lhasa apso) are smaller because they not only are born somewhat smaller than average but grow very slowly immediately after birth. In addition, they mature physiologically and skeletally before one year of age. In contrast, giant breeds (such as the Great Dane) are born slightly larger than average and, in addition, grow exceedingly rapidly after birth. The growth of great Danes slows down after the early postnatal period but continues well into the second year of life—as a result, Great Danes aren't fully mature until about two years of age. Other breeds often fall in between these two extremes of modified growth, as the first early dogs probably did. The gestation length for all dog breeds (large and small) is the same (sixty to sixty-three days), suggesting that *inherited differences in growth rates account for size and shape distinctions between various physical forms or varieties.*

Figure 5.4 Top, head shape of a one-week-old malamute puppy (a wolflike breed). Bottom, Great Dane and chihuahua mothers with their pups (from an Iams© dog food ad).

In contrast, small wild canids such as foxes (*Vulpes/Alopex species)*, which have post-natal growth rates similar to much larger wild canids like coyotes and wolves, appear to be smaller because their gestation period is about ten days shorter. A similar pattern of shorter gestation in small representatives of other closely-related animal groups has been reported by zoologist/palaeontologist Bjorn Kurtén (21). In other words, the dramatic shape change during maturation that occurs in the wolf is what has allowed artificial selection to fix so many skull shapes in modern dog breeds.

A similar situation also exists for the pig, where the dramatic shape change in the skull during ancestral ontogeny (from rounded to elongate) is reflected in a remarkable number of different physical types in modern pigs (Figure 5.5).

Figure 5.5 Pig breed variation in form and head shape (from Alderson 1979)

In contrast, animals such as the horse and cat (where there is only a small amount of skull shape difference between neonate and adult) would not be expected to show as much skull shape variation between modern breeds. As Wayne emphasizes, neither horses (Figure 5.6) nor cats exhibit anywhere near the amount of morphological variation as dogs and pigs. Archaeological material of early domestic horses and cats, therefore, might be expected to differ less in size and shape relative to their wild ancestors than do other animals (given that the same juvenilization process is at work), making the recognition of early domestic forms especially difficult in these animals—and indeed this seems to be the case (e.g., 127).

Figure 5.6 Head shapes in horses vary little, from the ancient-type Pzrewalski's horse to the modern pony (from Hemmer 1990).

Protodomestication of the dog

I suggest that dogs came to be when certain individual wolves decided to colonize human-occupied territory—those first early settlements that were established when humans finally stayed in one place year round. Only those wolves who were able to tolerate the physiological stress of having people close by would have even tried to move into these new human-dominated (*anthropogenic*) habitats—the rest would just have stayed in the woods.

As I described in the previous chapter, since patterns of thyroid hormone control the behavioural response of animals to this kind of stress (in addition to controlling all of those other characteristics, including growth), when only a few hormonally similar animals become isolated in a new habitat and breed together, there is very little hormonal variation present. Within only a few generations, a thyroid rhythm pattern becomes established that is different from the ancestral population (see Figure 6.2 in Chapter 6 for a useful colour analogy).

Without the full range of variability of thyroid hormone production, descendants of this invading population would almost certainly have been behaviourally, reproductively and morphologically different from the ancestral wolf population after relatively few generations, due to the extensive and pervasive influences of thyroid hormone. I suggest that this process could have produced a descendant population of early dogs with a well-coordinated suite of distinctive traits within about forty years, roughly the same short period of time as Belyaev's foxes, but adjusted for the longer generation time for wolves. Even if it took a bit longer (say eighty years), this extremely rapid change from one form to the other would be too fast to leave archaeological evidence for any intermediate stages.

Protodomestication would have been complete once a morphologically-distinct animal, the first dog, emerged with its own unique thyroid hormone production profile. Protodomestication must be irreversible after a new thyroid rhythm pattern is established because some essential variation of the ancestor has been lost. Occasional hybridization with wild wolves would have little effect because hybrid offspring with limited stress tolerance born in any generation would tend to flee a human-dominated environment, removing them from the breeding pool of early dogs.

Once wolves were biologically "domesticated," their distinctive traits (less fearful behaviour, increased tolerance to stress, smaller size, changes in reproductive seasonality, perhaps unique coloration) would have made them uniquely amenable to subsequent human subjugation and selection—whether that selection was deliberate or unconscious. In other words, once the speciation process of *protodomestication* was complete, *classic domestication* processes could begin.

Even the first early dogs, as described above, must have possessed behavioural and reproductive traits that encouraged them to mate only with each other (739) and thus qualify as a "true" species, distinct from their wolf ancestors from that point onward. This process constitutes speciation in every sense of the word. Wolves and dogs are a natural species pair, and there is simply no rationale for them to be considered "synonymous" entities, as some taxonomists have decreed (listing the dog as a subspecies of the wolf or *Canis lupus familiaris*), just because they are *capable* of interbreeding. I insist, as do many of my colleagues, that dogs deserve a distinct scientific name just as much as other descendants do that evolve via speciation from a closely related ancestor (262).

If protodomestication was a natural speciation process initiated by the animals themselves, what prevented subsequent interbreeding with the source population? In other words, how do so-called "isolating mechanisms" develop in early domesticates if people didn't deliberately isolate potentially interbreeding individuals from each other? I suggest that two kinds of changes contributed to the cohesion of the domesticated group: a significant shift in the timing of reproduction, similar to the reproductive timing shifts experienced by Belyaev's stress tolerant foxes (discussed in Chapter 2) plus changes in behaviour. Modern domestication experiments (50, 51, 325, 409, 754, 755, 758) provide concrete evidence that after several generations of isolation and selection for stress tolerant behaviour, really dramatic declines in fearfulness would have occurred concurrently with changes in reproductive timing.

The potential for such marked changes in behaviour during the early stages of protodomestication suggests that docile early dogs (and especially their young) might actually have sought out human contact. Coat colour differences that arose as a consequence of the process, such as whitespotting (piebald) and solid colours (*non-agouti*), would have clearly differentiated early dogs and other domesticates from their wild ancestors after relatively few generations of separation. Given these physical, physiological and behavioural changes, the shift to an active relationship between early domesticates and people seems almost inevitable.

On the rare occasions when breeding times overlapped, more docile behaviour would almost certainly have discouraged domestic males from leaving during the breeding season to mate with wild females, since wild female wolves are generally unwilling to accept domestic males as mates (in fact, they most often kill them). This has also been found to be true for other wild animals that have closely related domestic forms, such as bison and cattle, reindeer and caribou, wolves and coyotes (151, 241, 325, 442, 518, 597). (If and when hybridization does occur, the smaller or more docile (or domestic) species of the pair will be the female; therefore, to produce a bison/cattle hybrid you must use a domestic cow and a wild bison male—in part, because a domestic bull will be too intimidated by the aggressive behaviour of a wild bison female to attempt to mate but also, a wild female will not stand still long enough to mate with a domestic bull because he isn't sending her appropriate signals of dominance.)

Wild males, however, may still have approached early domestic females to breed, so that some movement of wild genes (*introgression*) may well have occurred in all animals on a continuous, if irregular, basis—although, as stated above, offspring of such matings are apt to leave the "domestic" environment and rejoin the wild group. As a result, protodomestication appears to be essentially irreversible after the initial physical changes have occurred, even with occasional hybridization.

Once protodomestication has occurred, the new species (with its more docile behaviour and longer juvenile imprinting period due to slower growth after birth) is one that could have been subjugated and managed by people with relative ease. This is the point at which all of the cultural influences traditionally described as "domestication" began to uniquely shape the history of early domesticates. Herbivores may have been corralled in order to keep them out of gardens and grain fields rather than to prevent them from "escaping," at least in the early stages of cultural control—it's easier to corral small groups of animals than to fence large grain fields (in contrast, dogs don't appear to have been confined on a regular basis until quite recently). A more intense symbiotic relationship would gradually become established as people took varying amounts of control over the living conditions of the animals, for example, supplying food, shelter and/or water.

Breed development in dogs

Slight individual variations in thyroid hormone rhythms must have continued to exist within early domestic populations (and probably increased over time as natural mutations in the genes controlling these rhythms accumulated) and this provided the raw material for both natural and human selection to implement further change. Some distinctive physical types (*morphotypes*) eventually developed as a result of continued selection for traits linked to differential growth rates and ages at maturity. Archaeological evidence suggests that large dogs were the first type to be preferentially selected, about 4,000 years ago, probably because large animals made the best guardians of property. Tethering guard dogs to keep them on the job would have seriously curtailed free-range mating, and as a result, encouraged owners to choose appropriate mates. Even without an understanding of genetics, the benefits of such controlled mating must have been readily apparent; by the Roman period, about 2,000 years later, many varieties of dogs existed. The intensely juvenile physical and behavioural features of some breeds (such as the chihuahua) reflect the continued impact that differences in growth rates and/or timing of sexual maturity has had on dogs subjected to artificial selection (136, 138).

The proposition that some breed variation is controlled through thyroid hormone metabolism is supported by recent evidence that some dog morphotypes (such as tall, long-snouted sighthounds) have distinctly lower thyroid hormone levels than the average for dogs in general (220, 221, 672). However, spontaneous mutations (such as dwarfism) in other genes have obviously occurred and have been perpetuated through controlled breeding (14, 745, 810).

Some traits, however, cannot be controlled by selection in this fashion, and a prime example is the extra digit (*dew claw*) in giant dog breeds. As Per Alberch (12) so eloquently explains, development of the extra digit is size-dependent. Continued selection against this trait by breeders of St. Bernard dogs has not eliminated the extra digit because breeders also select for large size. Alberch also suggests that small dogs, such as poodles, do not develop extra digits because they do not reach the critical size required (but see Box 5.1 for the classic genocentric explanation).

Some dog breeds with extremely "juvenilized"-wolf physical characteristics also possess more distinctly juvenile wolflike behaviour, with other breeds falling in between the two extremes (274). An explanation for such variation in breed behaviour is offered by Ray Coppinger and Richard Schneider (138), who emphasize that juvenile wolves are animals in transition: they are undergoing a metamorphosis of sorts from neonate (with one set of unique behaviour and motor patterns) to adult (with a new set of behaviour and motor patterns). Neonatal patterns gradually disappear and are replaced by adult patterns during this transition period, so that juveniles retain some motor patterns and behavioural attributes of the neonate while they add adult attributes (e.g., juveniles at a certain stage can both suck and chew). Juveniles are thus endowed with an especially wide range of motor skills and behaviour.

Coppinger and Schneider suggest that inherited breed-specific growth patterns that leave an adult dog maturing at an early juvenile stage would produce a different repertoire of behaviour and motor patterns than one that leaves the adult maturing at a later juvenile stage. They use this model to explain the differences in behaviour and motor patterns evident in the plethora of breeds that have emerged within the past one hundred years, and in crossbreed (i.e., hybrid) offspring of different dog breeds. They suggest that hybridization between breeds that possess unique inherited programmed growth (associated with both morphological and behavioural consequences) can create truly new patterns of inherited behaviour in the offspring, which in breed creation programs can be set by subsequent inbreeding and selection. My theory proposes that that such breed-specific growth patterns are controlled by breed-specific thyroid rhythms.

Most dog breeders now realize that selection for a single trait, like head shape, for example, results in other physical, physiological and behavioural changes whether the breeder likes it or not (745). Selection for both single and multiple traits that are thyroid-linked, and subsequent crossing of established breeds with different thyroid-linked traits, is probably responsible for a wide range of distinctive breed types. Traits controlled by chance mutations in genes not directly linked to thyroid rhythms further increase the variation available for artificial selection. The amalgamation of traits that are not thyroid-mediated with thyroid-mediated types could account for the existence of the four hundred or so distinct breeds of dogs known to have existed at one time or another (810).

It's clear that only a few critical genes must control breed characteristics—otherwise, we would not have been able to establish so many physically and behaviourally distinct types of dogs that breed true after as little as twenty years. I suggest that those "few critical genes" are in fact the clock and rhythm genes that control pulsatile thyroid hormone secretion.

As mentioned, by selecting a certain physical characteristic (such as head shape), we also select all of the other traits that are linked to it via its associated thyroid rhythm: behaviour, coat colour, size, etc. Mating together *only* animals with those characteristics isolates very similar thyroid rhythms into one breeding population, severely restricting the possible variation. As happens with the creation of a new species, this line should breed true as long as we keep it reproductively isolated.

In addition, even more combinations of thyroid rhythms can be created by mixing breeds of different types. When mixing breeds, their thyroid rhythms may blend cohesively together in the offspring, creating a new physical and behavioural type that is different from the two parental types but recognizably a *harmonious* combination. Alternatively, they may combine in an additive fashion, creating a very different type altogether. Other factors come into play, of course, like chance mutations at specific genes outside the thyroid hormonal system, but I believe this will prove to be the essential underlying mechanism of breed production and improvement.

Box 5.1 Genetic control over morphology in dogs?

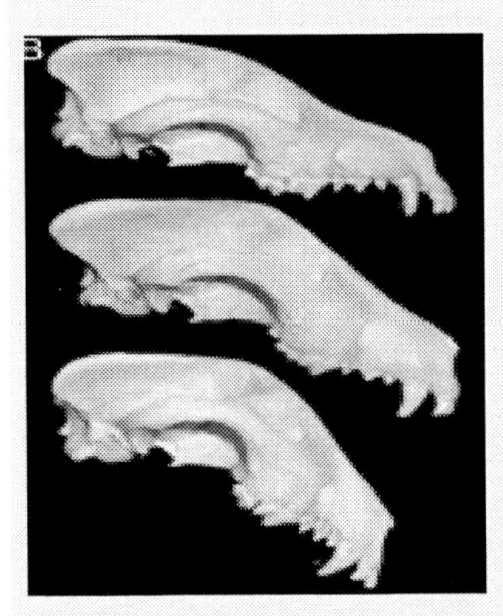

Left: Changes in shape over time in bull terriers (top to bottom, 1931, 1950, 1976) that seem to be correlated with a particular *Runx-2* tandem repeat sequence. Photo by Marc Nussbaumer, Bern Museum of Natural History. See also Pennisi, 2004 (*Science* 306:2172).

A recent study comparing genetic and physical variation in modern dog breeds by biologists John Fondon III and Harold Garner of the University of Texas Southwestern Medical Center (Dallas) addressed two issues discussed above: the extra digit in giant breeds and snout length (233). They looked particularly at slight differences in the lengths of certain nuclear genes—repetitive sequences of DNA bases that exist in multiple copies within a certain genetic region (called tandem repeats)—concentrating on genes known to be involved in fetal development. They found that patterns of tandem repeats in the Runx-2 gene correlated with differences in snout length and also with the relative amount of snout "deflection" in the bull terrier, while patterns of tandem repeats in the Alx-4 gene correlated well with an extra digit in Great Pyrenees dogs (of six dogs tested, four without a tandem repeat deletion had an extra digit, and a single animal of unknown size without the deletion had only five toes). While none of the other breeds tested had an extra digit or the deletion, none were breeds where extra digits are common, such as St. Bernard dogs.

This evidence may seem compelling, but remember that correlation does not mean causation! We really don't know yet how these tandem repeat sequences actually affect gene function, and in fact, these regions have long been considered "junk DNA" because they appear not to have any effect on normal gene function. I can think of at least two reasons that would explain why tandem repeats or deletions might be correlated with certain breed characteristics, even if they are effectively neutral. One is that the dogs in question had all been exposed to intense selective breeding (only purebred "inbred" lineages were tested), and the other is that the repeats are linked or otherwise associated with particular thyroid rhythms for reasons we don't yet understand (activated by specific levels of thyroid hormone at a particular time in developmemnt). If this last turns out to be true, it may mean that even if such tandem repeats have no effect by themselves, they may be useful "markers" for the thyroid hormone rhythm patterns that are actually generating the morphological traits we see. Only time will tell.

The first sheep, goats and cattle

While pigs probably originated in a very similar manner to that of dogs, because both are scavengers (363), protodomestication of sheep, goats and cattle would have proceeded somewhat differently. These herbivores likely adapted to human-dominated habitats through a process that mirrors the way some indigenous people handle their modern counterparts. Researcher Frederick Simoons (691) has described in some detail how modern domestic mithan stocks are handled in Bangladesh and Burma, which appears to perpetuate initial protodomestication of the *gaur* (a type of wild Asian cattle). Mithan, like their wild gaur ancestors, are attracted to salt. People today manage their mithan stocks in the most minimal fashion: they simply encourage the animals to return to the centre of habitation areas by providing salt. No attempts are made to handle or tame mithan, or to confine them—the regular proximity of the animals adequately facilitates occasional culling of individuals for ritual use (they are eaten only on rare ceremonial occasions). The animals feed on available wild forage in the forest and breed entirely at will, and yet mithan are classic domesticates: they are smaller than wild gaur, relatively docile in behaviour and often piebald.

While it's possible that previously domesticated mithan stocks are merely maintained by this system, there are compelling reasons to believe that protodomestication of wild guar occurred under these (or similar) conditions. This method provides both of the components necessary for precipitating protodomestication changes as described here: an attractive resource not readily available in the wild (salt) and the constant proximity of humans to provide the selection pressure. For other bovids such as cattle, goats and sheep, the attractant could have been either the newly domesticated grain crops under cultivation during what is called the Pre-Pottery Neolithic period (617, 733, 734; also see Appendix), deliberately supplied salt or the salt in human urine that would have accumulated naturally around settlements (human urine is known to strongly attract reindeer and seems to encourage domestic herds to associate closely with nomadic Sami tribes (846). Cultivated crops are especially likely as an attractant if stands of wild grain in the area had been depleted by previous centuries of human collection, as appears to be the case (307, 633); if so, accumulations of human urine may have been an added bonus.

Modern domesticates—farmed and hatchery salmonids

Now we move from considering the very first known domesticates to some of the latest: farmed salmon and trout. Modern "domesticated" rainbow trout (*Oncorhynchus mykiss*), now available in fish markets worldwide, descend from the wild eggs of coastal steelhead collected from a number of sources in western US states in the late 1800's (47). This founding population is known to have experienced an initial high mortality rate (perhaps as high as, or higher than, 80%) in response to the crowded conditions experienced during their initial captivity (Gary Thorgaard, p.c., Nov. 1999). Thus, all domestic trout are descendants of the few animals that were capable of surviving the stressful conditions of initial captivity. Such selective mortality actually works the same way as stress tolerant animals leaving their ancestral population to colonize a new human-dominated habitat, as I proposed for protodomestication—it's just that in this case stress intolerant animals simply die.

Marked differences in behaviour exist between modern domestic trout and wild rainbow trout (47), a phenomenon also seen in other captive-reared salmonids (290, 519, 544, 780). For example, as Zoology professor Gary Thorgaard of Washington State University (WSU) showed me several years ago at their trout raising facility, a tank in which twenty or more domestic trout thrive can hold only two wild-caught fish for any length of time. This suggests that most wild fish need considerable inter-individual space, often more than is economical for raising captive trout into adulthood.

In addition, while wild-caught trout kept at the WSU facility retreated to the bottom of the holding tank when the lid was raised (attempting to hide in the shadows), domestic trout were so vigorous in their rush to the surface that some leaped clear of the water. This behavioural difference suggests that most wild trout have a natural fear of activity at the surface of the water that compels them to dive and seek shelter, a predator-avoidance response that may be associated with, or linked to, the stress response to crowding. Therefore, the high mortality of individuals in the original captive population that were sensitive to the stress of crowding may have simultaneously eliminated those individuals with a high predator-avoidance response, which means that the facilities used for raising wild-caught trout eggs in captivity may have unintentionally selected for those few animals that lacked extreme stress responses of all kinds. As a consequence, the descendants of this population are now amenable to being fed and otherwise tended from the surface of holding tanks without stress and can be held in concentrations high enough to make trout farming economically viable (290).

Recent study has also shown that the brain morphology of farmed rainbow trout differs markedly from that of wild fish, suggesting that protodomestication has affected brain development after little more than one hundred years (492). There is evidence that changes may appear even sooner, however: after less than twenty-five years in captivity, farmed Atlantic salmon (*Salmo salar*) show faster growth rates and increased levels of growth hormone than wild fish (230, 231, 472). Faster changes still are seen in hatchery-raised Chinook salmon (*Oncorhynchus tshawytscha*), which, after only a few generations, lay more eggs that are smaller in size than wild salmon (320, 232). Smaller eggs are more vulnerable to predation than larger ones and some hatchery fish with these traits, when crossed with wild fish, lay eggs that are intermediate in size, indicating that these changed traits are inherited in some fashion.

Perhaps most surprising and significant is that similar biological effects have also been documented between wild Chinook salmon and hatchery-raised fish within a *single* generation. Hatchery Chinook are raised to the fry stage from eggs taken from artificially spawned wild adults, and then released back into their natal river as a method of supplementing wild stocks. A report on levels of predation by "plunge-diving" Caspian terns (*Hydroprogen caspia,* which forage at or near the surface of the water) on juvenile salmonids in the Columbia River estuary, Oregon (US), concluded that hatchery-reared smolts of Chinook salmon are more vulnerable to tern predation because they travel in the upper water layers of the river, while wild smolts are more evenly distributed in the water column (132). Hatchery-raised fish of many salmonid species have been shown to be less responsive to all forms of predation than wild fish (290, 373), and this trait has been shown to be passed along to offspring of crosses with wild fish (763). Clearly, simply raising salmon eggs to the fry stage in captivity, with no intention of changing the population, can initiate biological changes that are detrimental to survival in the wild. This suggests that significant and irreversible protodomestication changes can indeed be manifested within a single generation.

Protodomestication vs. speciation

Protodomestication as a biological process defines both early species and early domesticates as equivalent population subsets of wild source populations that colonize new habitats. However, there are differences in the nature of the new environments they invade. For early domesticates, these environments are physically altered by people and dominated by their presence; for early species, the new environments they colonize may be either a previously unoccupied adjacent niche or a niche newly created by climatic change.

In both cases, the new habitat offers resources (such as food and breeding sites) that are either unavailable or scarce in the original or source territory, making it highly attractive. In both cases, those individuals within the source population that have the highest tolerance to stress (or, which possess a *particular* thyroid hormone rhythm) are the most likely to colonize a new territory.

The division of the source population during colonization of a new habitat (via what biologists call *allopatric* and *peripatric* speciation), according to my model, is distinctly *non-random.* Non-random subdivision of the source population according to existing variation in thyroid rhythms (encompassing the underlying clock and rhythm genes that control them) makes this model distinct from speciation events described by purely genetic models (323, 507, 511, 571, 607, 670). Geneticists, because they use mathematical models to predict the effects of selection over time in a population, assume that mating is random among individuals (this is stated explicitly). They also appear to assume a process of random division of populations during a split, which I conclude partly because they do not explicitly state otherwise and partly because they do not give a basis for suggesting that certain *alleles* (genetic variants) might preferentially end up in one population or another. This distinction between my model and theirs (non-random vs. random population division) is significant and stems largely from the fact that my model is *not* mathematical.

In my model, interbreeding within a small, isolated population of only stress tolerant individuals rapidly establishes a new mean pattern and range of variation for thyroid hormone rhythms in the colonizing group. This shift precipitates significant changes in descendants because of the intimate connection that exists between the adrenal stress-response system, fetal growth and thyroid hormone. Descendants of founders are recognizably different from their ancestors within relatively few generations because the reduced variation of thyroid rhythms within small populations has morphological, behavioural and reproductive repercussions. Viewed from this perspective, it is apparent that in both speciation and protodomestication, the initial populations of ancestors are equivalent subsets of wild source populations and the selection mechanism is the same. Therefore, descendants of both processes are equivalent entities that must qualify as equally *real* species.

It is also apparent that the point at which gradual adaptation to the new habitat occurs is the first point of departure between the two processes, although the magnitude of difference is small. While some might still argue that deliberate human selection played a significant role in the changes that affected early domesticates during this stage, the prevailing view now concedes that natural selection was probably more significant (128). If human selection did occur, it was probably *unconscious* (165). Unconscious selection refers to, among other examples, choosing to slaughter individuals with intractable temperaments (thus removing them from the breeding pool), or sparing from slaughter individuals that are good mothers or produce large litters.

In other words, adaptation of early domesticates to the conditions within their anthropogenic environment proceeded primarily via the same mechanisms of natural selection that adapt early wild species to their new environments. However, anthropogenic environments do possess some inherent properties that may not be present in the wild, including the constant psychological stress of close association with humans, the immunological stress of very close association with conspecifics, the inbreeding stress imposed by limited choices of mates, and the nutritional stress imposed by severely limited food resources, especially for confined populations. Therefore, the selective forces operating on early domesticates will be different from those acting on populations of wild species, even if the actual process of adaptation is the same.

Ultimately, geographic radiation of established wild species will occur as their population size increases; the subsequent adaptation of geographically distant subpopulations to their local environments leads to the development of geographically distinct *subspecies*. For established domestic species, whose geographic radiation is limited by associated human radiation and migration, population increases and adaptation must take place within the limits of each anthropogenic environment, leading to the development of what are usually called *local regional varieties*.

It is not until humans assume absolute control over domestic species that the greatest difference in the history of the two types of descendant species occurs. Through confinement and deliberate selective breeding, we have developed lines of animals with distinct physical and behavioural characteristics that bred true generation after generation. We have been very successful, I would say, in first discerning, and then artificially isolating, subsets of individuals that share behavioural, morphological and reproductive traits that are controlled by similar thyroid rhythms.

Artificial isolation of particular thyroid rhythms allows us to initiate and perpetuate juvenilization processes decoupled from nature—whether or not people realize this is what they are doing. This kind of selection, along with fixation of rare mutant alleles through inbreeding and hybridization of existing breeds, has resulted in a truly astonishing number of domestic variants in some animals, including more than four hundred recognized breeds of dogs and almost as many pig breeds.

However, the diversity of breeds resulting from human manipulation stay distinct only if artificial isolation is maintained. In contrast, the diversity of wild forms shaped by natural selection are true species that stay distinct because they are uniquely adapted to specific habitats and have mechanisms that encourage them to mate only with each other. I'll go over some particular examples of this in the next few chapters.

This model emphasizes that the changes in growth rates and timing of sexual maturation associated with protodomestication are significant consequences of the evolutionary process involved. Animals that do not manifest the distinctive biological changes associated with protodomestication, such as Asian elephants (561, 617), the bottle-nosed dolphin, *Tursiops truncatus* (656), and reindeer (694), have not undergone this process as described and are therefore more appropriately referred to as *managed species*; they are not true domesticates.

I thought initially that managed species would also include those domesticates for which historical records indicate humans deliberately removed randomly selected animals from the wild and subsequently maintained populations in captivity, such as the Syrian or golden hamster (146, 148). However, I've since come across details on founding conditions for the domesticated golden hamster (640) that suggest initial mortalities suffered in the first generation (an adult female and eight young reduced to one surviving male and two females, in 1930) could actually have been a "selected mortality" event, such as occurred with the first generation of captive-raised rainbow trout, described in the previous section. This would make the process entirely analogous to the process of protodomestication in other animals. In selected mortality cases, rather than stress tolerant colonizers leaving intolerant cohorts behind, the stress intolerant portion of the population leaves via death, so that stress tolerant colonizers alone remain in the new habitat. There is still a non-random subdivision of the source population according to hormonal makeup that leaves only stress tolerant individuals as founders, but the factor of choice present in the colonization case does not exist for populations undergoing selective mortality.

The somewhat ameliorated behaviour and many coat colour variants of domesticated hamsters, as compared to wild conspecifics, would tend to support this interpretation (419); descendant populations have indeed changed in some physical and behavioural respects as a result of being raised in captivity, and so probably qualify as valid examples of protodomestication.

In some domesticates, however, significant biological differences from wild forms cannot be substantiated. Obviously, when the ancestral animal has not been clearly established, as for the horse (74), it's not possible to compare biological features. These animals require further investigation to establish their taxonomic status. In other animals, however, morphological, physiological and behavioural differences between ancestors and descendants are present but just not so definitive, as occurs for example in all camelid species, in the turkey, and in the cat (143, 238, 321, 501, 639, 698, 778). These animals are simply less convincing examples of the protodomestication process under discussion.

The genetics of domestication

In contrast to morphological, ecological and behavioural differences, the genetic distinctions between wild and domestic animals are much less clear. Modern domesticates are generally very similar genetically to living species of their wild ancestors. It must be emphasized, however, that living species of some animals (such as modern wolves) are not the ancestral species of the domestic form (530), but simply their closest living relatives (just as the chimpanzee is our closest living relative). Similarly, modern domestic breeds cannot be studied as genetic equivalents of early domestic ancestors, especially for those animals under consideration here that have been subjected to artificial selection, crossbreeding and human-mediated migration for thousands of years.

Nevertheless, there have now been comparative studies examining genetic diversity and ancestral relationships (see Table 5.2 near the end of this section for summary) for dogs, cattle, water buffalo, sheep, horse, goats and pigs, as well as for the South American camelids (90, 268, 334, 359, 375, 410, 435, 467, 485). Comprehensive analyses have been undertaken so far for dogs, cattle, pigs and horses: these involve utilizing tissue samples from ancient (extinct) as well as living individuals. All of these studies look at patterns of variation within *mitochondrial DNA*, the genetic material that's found within each mitochondria of an animal's cells (see also Figure 6.3 in Chapter 6). Mitochondria produce the energy needed for each cell to live and divide, and there may be thousands of them in any single cell. The genes in mitochondria govern the machinery of this little "powerhouse" organelle only; they do not appear to affect any physical, physiological or behavioural traits that might be subject to natural or artificial selection (only genes in the nucleus of the cell have the power to control selectable traits). Within the entire mitochondrial genome (which is only a fraction the size of the nuclear genome) there are parts that don't appear to have responsibility for manufacturing any particular enzyme or protein—these are known as "non-coding" regions and seem to be particularly useful for unravelling relationships between closely related animals. The idea is that mutations accumulate randomly within these non-coding regions as time passes; because they don't get repaired, the longer two species or populations have been reproductively isolated from each other, the more distinct their mtDNA will be. Large genetic differences don't really mean anything except that the populations in question haven't been interbreeding for a long period of time.

The other unique attribute of mtDNA is that it's passed along unchanged by a mother to all of her offspring, male and female. This means that if a domestic dog female happens to mate with a wild wolf male, all of her pups will have *dog* mtDNA; and if one of those female pups happens to prefer living a wild existence and mates with a wild wolf male herself, all of *her* pups will also have dog mtDNA. You can see then the danger of using mtDNA: it masks what's going on from the male side of the equation and this may have important implications. However, the general utility of mtDNA analysis supersedes this drawback and for most researchers who want to examine the relationships of closely-related species, it's still the best tool at their disposal.

Of the comprehensive studies that used mtDNA samples from ancient animals as well as from living ones, all generated results that support the suggestion that more than one domestication event must have occurred. For example, the traditional explanation for the domestication of cattle—as a single domestication event followed by human-mediated dispersal, with humped varieties derived later as local adaptations to arid conditions (207)—was not supported by the genetic evidence (462). Evidence from mitochondrial DNA suggests that at least two distinct domestication events occurred, one in Europe that produced cattle varieties without humps (presumably from the aurochs subspecies *Bos primigenius primigenius*), and another in India that produced humped cattle varieties (from *B. p. namadicus*). The study produced no clustering of mtDNA haplotypes according to breed in either type of cattle.

This evidence supports the taxonomic distinction that has long been made between zebu-type (humped) cattle as *Bos indicus* vs. taurine-type (those without humps) cattle as *Bos taurus*, except for African zebu-type breeds. African zebu-type breeds were found in the genetic study to have taurine-type mtDNA haplotypes, a pattern usually explained by proposing European origins for African cattle populations that were subsequently modified by hybridization of taurine females with Asian zebu males. Males would have passed along the zebu physical and physiological type (the hump and heat tolerance), presumably through their nuclear genes, while offspring would have retained the taurine mitochondrial DNA of their mother. The large sequence divergence between the two *Bos* lineages was interpreted by the researchers as evidence for at least two domestication events for cattle, from two subspecies of aurochs which had been geographically and genetically distinct for thousands of years prior to protodomestication events. Subsequent work (90, 98, 485), which included sequences from extinct individuals as well as nuclear DNA from males, further strengthens the case for a minimum of three events involving genetically distinct subspecies of aurochs in Europe and western Asia.

Horse mtDNA analysis is a bit more complicated, and requires more study. At present, the results are still consistent with archaeological evidence of a single protodomestication event followed by human-mediated dispersal of the animals. However, the results also support the notion that newly arriving domestic stock throughout Eurasia were mated with local wild females (359).

Ben Koop and colleagues from the University of Victoria in Canada (410) similarly presented mtDNA evidence for at least three, and perhaps as many as five, protodomestication events for the dog. Significantly, as for cattle, none of the mtDNA sequences grouped according to breed in this study, refuting the suggestion that certain dog "morphotypes" (such as sighthounds, spitz, etc., as discussed previously) derived from particular regional wolf subspecies (127, 533). This was the first study on dogs to use prehistoric samples as well as modern ones, although we now have additional data of this type (444, 588). Koop and colleagues comment that the presence of wolf sequences in several "dog" sequence clades generated by previous researchers dating the divergence of dogs from wolves (779) precludes the use of "molecular clock" estimates—that evidence in any study of hybridization between ancestor and descendant violates one of the basic assumptions required for use of this technique. I've discussed other problems associated with unravelling dog/wolf genetic relationships in another publication (147).

Table 5.2 Number of protodomestication events determined from mtDNA evidence

Domesticate (Common Name)	**Domesticate (Latin) (ref. 262)**	**Wild Ancestor (Latin) (ref. 262)**	**Protodomestication Events**
Dog	*Canis familiaris*	*Canis lupus*	at least 3, perhaps 5 or more (refs. 410, 444, 530, 779)
Sheep	*Ovis aries*	*Ovis orientalis*	at least 2, perhaps 3 (refs. 98, 485, 334)
Goat	*Capra hircus*	*Capra aegagrus*	at least 3 (refs. 98, 467, 485)
Pig	*Sus domesticus*	*Sus scrofa*	at least 6 (perhaps more) (refs. 268, 435, 787, 788)
Cattle	*Bos primigenius/ Bos namadicus*	*Bos taurus/ Bos indicus*	at least 3 (refs. 90, 98, 485, 489, 751)
Water buffalo, swamp & river	*Bubalus bubalis*	*Bubalus arnee*	only 2 (refs. 98, 485)
Horse	*Equus caballus*	*Equus ferus*	at least 1 (perhaps more) (refs. 98, 359, 485)
Chicken	*Gallus gallus*	*Gallus gallus*	only 1 (ref. 245)

The studies described above, which present evidence that the genetic history and presumed historical relationships of most domestic animals do not correspond, is not a situation unique to domesticates. Studies of several wild species that were expected to show genetic support for their ecologically divergent histories, or their ecological and historical similarities, did not show the anticipated mtDNA patterns (e.g., East African chimpanzees, ref. 269; Southern Hemisphere fur seals, ref. 443; Lake Tanganyika cichlid fishes, refs. 396, 653; East African black-backed jackals, ref. 795; Pacific harbour seal, ref. 427).

In general, because natural domesticates and their extant ancestral species are so close genetically, hybridization between them almost always produces viable offspring (for sheep and goats, ref. 325; wolves and dogs, ref. 350; llamas and alpacas, ref. 375). However, the ability to hybridize with systematically close relatives is not by any means restricted to domesticates and should not (by itself) indicate that domesticates are not valid species. In fact, one of the unexpected results of recent molecular genetic studies has been the ability to identify previously unrecognizable hybrids between wild species, as I discuss more fully in the context of human evolution in Chapter 8.

If apparent genetic differences are so small, however, what accounts for the significant biological differences between wild and natural domestic animals? As previously stated, the answer appears to be that significant shifts in either developmental growth rates or timing of developmental events have occurred. The biological evidence thus suggests strongly that the essence of protodomestication is heterochronic speciation (a process of juvenilization), while the archaeological evidence suggests this speciation was implemented exceedingly rapidly.

Summary and discussion

Research on dogs and wolves suggests that evolutionarily-significant growth and developmental differences account for all of the physical, physiological and behavioural differences between them—and for the differences between dog breeds subsequently developed through artificial selection. In addition, the same patterns evident in dog evolution are seen in the other domestic mammals included in this discussion. As natural domesticates constitute some of the best-known and important examples of evolutionarily-significant growth and development changes, understanding how and why protodomestication occurs may shed significant insight into what drives this evolutionary process.

In the case of protodomestication, however, growth and development changes have a specific directionality. Early domesticates are all distinctly juvenilized because the anthropogenic stress component is the same in all cases, for all animals. Human settlements created unique environments in evolutionary terms, ones which had never been available to wild animals before, and the proximity of humans constantly operating within these new habitats was also a new kind of environmental stress. The selection pressure imposed on founders by the stress of constant human proximity would have become increasingly more intense as time passed, driving out individuals with only marginal tolerance because as the population of both people and early domesticates increased in numbers, the frequency of encounters between them would also have increased.

The underlying physiological interactions between the traits which change predictably as a result of protodomestication suggest that an entirely natural, biological explanation is both plausible and testable. If "non-fearfulness" or "low anxiety" in relation to people can be considered a specific manifestation of the more general characteristic of "high stress tolerance," then the process of protodomestication can be seen as an entirely natural process; it did not necessarily involve direct interference by humans in the early stages, just their proximity. In other words, while the hypothesis does not preclude the deliberate actions of people in the initial stages of protodomestication, it does not *require* such interference.

One could propose, for example, that people living during Neolithic times might have observed a connection between stress tolerant behaviour and certain physical characteristics in wild progenitors, such as specific coat colours or the position of facial hair whorls (which have been shown to be strongly correlated with stress tolerant behaviour in domestic cattle) (431). Noting such connections between physical characteristics and behaviour would have allowed people to deliberately select stress tolerant individuals out of large groups of animals, even newborns, without requiring any specific individual to demonstrate stress tolerant behaviour at the time the selection was being made. If this were the case, a founding population of only stress tolerant individuals could conceivably be created by deliberate intent that would be indistinguishable in result from natural protodomestication.

I propose that individual variation in thyroid rhythms among the wild ancestors of domestic animals provided the raw material for selection pressure to implement directed growth and development change in a process not significantly different from that which changes any wild species into a new one. It is only *after* the protodomestication event occurs that cultural influences impart a distinctive history to early domesticates that distinguishes them from other species.

While this hypothesis doesn't preclude the possibility that protodomestication of any single species or subspecies could have happened more than once, it is doubtful that it happened often. The environmental and population pressures on the animals would have to have been quite severe, and the attractions of the human-dominated habitat very strong, to encourage wild animals to expand into an anthropogenic environment. In addition, the stress tolerant (i.e., less fearful) behavioural types must have been present as natural variations in the ancestral population to start with (populations of animals which all tolerate this kind of stress well—or very poorly—do not lend themselves to protodomestication-type speciation). This combination of necessary factors could explain why colonization of anthropogenic environments by wild animals happened rarely overall in relation to the number of potential animal species available for domestication.

Recommended reading

Alberch, P. 1985. Developmental constraints: why St. Bernards often have an extra digit and poodles never do. *American Naturalist* Vol. 126:430-433.

Alderson, L. 1978. *The Chance to Survive: Rare Breeds in a Changing World.* Cameron and Tayleur Books Ltd., London.

Coppinger, R. and Schneider, R. 1995. Evolution of working dogs. In J. Serpell (ed.), *The Domestic Dog: Its Evolution, Behaviour and Interactions With People*, pp. 21-47, Cambridge University Press, Cambridge.

Fagan, B.M. (ed.) 1996. *The Oxford Companion to Archaeology.* Oxford University Press. [highly recommended, dictionary-style reference to archaeology of the world]

Thomson, K.S. 1996. The fall and rise of the English bulldog. *American Scientist* 84:220-223.

Wilcox, B.W. and Walkowicz, C. 1989. *The Atlas of Dog Breeds of the World.* T.F.H. Publications, Neptune City, New Jersey.

Chapter 6

How Speciation & Adaptation Actually Work: Some Examples

Seasons, stress and food

As the seasons change, so do light, temperature, rainfall, and perhaps vegetation (depending on where animals live)—and many of the essential body functions that must change on a seasonal basis to adjust to these environmental changes are known to be controlled by thyroid hormone. The physiological systems that change most dramatically along with seasonal environmental shifts include:

- basic metabolism (including the initiating and maintenance of hibernation in some animals that are exposed to severe winter conditions)
- reproduction (especially synchronization of mating times between males and females, which usually results in production of offspring at optimal times)
- hair colour and hair growth (especially the shift to white winter coats or the growth of thick winter undercoats, and simple annual hair replacement)
- metamorphosis from one body form to another (such as tadpole to frog), allowing the animal to move from a wet habitat to a dry one
- migration initiated by changing light, temperature and/or rainfall

While some environmental conditions change routinely and predictably on a seasonal basis (this is especially true for light), less predictable changes also occur. One example is the recurring but unpredictable shifts in rainfall that accompany alternating El Niño/La Niña conditions in the eastern South Pacific. Although such conditions usually last a single year, they can occur three or more years in a row—with devastating results on habitats along the west coasts of South and Central America. Amounts of rainfall may swing between drought and deluge conditions, dramatically affecting the dominant vegetation that can survive and, as a consequence, what foods are available to herbivorous birds and mammals (later I'll describe in detail an example of how this drives short-term adaptation in birds). Another example is the extent of pack ice in subarctic environments, which can vary from year to year or remain stable for decades at a time, depending on prevailing winds and temperature changes.

The essential point is that environments are constantly shifting, on scales that run from highly regular daily and seasonal changes in sunlight to wildly unpredictable seasonal and decadal fluctuations in temperature, rainfall and snowfall. Individual animals must cope with these changes as best they can, but at some point change will surpass the ability of individuals to adjust, and this is where population-level variation becomes really essential to species survival. Species are simply groups of individuals living in interbreeding populations. If a species is to survive over the long haul, a biological mechanism must exist that allows the *population* to survive extreme but short-term changes, even if some individuals do not.

I contend that thyroid hormone rhythms are the biological control mechanism that orchestrate the many subtle biological changes that all individuals within a population must make in adjusting to minor changes in environmental conditions. As a consequence, when environmental conditions change in more unpredictable and extreme fashion, individual variations in thyroid hormone rhythms become essential to species survival: unless conditions become absolutely intolerable, at least some individuals are likely to survive. And as long as some individuals survive, so does the species. (Although most conservation concepts (e.g., 16) suggest that very small populations of animals are "doomed to extinction" because of low levels of measurable genetic variation, several studies have shown this is not true (northern elephant seals, ref. 796; Serengeti cheetahs, ref. 384; California island foxes, ref. 8): not surprising, considering that virtually all new species must start with very small founder populations.)

In this way, the role of thyroid hormone in the adaptation of individuals to environmental conditions that change on a daily and seasonal basis can be expanded to the permanent adaptation of populations of animals (*species*) to environmental changes over evolutionary time. Think big-picture changes here, such as those that occurred during past ice ages: large-scale alterations in vegetation and climate resulted, to which species of animals had to adjust—even if this meant transforming into a different kind of animal. Those that did not adjust became extinct.

I believe that the intimate role that thyroid hormone plays in the response of animals to stress is pivotal to this concept. As outlined in Chapter 2, stress can take many forms and can originate from the environment (e.g., from cold or lack of protective cover) or from within (e.g., the physiological stress of starvation or the mental stress that comes from avoiding predators or competitors). Life can become stressful in the habitat a species regularly occupies simply because the population has grown too large to support all individuals optimally. When these stresses reach a critical level, some individuals are apt to start looking beyond the established species territory for somewhere else to live.

For individuals expanding beyond their usual boundaries, new habitats offer many attractive benefits, but they also present stressful conditions. As for protodomestication, novel habitats of all kinds are more likely to attract physiologically stress tolerant individuals—those with all of the physiological and behavioural manifestations of a few particular thyroid hormone rhythms—than less stress tolerant animals. As only those individuals with a stress tolerant physiological and psychological nature become colonizers, founder populations possess a specific subset of the thyroid hormone rhythms that were present in the original population. Because of this, the range of variation of thyroid hormone rhythms within small founder populations will always be much smaller than that which existed in the ancestral population as a whole (see Figure 6.2 in the next section for a colour analogy).

Thyroid rhythms have a particularly strong effect when divided in this fashion, however, because of the essential developmental role of thyroid hormone: there would almost certainly be immediate consequences to offspring that result from mating within small and isolated founding populations which have limited variation in thyroid rhythms. As a new thyroid rhythm average for the group becomes established in descendant populations, a new physical form also becomes established, one with much less variation than that found in the ancestral population. Most importantly, changes to behaviour and reproduction would occur at the same time.

Under these circumstances, colonizers of new habitats become physically, reproductively and behaviourally distinct so rapidly that they retain a very close genetic relationship to their ancestors. Nevertheless, a descendant population resulting from such a colonization event represents a new species as soon as a new thyroid hormone rhythm average becomes established. Once this speciation change occurs, it is permanent as long as reproductive isolation is maintained. Behavioural and reproductive timing changes that occur as a consequence of the thyroid rhythm shift, in addition to the ecological differences that precipitated the event in the first place, tend to keep the new population reproductively isolated.

What DNA is and what it can tell us about evolution

DNA is a string of *genes* composed of particular combinations of four basic chemical building blocks—usually referred to by their initials (A,C,T,G)—that form the basis of the *genetic code*. The composition of these chemical building blocks can change (e.g., A to C, T to G) due to *mutation*, which sometimes (but not often) changes how well the gene itself works. Most genes produce a protein (often an enzyme), which has a specific job in the body's complex chemical machinery; while sometimes a mutation will completely disable a gene (producing no enzyme), in most cases mutations cause a gene to over- or under-produce the enzyme (like the rheostat on a light switch), which then affects the overall efficiency of the chemical machinery that depends on it, and down the line, specific body functions.

mtDNA mitochondrial DNA (mtDNA) is a small strand of genetic material found in the *mitochondria* of cells (mitchochondria are tiny *organelles* found in the material that surrounds the *nucleus* of a cell); a cell may contain thousands of them, each with a string of mtDNA—picture a bowl of clear soup (the cell) with lots of rice (mitochondria) and a single egg yolk (the nucleus). The genes found within mtDNA produce enzymes necessary for each cell to make the energy it needs. It is currently assumed that mtDNA has nothing to do with inherited characteristics of form or behaviour in an animal. However, comparing the *sequences* (any string of mtDNA) between species is a way of unravelling particularly tricky evolutionary relationships. Put most simply, the technique assumes that *amount of difference = time spent apart*. Sequences that are most similar are most closely related. A mother's mitochondria is passed along to each of her offspring unchanged in her eggs—thus maternal mtDNA does not mix from one generation to the next, although it may accumulate mutations.

nDNA nuclear DNA is the genetic material found in the nucleus of each cell, which is arranged into physically discrete groups called *chromosomes* and, within those chromosomes, specific genes. After an egg is fertilized, nDNA from both the mother and the father can combine in a huge variety of ways. nDNA produces enzymes that may have wide-ranging effects throughout the body, and thus appear to be involved in producing particular inherited physical, physiological and behavioural characteristics. Because of the potential for mixing at conception, interpreting evolutionary relationships from nDNA is trickier than with mtDNA, but it's much better for distinguishing short-term relationships, like that between siblings.

Figure 6.1 DNA: what it is and what it can tell us about evolution

There is no evidence so far to suggest that any particular *mitochondrial DNA* sequences are associated with particular thyroid hormone rhythm characteristics (see Figure 6.1 for an explanation of various types of DNA). Thus, by chance, even a small colonizing population of stress tolerant individuals may possess all (or nearly all) of the mtDNA variation that existed in the ancestral population. This explains numerous examples of the extremely close mtDNA relationship between many taxa that by all other measures are considered fully distinct species. Such is the case for the polar bear, discussed in more detail in the next section.

Origin of the polar bear from a brown bear ancestor

One of the most spectacular examples of this type of speciation is demonstrated by the evolution of the polar bear. Recent molecular genetic research by Sharon Talbot and Gerry Shields (729, 730) on brown bear (*Ursus arctos*) populations in Alaska has revealed surprising similarities between several mitochondrial genes of the polar bear (*Ursus maritimus*) and one particular coastal subspecies of brown bear. While polar bears had previously been proposed as descending from brown bear stock, based on skeletal similarities and other physical criteria (421, 422), the results of the genetic study place *one particular population* of brown bear as either the ancestor of all polar bears or its closest living relative—this means that brown bears from three islands in southeast Alaska (Andreanov, Baranoff and Chicagoff) are more closely related to polar bears than they are to brown bears that live on the mainland of Alaska. This surprising result led the authors of one study (see ref. 730:574 and Figure 6.3) to conclude that "the morphological features distinguishing polar bears from brown bears have evolved rapidly in response to selective pressures of adapting to a new environment, *prior to the emergence of distinguishing molecular features* [italics mine]."

We know that brown bears existed on the islands off southeast Alaska during the late Pleistocene (the end of the last major glaciation event) as recently as 40,000 years ago and that these brown bear populations must have been compressed into ever diminishing territories by ice advancing from mainland ice sheets as the glacial maximum approached at about 20,000 years ago (322). Such small areas would soon become saturated with animals as large as a brown bear, encouraging some to seek other options. These conditions almost certainly encouraged a few stress tolerant individuals to colonize the shore-fast ice and pack-ice environments that would have surrounded these island refuges.

The ice at the edges of glacial refuges would have posed some distinct advantages. Such an environment would have been rich in high-fat prey resources—especially seals—foods rarely utilized by terrestrial bears in the original source population, although some brown bears in the northern extremes of their range are known to kill and eat seals (469, 724). But a steady diet of seals presents a different thyroid hormone composition than a diet of salmon mixed with berries, and this dietary shift would have contributed to the outcome of the colonization of the new pack-ice environment.

I propose that a few stress tolerant brown bears (those with particular thyroid hormone rhythms), colonizing pack ice where only foods high in dietary thyroid hormone (seals) were available, were exposed to virtually perfect conditions for rapid speciation to occur (see Figure 6.3). A few stress tolerant bears, out on the ice hunting seals further and further away from the islands, would have had only each other to breed with when mating season rolled around. Any offspring born to these founders would be exposed to the same ice habitat and carnivorous diet; if it didn't suit any of these offspring (because they were slightly less stress tolerant than their parents), they could always return to dry land and try to survive there.

But how, you may ask, did polar bears end up completely white when brown bears are dark all over? Polar bears are not albino: albino coloration is caused by specific mutations to pigment-producing genes within skin and hair cells. Although albino bears are known in many populations (40, 469), polar bears have dark eyes and noses (674), a colour scheme that is a prime example of extreme *piebaldness,* otherwise known as whitespotting (42). Such whitespotting, remember, is a common consequence of protodomestication. Within only a few generations of colonizing ice-dominated habitats, founder populations of brown bears destined to become polar bears must have included some individuals with a white-spotted coat. It may have taken many generations for the extreme piebaldness characteristic of modern polar bears to dominate the population, but a less extreme black-and-white coat must have been present very early.

If being completely white gave polar bears a survival or reproductive advantage (because animals that are all white might be more successful at sneaking up on the seals that make up the bulk of their diet), partially piebald animals may have been selected against until only all-white ones remained. A similar selection process occurred with the Samoyed dog, a breed recently developed from a black-and-white dog native to Siberia in the late 1800's. The Samoyed was transformed into a completely white dog within a few decades as a result of deliberate selection by North American breeders in the first half of the twentieth century (810).

A colour analogy for thyroid rhythm theory

Think of the original population as being composed of individuals whose variations in colour, size and temperament represent different colours of the rainbow. Aggressive individuals might be blue, timid ones red, and curious ones yellow (reminiscent of the simple temperament test used on the Russian foxes described in Chapter 2; remember that these traits were associated with several other physical and physiological characteristics). Among individuals classified as "aggressive," for example, there are probably degrees of intensity, so we could characterize them as being various shades of light or dark blue. Similar degrees of intensity would apply to animals classified as either timid or curious, giving us a population characterized by various hues and shades of blue, red and yellow. Some animals might better be described as aggressively curious, and so we might characterize them as a shade of green, while others might be so timidly curious we'd characterize them as various shades of orange. We now have a veritable rainbow of physical and behavioural characteristics.

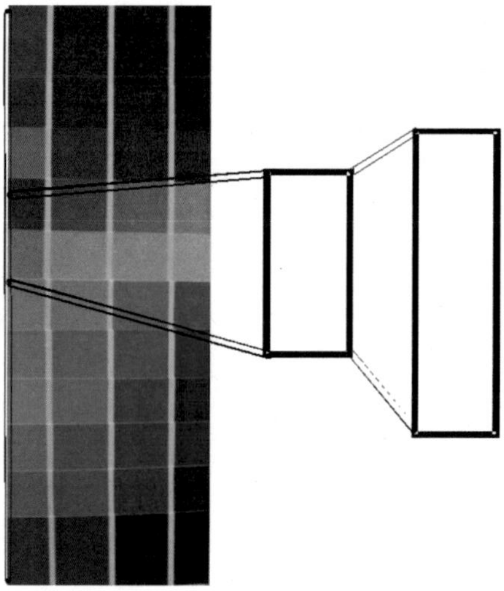

Now think about mixing paint: if only animals characterized as yellow colonize a new habitat (or survive under adverse environmental conditions), then not only are the traits represented by blue and red not available to future generations, but neither are most green or orange variations. Limited variation constrains the future options of the founding or surviving population—at least until new genetic variation develops.

Figure 6.2 A colour analogy for thyroid rhythm theory

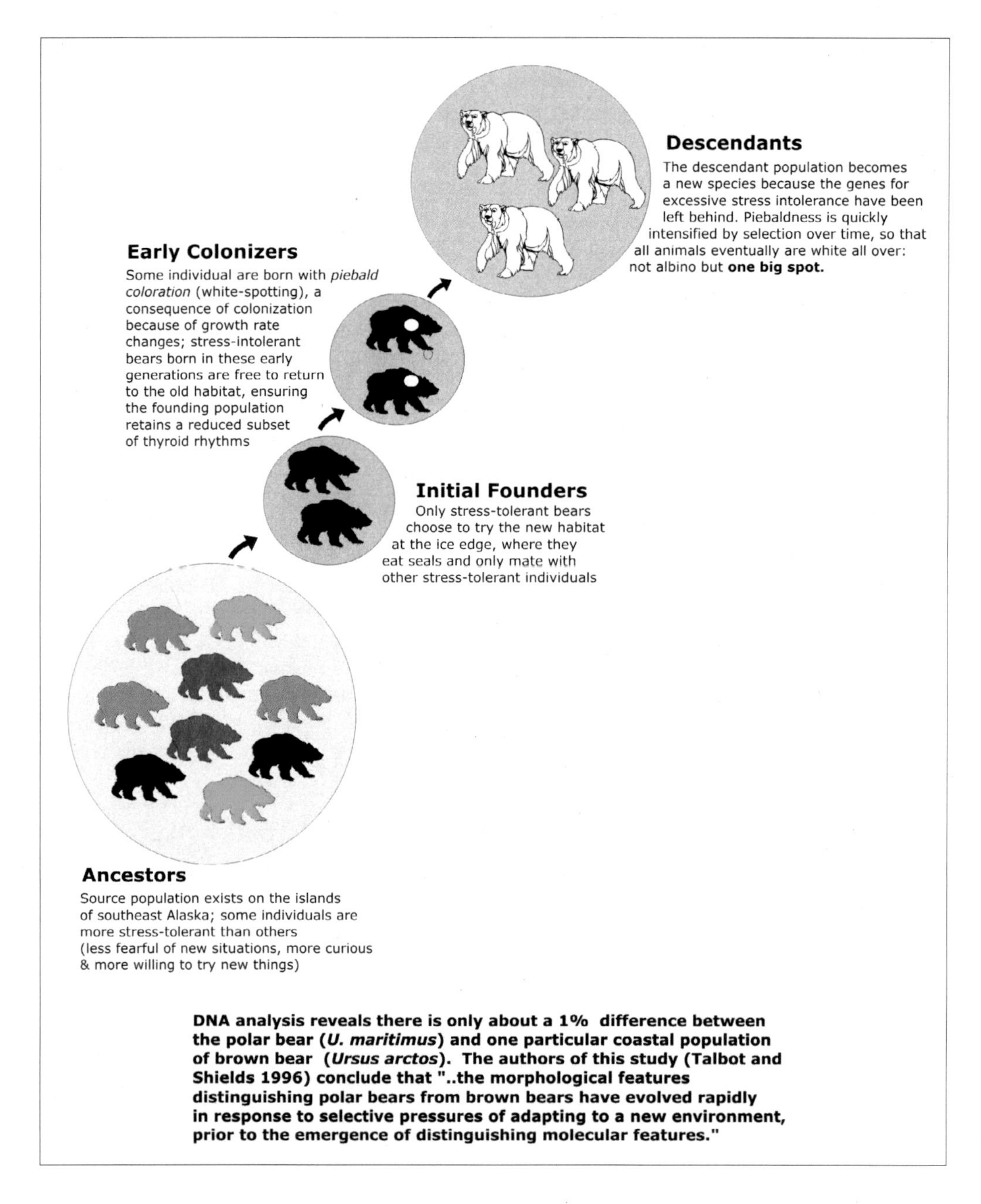

* *Figure 6.3 A model for the speciation of the polar bear via thyroid rhythm theory*

If polar bears possess extreme piebaldness as a consequence of the rapid speciation changes precipitated by colonization of a radically new environment, then perhaps other white arctic animals (such as hares, foxes and wolves) and alpine species (such as mountain goats) are white for this reason as well (40, 118, 371, 259, 260, 674, 688). Being completely white in arctic or alpine environments may have conferred such an immediate survival advantage that, as for primitive polar bears, partially piebald animals may have been heavily selected against until only extreme piebald (all-white) animals remained.

The essential point, however, is that piebaldness could only have become subject to natural selection pressure in any new species because the changes associated with the speciation event itself made piebaldness an available option. Piebald animals are very rare in natural wild populations, much more rare than albino forms. Piebaldness is normally far too rare in virtually all wild populations for selection alone to have created a whole population of extreme piebald animals (even over a very long period of time) unless something occurred to increase the natural incidence of the anomaly substantially.

This suggests that animals whose normal colour is a striking black-and-white pattern (such as skunks, giant pandas and killer whales) probably have not been exposed to selection pressures as extreme or intense as animals that are completely white (such as polar bears, beluga whales and mountain goats). Just because some animals have successfully dealt with being conspicuously piebald does not mean that piebaldness in and of itself confers any kind of adaptive advantage, as is usually suggested (110, 539, 674). On the contrary, it's very likely that the stress tolerant behavioural traits associated with becoming a new species are what conveyed the initial adaptive advantage for the ancestors of piebald animals. Piebaldness in wild animals, as in domesticates, was almost certainly an inevitable consequence, an *artefact*, of small populations of physiologically-similar individuals colonizing radically new habitats. If so, piebaldness as a normal species trait may be an important indicator that a past speciation event in the history of those animals involved a major habitat shift.

Multiple speciation events

As stated previously, stress is an essential feature of virtually all speciation events. Stress is an ever-present factor that can have purely physical manifestations (such as reaction to light or temperature), or involve psychological or behavioural components (such as fear of predators, finding new food sources or competing for mates or breeding sites). Existing variation in physiological or behavioural tolerance to stress among individuals in a population can lead to a predictable split if there are changes in stress-inducing conditions.

After a colonization event splits a few similarly stress tolerant individuals away from an existing population, the particular circumstances of each new habitat present specific challenges to the founding population. Such challenges would impact the first few generations of founders heavily, leaving only those able to stay healthy enough to reproduce successfully as contributors to the next generation. The resulting changes in the descendants will vary according to the physiological makeup of each colonizing population, its size and the severity of the selection pressure. However, this concept supplies the first really plausible explanation for how multiple speciation events occurred, as seen in taxa demonstrated to have *polyphyletic origins,* also called *sibling species* (506, 508, 511). Given virtually identical ancestral populations (such as populations of geographically isolated subspecies) and similar or identical new habitats available for colonization, virtually identical biological changes would be expected to occur. If the geographic subspecies in question have been isolated long enough for them to have developed some level of genetic distinctiveness, the new species each population generates will be genetically distinct despite the extreme similarities in morphology, behaviour and life history traits.

The concept also explains how *convergent* evolution could easily occur, generating animals with very similar form from different ancestral species. When similar but distantly related species are exposed to nearly identical climatic or habitat changes, we would expect similar kinds of changes to result because the thyroid control mechanisms are the same (and would respond in a like manner). The resulting descendants should be genetically distinct but similar in physical and behavioural form—in some cases, the similarities may be so strong that they make sorting out evolutionary relationships extremely difficult.

Numerous examples of convergent evolution are now known, including early hominids and other bipedal apes (which I discuss in detail in Chapter 8); indeed, the list is expanding rapidly now that comparative genetic studies are more common. However, an interesting example of this phenomenon is seen in the true seals of the Arctic, family Phocidae (27, 619). All of the seals in this family (except the harbour seal, *Phoca vitulina*) generally live in close association with ice and have been economically important to humans for tens of thousands of years. They include the ringed seal (***Phoca**/Pusa hispida*), harp seal (*Phoca/**Pagophilus** groenlandica*), ribbon seal (***Histriophoca**/Phoca fasciata*) and spotted seal (***Phoca** largha*). Genetic analysis confirms that ringed and spotted seals are much *less* closely related to each other than their similar body forms, habitat preferences and reproductive characteristics would suggest. Such results suggest these seals evolved from a common ancestor under similar conditions but at widely distinct times in the past (I've given the most recent suggestion of appropriate generic names in bold). Ironically, the same studies show that two Antarctic seals (the Weddell seal, *Leptonychotes weddelli*, and the leopard seal, *Hydrurga leptonyx*), which are much more distinct physically from each other than any of the Arctic seals mentioned above, are nevertheless *very similar* genetically—which suggests both species evolved from a common ancestor at about the same time but in response to distinctly different conditions (other examples are described in refs. 28, 29, 333, 346).

Similar growth and developmental responses, such as the changing body sizes of mammals during the Pleistocene and Holocene (420-422, 697)—first in response to overall cool conditions and then to warming trends—could have occurred across many unrelated groups because both the stressor and the stress response system would have been the same for all (in much the same way that protodomestication produces similar changes in every instance).

As a consequence of the interactions between thyroid hormone, reproductive hormones and growth hormones (see discussion in Chapter 3), species-specific rates of physical and sexual maturation determined by species-specific thyroid rhythms can also generate sex-specific size, shape or colour patterns. While the concept does not explain *why* females are the larger sex in some animals while males are larger in others, it does explain how such a pattern can be maintained as a species-specific trait, generation after generation.

In addition, because growth in vertebrate animals involves embryonic, fetal and postnatal programmed maturation that is thyroid hormone dependent, shifts in rates of physical and sexual growth can be implemented at any stage from conception onward—changes that occur only during embryonic stages (very early development) will have a very different result than shifts that affect only growth after birth or hatching.

Finally, colonization of radically different environments may often have necessitated quite dramatic dietary shifts that involved adding or removing sources of dietary thyroid hormone, factors that appear to be significant in the generation of the many new species, including the polar bear (as discussed in the previous section), whales (742) and many early hominids—our own ancestors (which I discuss in Chapter 8).

Adaptation in birds and fish

I've so far addressed microevolution and macroevolution almost exclusively. However, I believe that thyroid rhythm theory also explains both the temporary and permanent adaptation of established species over extended periods of time. Because new conditions can be imposed on an entire population—by sudden climatic shifts, for example—I suggest that only those individuals who already possess the particular traits that allow them to survive such changes and reproduce successfully despite them, pass their genes on to future generations. Since I maintain that such variable traits are simply manifestations of individual differences in thyroid hormone rhythms, the entire population may be reduced to only a few particular rhythms in one generation. This reduced variation occurs because most individuals that have the "unselected" rhythm pattern either die or do not reproduce, leaving only one or a few similar thyroid rhythm patterns to contribute to the next generation.

An example of how rapidly adaptation can occur within established species has been reported recently by Peter and Rosemary Grant (284). I think this might represent an especially good example of how individual variation in thyroid rhythms give populations the flexibility they need to respond rapidly to even short-term changes in environmental conditions so that overall population survival is ensured. Between the years 1972 and 2001, the Grants undertook a rare long-term investigation of two species of finches from the island of Daphne Major in the Galapagos: the medium ground finch (*Geospiza fortis*), which has a broad, deep beak, and the cactus finch (*Geospiza scandens*), which has a long pointed one (Figure 6.4).

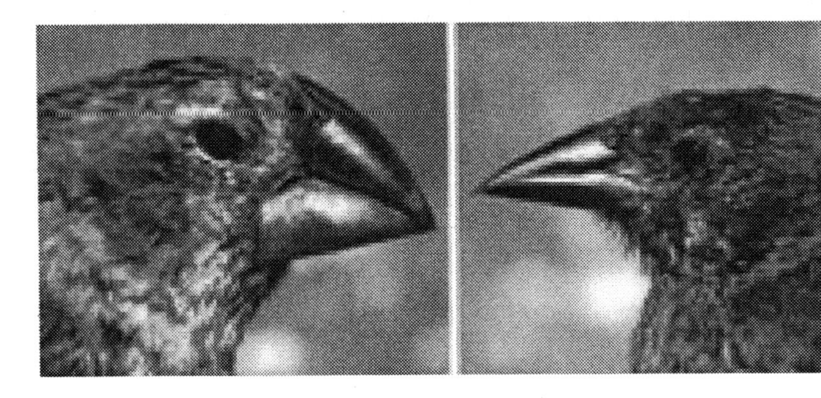

Figure 6.4 Finch beak shapes studied by the Grants: medium ground finch is on the left, cactus finch on the right (photo by Arhat Abzhanov; see refs. 2, 589)

In their study, changes in the size of readily available plant seeds over the thirty-year period (brought about by alternating 3–5 year periods of drought and flood) resulted in changes in the size and shape of finch beaks that were sometimes accompanied by changes in overall body size. If beak shape and size, as well as body size, are traits set during early development and thus controlled by species-specific but individually variable thyroid rhythms, as I propose must be the case, shifts in abundance of different seed types could alternately select for beaks of one size and/or shape according to which of the various seed-producing plants is the most successful in any given year.

In the final ten years of the finch study, the Grants also found evidence of hybridization between the finch species—hybridization that appears to have increased the variation in beak size and shape available to both species (or replaced variation lost through adaptation processes) without negatively affecting the fitness of either. These effects of hybridization raise an important question: if variation becomes much reduced in an adapting or colonizing population, how does it later increase so that adaptation can continue in the future? Obviously, hybridization with a closely related species is one way to add variation back into an otherwise restricted gene pool.

Box 6.1 Genetic control over beak shape changes?

Not surprisingly, a gene that alters finch beak shape in precisely the ways described by the Grants was identified in 2004 by Arhat Abzhanov and colleagues at Harvard Medical School's Department of Genetics (2). The gene is known as *Bmp-4* (because it produces *bone morphogenetic protein-4*), the same gene shown to control the shape of chicken and duck beaks and mammal teeth (380, 835). *Bmp-4*, along with several other genes that moderate its function (such as *sonic hedgehog, Shh* and *fibroblast growth factor-8, Fgf8*), are known to respond to thyroid hormone in a time- and dose-dependent fashion (41, 348, 353, 719). In fact, alteration of *Shh* gene function, depending on the timing of disruption during embryonic development, is known to generate a variety of effects on brain and skull conformation in all animals, including humans (140). Such genes may be especially responsive to shifts in thyroid rhythms during early development..

Another way to add variation back into a founder population may be through chance genetic mutation. It may be that the genes controlling thyroid rhythms that reside in the clock neurons of the SCN (as described in Chapter 4), accumulate mutations at a fairly rapid rate, quickly replacing variation lost during a population crash or speciation event. There must be many genes whose combined actions generate thyroid rhythms. Each individual gene within that SCN clock complex cannot be subject to selection directly: only the *entire* neurohormonal output (which generates the unique thyroid rhythm) can be selected—particular physiological and physical traits, as manifestations of that rhythm, are all selectable "end-products" of this output. The individual genes that produce thyroid rhythms probably undergo slight mutations almost continually, with no single genetic variant either advantageous or disadvantageous on its own. Such constant slight mutations would create a multitude of rhythm variants in every generation. These mini-mutations would ensure that individual differences in thyroid rhythms are constantly replenished. This is pure speculation, of course, but worth checking once the suite of genes that produce thyroid rhythms have been confidently identified.

Alternatively, given the critical requirement for thyroid hormone in all vertebrates in generating fine-scale neural architecture of the brain in a dose- and time-dependent manner, and given the evidence that an intact set of maternal clock neurons is required for normal development of SCN rhythms in newborn rats (625), another explanation for continually renewed variation presents itself. Rather than the genes of pertinent SCN oscillating clock neurons changing, perhaps what changes is the physical relationship of those cells to others in the clock complex.

I suggest the possibility exists that minute shifts in such physical inter-relationships—cell-to-cell positional differences—might ultimately result in minor differences in thyroid rhythm output, due to slight disruption of maternal thyroid rhythms during fetal brain development of offspring. If the individually-unique thyroid hormone rhythms of a mother strongly influence the neural cell migration, proliferation and maturation (at least in placental mammals), as I suggest must be the case, disruption of the maternal signal (due to dietary change, stress, etc.) may result in slight differences in the relationship of critical clock complex neurons to each other. For offspring of mothers exposed to such stresses, such slight differences in clock complex architecture should produce a new thyroid rhythm that is not the result of genetic mutation. What is significant and exciting about this proposition is that it provides a mechanism for some of the variation that's needed for evolution and natural selection to be generated by means that are not genetic in nature but *are* eminently heritable by future generations (see also Box 8.1 in Chapter 8).

Speciation, adaptation—or something else?

In some situations it may be hard to determine whether permanent speciation or temporary adaptation has occurred. Certain animal populations may have transformed in significant ways but not to the point of becoming genetically distinct. One example that springs to mind comes from my part of the world and involves Pacific salmon. All Pacific salmon (six species, all classified to the genus *Oncorhynchus*) are generally *anadromous*: that is, they hatch into fry from eggs laid in fresh water, transform into another form that moves downstream into the sea to feed, and eventually return to their natal stream or river system to reproduce as sexually mature fish. Pacific salmon reproduce just once and then die, in contrast to Atlantic salmon, some of which spawn several times.

One of these Pacific salmon species, the sockeye (*Oncorhynchus nerka*), prefers an inlet stream to a freshwater lake for spawning. The young fish spend at least a year growing in these lakes before transforming into what are called *smolts,* which then head downstream to the sea (84). Interestingly enough, the change to this critical smolt stage, which prepares the fish to live in salt water, is known to be initiated and controlled by thyroid hormone (189, 200, 707).

Some sockeye individuals, however, remain in the nursery lake their entire lives and are referred to as *kokanee.* Both forms return to their natal stream to spawn at sexual maturity. Kokanee have more gillrakers than anadromous sockeye and remain quite small: kokanee are never larger than 22 cm while sockeye are never smaller than 38 cm (both size and gillraker count are inherited traits, see Figure 6.5). In the past, these two salmon were considered to be distinct species due to their markedly different size and habitat preferences, but we now know this is not the case.

Kokanee spawn at virtually the same time as sockeye and often in nearby locations, but the two types prefer different gravel size and stream current velocity. While a few crosses do occur between the types (with gene flow estimated at <1%), in general both kokanee and sockeye prefer to mate with their own kind.

Figure 6.5 Dramatic size difference of kokanee (top) compared to sockeye (bottom), photo by Chris Wood

As a consequence, the two types are considered to be reproductively isolated. When crosses do occur, mating is between small kokanee males and large sockeye females. A recent genetic study by fisheries biologists Chris Wood and Chris Foote compared samples of kokanee and sockeye from several lake systems. If each were a distinct species, we would expect all kokanee to be most closely related to each other, and all sockeye to other sockeye. However, it turns out that, although kokanee are indeed genetically distinct from sockeye, the kokanee in each lake tested were actually more closely related to the sockeye that spawned in their own lake than they were to kokanee from other lakes (234, 732, 826).

This suggests that kokanee must be generated in each lake system independently, time and again, from the specific stock of sockeye that spawn there. Once established, the two forms become reproductively isolated. How could this occur? It's been noted that on occasion, even for stock transplanted to New Zealand, large anadromous sockeye produce small, non-anadromous (non-migrating) offspring, called "residual sockeye" (610, 826). These residual sockeye are almost always male, do not go to sea, and are smaller at maturity than anadromous sockeye. Mating of small residual males with normal-size females produces hybrid females of intermediate size. Such hybrids, if they continue to interbreed with small residuals, would eventually produce females of a small form that would then choose small males as mates.

This raises a taxonomic dilemma of some significance. If the kokanee from a particular lake are reproductively isolated from other kokanee, as well as being genetically distinct from the sockeye stocks in their particular drainage, doesn't that make them a distinct species in their own right (by virtue of being the result of a distinct speciation event *and* reproductively isolated)? If so, would we also have to give species status to each lineage of domestic pig, dog or goat that resulted from a distinct speciation event (as described in Chapter 5)? While either outcome seems unlikely, it does raise additional questions about what we expect a species to be.

Along a similar line of thought, we might consider the taxonomic plight of the right whale (Figure 6.6), which has populations in the Atlantic, the Pacific and the Antarctic. Until recently, the right whale was simply considered a globally distributed species, with no documented geographical variation in external physical features or behaviour. Prior to the genetic revolution, some researchers thought that right whale populations in the Northern Hemisphere were probably reproductively isolated from Antarctic ones, suggesting that distinct Northern and Southern species, or at least subspecies, might be valid taxonomic entities. Genetic analysis, however, has provided exceptionally strong evidence (based on both mitochondrial and nuclear DNA) that *none* of the whales from the three distinct ocean basins interbreed.

It's now clear that there are three genetically distinct populations of right whale, suggesting that each deserves elevation to full species status (247, 648). Such a new taxonomy would give us the North Atlantic right whale, *Eubalaena glacialis*, the Southern right whale (of Antarctic waters), *Eubalaena australis,* and the North Pacific right whale, *Eubalaena japonicus*. It's interesting to note that a similar pattern of genetic distinction without physically diagnostic characteristics among the three ocean basins has also been documented in the minke whale, one of the smallest baleen whales (619). Although a taxonomic distinction has been made between the Northern population, *Balaenoptera acutorostratus*, and the Antarctic population, *Balaenoptera bonaerensis*, recent evidence suggests that North Atlantic minke are genetically distinct from those in the North Pacific (247).

Figure 6.6 North Atlantic right whale Eubalaena glacialis (photo by Moira Brown, New England Aquarium 2005)

I suggest that these genetic distinctions, which exist despite any appreciable physical or behavioural diagnostic characteristics, are indications that fairly large, hormonally well-mixed founder populations of these species colonized adjacent ocean basins millions of years ago; the founder populations must have possessed a good representation of all thyroid rhythms present in the original population. Over time, because populations in each ocean basin did not mate with each other, distinctive genetic mutations accumulated in each population. However, the large number of founders involved in the colonization events, in contrast to what occurs in protodomestication and colonization of distinctly different habitats, prevents each of the populations from becoming physically and behaviourally distinct despite several million years of isolation. The likely reason: big, deep-water ocean basin habitats are virtually identical.

E*cotypes*: neither species nor subspecies

Another intriguing phenomenon of nature is what is generally called an *ecotype*. Ecotypes are usually found only in species that occupy several distinct types of habitats. One particular physical variant then occupies only one particular ecological habitat within the total species range. Ecotypes are sort of evolutionary fence-sitters, neither true species nor a true geographic subspecies. One habitat that seems capable of spawning such population divergence is polar ice. Some of this ice is permanent, unattached and constantly moving (*pack ice*); the rest is seasonal and attached to shorelines (*shore-fast ice*). In northern polar seas, ringed seals (*Phoca hispida*) are probably the most successful ice-adapted species (Figure 6.7).

Ringed seals are among the smallest of the Arctic seals and are the only ones that maintain and defend breathing holes in the ice throughout the long arctic winter. Ringed seals also construct snow "lairs" above the ice, both for resting (*hauling out*) and for giving birth, which affords them some protection from their primary predators, polar bears and Arctic foxes.

It has been assumed, perhaps because most hunting of seals is a land-based activity often assisted by dogs (339, 587), that all ringed seals breed on shore-fast ice along coastlines (809). However, there is strong evidence that a distinct ecotype of ringed seal exists that lives and breeds only within the shifting pack ice.

Figure 6.7 Ringed seal pups a few months old (photos by US NOAA)

Pack-ice breeding ringed seals are small: they are reportedly both smaller at birth and as adults than shore-fast breeding ones. Sexual maturation appears to occur at least a year or two earlier in pack-ice seals than in seals that breed on shore-fast ice. Evidence that substantial breeding populations of ringed seals live within the pack-ice environment has been documented for the ice-filled waters adjacent to the Bering Sea (the Sea of Okhotsk and South Chuckchee Sea, ref. 219), Davis Strait in the Canadian Arctic, ref. 225) and in the Barents Sea off Norway, ref. 809). Measurements of skulls from a small sample of each of the two ecotypes in the Canadian Arctic (225) lends support to the observation that pack-ice breeding ringed seals are significantly smaller.

How and why has this phenomenon occurred? I suggest an answer may lie in the connection between thyroid rhythms, body size (size and age at maturity) and territorial behaviour. Anecdotal evidence from Inuit hunters suggests that one of the factors contributing to the division of ringed seals into two separate breeding populations is individual variation in degree of territorial behaviour. That is, shore-fast breeding seals are extremely territorial in defence of their breathing holes: indeed, it's probable they *must be* strongly territorial in order to maintain personal breathing holes within the limited amount of shore-fast ice available. If intensely territorial individuals are also larger than non-territorial seals, it's likely both of these traits are controlled by individually-unique thyroid rhythms.

In a competition for limited amounts of shore-fast ice habitat, smaller and less aggressive ringed seals are likely relegated to the extensive offshore pack-ice. However, being non-territorial in pack ice may actually be advantageous, especially if it means seals willingly re-establish breathing holes whenever ice movement takes old holes over suboptimal feeding grounds.

If this is indeed how the two ringed seal ecotypes became established and maintained, it implies that populations of land-fast ice breeders must continually contribute their non-territorial individuals to the pack-ice population. However, it also suggests that few, if any, pack-ice seals join the shore-fast ice population. As a consequence, if a genetic survey were done, we would expect that any population of shore-fast breeding seals would show a distinct genetic signature but that pack-ice populations would not. I also suggest that such continued addition of shore-fast ice individuals (what biologists call *introgression*) would prevent pack-ice populations from establishing the unique thyroid rhythm necessary to transform it into a new species. Although situations like this are rarely mentioned in the scientific literature, they may be more common than we realize. While examples of this phenomenon, plus a related but not identical one, *resource polymorphism,* have been documented in fish, amphibians and birds (700), it appears to have been studied less in mammals.

Summary and discussion

If thyroid rhythms are indeed responsible for keeping the individual adapted to environmental conditions that change on a daily and seasonal basis, as I propose must be the case, this physiological system has the potential to initiate and implement the changes that allow species to adapt over longer periods of time. The range of individual variation for thyroid rhythms possessed by any species undoubtedly influences its evolutionary adaptability. We would therefore expect some species to be better able than others to adapt to specific kinds of environmental change because their populations are composed of individuals with more physical, physiological and behavioural variation.

Some sets of quite distantly related species could all change in a similar fashion on exposure to identical shifts in environmental regimes over evolutionary time (such as extreme low or high temperatures). Similar ontonogenic responses, such as the changing body sizes of mammals during the Pleistocene and Holocene, could have occurred across families and orders because both the stressor and the stress response system would have been the same for all taxa (in much the same way that protodomestication produces juvenilized descendants in every instance).

In addition, because growth in vertebrate animals involves embryonic, fetal and post-natal programmed maturation that is thyroid hormone dependent, shifts in rates of physical and sexual growth can be implemented at any stage from conception onward. Finally, the colonization of radically different environments often necessitates quite profound changes in food supply that can either add or remove sources of dietary thyroid hormone. The added effects of shifting dietary sources of thyroid hormone on growth and development appear to be significant in many lineages, including those of our ancestors, which I discuss in Chapter 8. But before I tackle human evolution, in the next chapter I discuss a few more fascinating examples that should help clarify the usefulness of thyroid rhythm theory in explaining how various animal populations adapt to different conditions.

Recommended reading

Pelly, D. F. 2001. *Sacred Hunt: A Portrait of the Relationship Between Seals and Inuit.* University of Washington Press, Seattle.

Reeves, R. R, Stewart, B.S., Clapham, P.J. and Powell, J.A. 2002. *National Audubon Society's Guide to Marine Mammals of the World.* Alfred A. Knopf, New York.

Struzik, E. 2003. Grizzlies on ice. *Canadian Geographic* (November/December):38-48.

Chapter 7

More Examples: Island Dwarfs & Giants

The unique and the bizarre—island species

Most animal populations on islands show similar trends over time: small mammals get larger (as do some birds), large mammals get smaller (as do some small animals, like rabbits and shrews), and birds often become flightless; see Figure 7.1. Many examples are known, most only from fossils, including recently discovered hominid remains that have been designated as a new dwarf species, *Homo floresiensis*. In this chapter, I apply thyroid rhythm theory to the phenomenon of island dwarfs, giants and flightless birds. The concept explains how and why such changes could occur in all animals, including ancient hominids. As for domesticates and other colonizers of new habitats, thyroid rhythm theory predicts that changes to the growth programs and other life history traits of island populations will vary depending on the particular thyroid hormone rhythms possessed by colonizing founders.

Animal species found on remote and continental islands are often unique (*endemic*) from their mainland counterparts; occasionally, they are even bizarre. Sometimes, island-dwelling animals are so different in size or shape that determining their closest relatives was a real challenge before DNA technology came along. Many are, or were, especially vulnerable to rapid extinction, particularly from the hunting practices peculiar to humans: the recently exterminated dodo of Mauritius and great auk of Newfoundland and

Britain are recent examples, while the pygmy hippo of Cyprus likely fared the same fate in prehistoric times (609, 690).

The intriguing phenomenon of island endemism has captured popular interest for decades, but most island oddities vanished so long ago they are poorly known outside the scientific world. However, within that specialist realm, island-dwelling animals have been the focus of study for evolutionary biologists and ecologists for more than a century, beginning in 1880 with Alfred Russell Wallace's fascinating account of his studies in Indonesia (609, 784). Much of this scientific interest is fuelled by the fact that animal populations on islands develop under such controlled conditions they come close to qualifying as natural experiments. As a consequence, in-depth studies of island endemics are particularly well suited to answering questions of how adaptive evolutionary processes actually work. In principle, understanding how these processes work in island situations should help enormously in unravelling how and why giants and dwarfs occur in other circumstances (85, 282, 747).

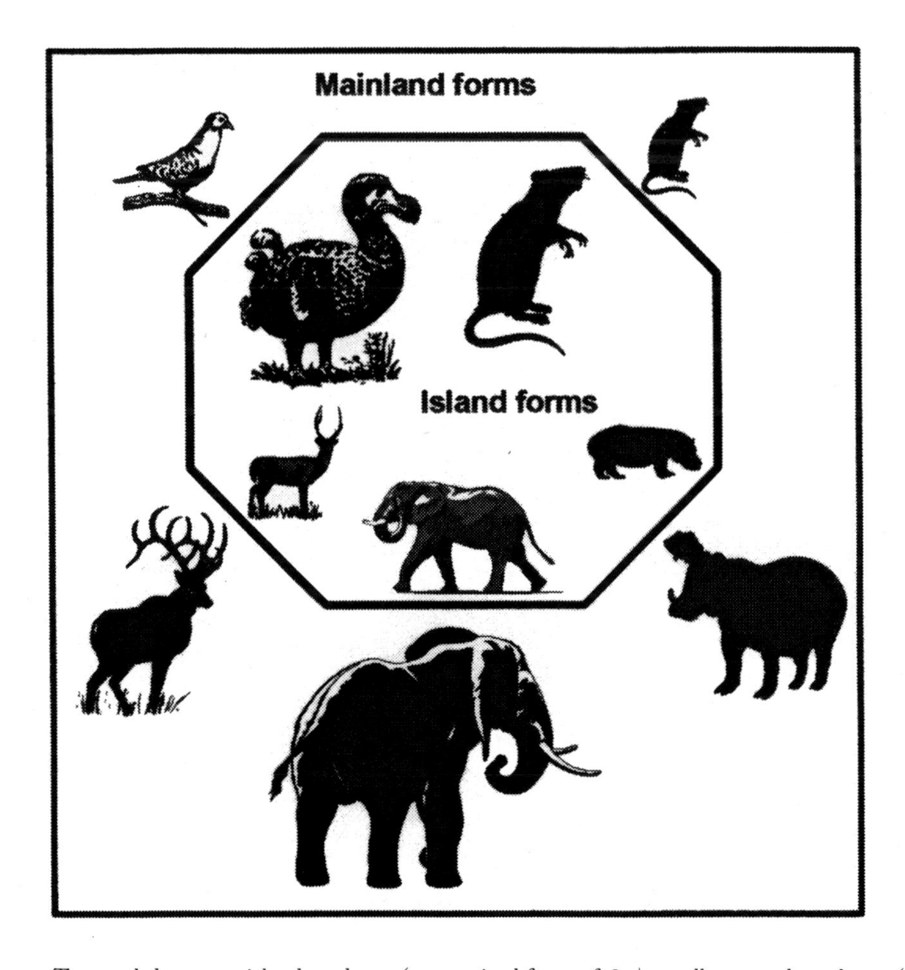

Figure 7.1 The trends known as island syndrome (summarized from ref. 811): small mammals get larger (as do some birds); large mammals get smaller (as do most rabbits/hares and shrews); many birds become flightless, with enlarged bills and well-developed hind limbs.

Attention to this topic has peaked again in all quarters recently, driven by a contentious debate on whether remains of an extinct hominid from Flores Island in Indonesia should be added to our list of island dwarfs (96, 159). These hominid finds have generated renewed calls to explain what has become known as *island syndrome*, the phenomenon of dramatic change in size, shape and behaviour characteristic of isolated island animals as compared to their mainland ancestors. The specific question that now demands an answer is this: how could one of our closest ancestors (*Homo erectus*), a species similar in size to ourselves and clever enough to manufacture an arsenal of stone tools, be reduced to a small-brained creature one metre tall just because it inhabited an island (37, 824)?

I believe I can answer this question, not just for these hominid curiosities but for all island vertebrates, by applying the concept of thyroid rhythm theory to the phenomenon. Island colonization considered in this light explains not only the parallels that have long been noticed between island endemics and domestic mammals (282, 405), but also the strong trend toward flightless gigantism (such as seen in dodos and great auks) and other anomalies of limbs and body shape found among island birds and reptiles. Thyroid rhythm theory provides a more comprehensive explanation for island syndrome than previously proposed hypotheses because it describes a specific biological mechanism to account for these changes that is widely applicable and testable.

Island syndrome

Not all marine islands are equal. Some islands—such as those off California in southwestern North America and many islands in the Mediterranean—lie relatively close to continental land masses and were connected to them for varying lengths of time during periods of lowered sea level. Others—such as the islands of Hawaii—have always been remote and isolated. As a consequence, the ability of animals to colonize islands varies with the kind of island and its location, as well as the dispersal ability of the animal itself. Regardless of these differences, however, many animal populations on once-connected continental islands show similar trends over time as those on remote oceanic islands (283), a pattern that certainly begs for an explanation.

Obviously, remote oceanic islands are difficult to colonize because they're hard to reach. Small terrestrial mammals and weak flying birds generally end up on oceanic islands by accident (this is often referred to as *sweepstakes dispersal*). Examples of such accidental colonization include rodents becoming marooned on the vegetative flotsam generated by storms (uprooted trees, etc.) and small forest birds and bats getting blown far out to sea by fierce winds (811). Migratory birds that are powerful enough to withstand storm winds may nevertheless colonize remote oceanic islands, perhaps out of curiosity or as a result of navigational error.

Large mammals are much less likely to colonize isolated oceanic islands, quite simply because they're too far away from continents for them to reach. But islands just offshore from continents, or islands that are part of archipelagos connected in stepwise fashion to large land masses (such as Indonesia), are often close enough to allow strong swimming large mammals to establish successful island populations.

Island-dwelling animals differ from their mainland ancestors in surprisingly similar ways. The most prominent trends seen in island syndrome have already been mentioned (Figure 7.1): small mammals get larger (as do some birds and many reptiles); large mammals get smaller (as do some small animals like rabbits/hares and shrews, as well as some reptiles and birds); many birds become flightless. Some island-dwelling birds with reduced wing bones also possess enlarged bills and well-developed hind limbs; among island mammals, pronounced changes are also observed in dentition, skeletal proportions, antler development, behaviour, and life history characteristics such as age at sexual maturity (6, 155, 261, 481, 574, 613). Some examples of island dwarfs and giants are listed in Table 7.1, where I've included a notation that emphasizes the range of variation in scientific naming among them. Some island endemics are considered to belong to the same species as mainland forms—sometimes (but not always) distinguished by a distinct *subspecies* name—while others are considered distinct species (there doesn't appear to be any rule of scientific nomenclature that governs the official naming of island forms, and so scientific names vary according to preference or the opinion of individual scientists).

Regardless of how they are named, several explanations have been offered over the years to account for these trends among island-dwelling animals. These explanations typically focus on a combination of three main factors that differ between mainland and island environments: lack of competition between species, lack of predation, and either restricted or abundant food (574). A typical account suggests that a founder population of large animals will eventually become small (sometimes *very* small) because as their population grows they deplete limited food resources, allowing natural selection to favour the smaller body sizes that require less food. Alternatively, founder populations of small animals (such as rodents and birds) that colonize an island often arrive to find not only abundant food but no predators or competitors for the food they require for optimal growth—in this case, the story goes, there is virtually no restriction on how large animals can become, and they grow to larger and larger sizes with each passing generation.

Some researchers (574, 613) believe that these selective factors alone aren't adequate to explain the phenomenal range of body size changes that have been documented for various populations of island-dwelling animals. They suggest instead that shifts in *life history traits* (such as rates of mortality or the age when individuals reach sexual maturity) offer a better explanation for why island endemics change as they do.

Table 7.1 Examples of a few insular birds and mammals, living and extinct

Common name (*extinct) Mammals	Scientific name[1-4]	Change	Reference
Skomer bank vole (Skomer Is., Scotland)	*Clethrionomys glareolus skomerensis*[1]	giant	6
hutia (Bahaman islands)	*Geocapromys ingrahami*[4]	giant	85
giant rat (Lesser Sunda Islands, Indonesia	*Hooijeromys nusatenggara*[4]	giant	766
*straight-tusked elephant (Sicily)	*Elaphas falconeri*[2]	dwarf	576
*mammoth (Santa Rosa Is., CA)	*Mammuthus exilis*[2]	dwarf	7
*mammoth (Wrangel Is., Siberia)	*Mammuthus primigenius*[3]	dwarf	771
*hippopotamus (Cyprus)	*Phanourios minutes*[2]	dwarf	690
*cave goats (Majorca), five species	*Myotragus spp.*[2]	dwarf	405
*red deer (Jersey, English Channel)	*Cervus elaphus jerseyensis*[1]	dwarf	455, 456
reindeer (Spitsbergen)	*Rangifer tarandus platyrhynchos*[1]	dwarf	261

Birds

Common name (*extinct)	Scientific name[1-4]	Change	Reference
*dodo (Mauritius)	*Raphus cucullatus*[4]	flightless	609
*giant goose (Hawaii, USA)	*Branta sp.* undescribed	flightless	583
*Laysan rail (Hawaii, USA)	*Porzanula palmeri*[2]	flightless	109
*giant auk (Newfoundland, Britian)	*Pinguinus impennis*	flightless	543, 254
Auckland Island teal (New Zealand)	*Anas aucklandica*[2]	flightless	461

Notes:

1) Inclusion of a subspecific designation indicates the island form is considered conspecific with mainland forms (if they still exist), not a unique species.
2) These are definitely considered distinct species derived from mainland ancestral forms.
3) These are considered conspecific with mainland forms, but no subspecific designation exists.
4) The ancestral species is not known, but the island form has a unique species name.

But while there may indeed be a significant connection between life history traits and island syndrome in a general sense, that doesn't address the question of how the diversity of changes we see are actually implemented biologically. In other words, I could amend my original question about dog protodomestication (the one that started me along this research path in the first place) and ask: what has to happen to a pigeon, in strictly biological terms, in order for it to become a dodo? Until that question is answered, we really won't have a clear understanding of how and why islands are so often home to populations of bizarrely-shaped animals.

Parallels with protodomestication

Let's deal with one kind of change at a time, starting with the most common and dramatic trend, that of size reduction, or *dwarfing*, of large animals. The dwarfing associated with island colonization is achieved, not by *achondroplasia* (a specific genetic mutation leading to disproportionately stunted growth, which occurs occasionally in all kinds of animals, including humans), but by a shift in the rates of growth. This process is known scientifically as *paedomorphosis*, or less formally as *juvenilization*, in which altered growth rates result in individuals attaining smaller sizes at sexual maturity than their ancestors. Domestic animals, as discussed earlier in this book, are prime examples of this process, characterized as they are by distinctly juvenilized appearance and behaviour as compared to their wild ancestors. Indeed, for decades the same sorts of size and shape differences documented between wild ancestors and domestic descendants have been noted between island endemics and their mainland ancestors (261, 405, 650, 576). Not only that, but many island endemics are also relatively tame or show markedly less aggression than mainland forms (6, 282).

I contend that the similarities between island-dwelling animals and domesticates are not coincidental but in fact represent evidence that an identical evolutionary process is at work. Just as the common physical and behavioural changes seen in different domestic mammals can be effectively explained as being the result of deliberate colonization of a particular habitat type by stress tolerant individuals of wild ancestors (as explained in Chapter 3), so can the changes seen in island species.

I maintain that animals which consciously (i.e., deliberately) colonize islands are a hormonally-similar subset of the original population—that if we were to measure thyroid hormone secretion patterns in colonizers, we would find the particular thyroid rhythms associated with stress tolerant behaviour in those animals. As a consequence, similar trends in outcome are inevitable every time an island is deliberately colonized. Stress intolerant animals simply choose not to participate in deliberate colonization events. Why choose to colonize an island in the first place? Large herbivorous mammals in particular can probably smell vegetation on offshore islands. Stress tolerant individuals of species that are good swimmers (such as deer, elephants and hippopotamus) may occasionally cross long stretches of water to explore such offshore resources and simply decide to stay. Although colonization of islands happens under rarely occurring circumstances, once undertaken it seems not to be reversed—which means that once a few colonizers occupy an island, the ocean is as effective as a fence in keeping them there.

All new habitats are associated with novel stressors, and islands are no exception. Island colonization presents both limitations and opportunities that not only are potentially stressful but require behavioural responses:

1) Predator vigilance is still required—even if predators are not present, their absence can never be assumed.

2) Decisions are required regarding the many unoccupied habitats available for use—if no other animals are present on an island to prevent the colonizer from selecting one new habitat over another, a choice will have to be made somehow (109).
3) Former seasonal movements between resource areas (such as pastures around water sources that don't dry out in the summer) are no longer possible—movement is restricted by the small physical space and oceanic boundary (613).
4) The small number of initial founders severely restricts mate choice for many generations.
5) Above all, there is—or appears to be—no escape.

To the initial stress tolerant founders, these factors (predator vigilance, opportunities that require decision-making, restricted movements, limited mate choice and virtual imprisonment) may not be especially disturbing, but they may prove very stressful for individuals born in subsequent generations who by chance inherit a thyroid rhythm that is less stress tolerant. In contrast to protodomestication and other examples of colonization, on islands such stress intolerant descendants are not free to leave. This inability to leave the new colony presents the largest difference between island colonization and proto-domestication. Even if there is still a swimmable distance between a continental island and the mainland, a stress intolerant individual is as unlikely to attempt an escape from the island as to have chosen to swim there from the mainland in the first place. This is one reason I suggest that original founders must be particularly stress tolerant, at least when colonization is by deliberate choice.

So what happens when one of these less stress tolerant second or third generation founders becomes pregnant? We know that absolute levels of thyroid hormone rise significantly during pregnancy, often to double pre-pregnancy levels (115), because T_4 is essential for fetal development and growth at all stages. Studies in both humans and experimental animals have shown that when T_4 levels do not rise to appropriate levels during pregnancy, premature birth or reduced uterine growth are almost inevitable, resulting in smaller than usual newborns (392, 710). Because stress of any kind reduces T_4 production immediately, and often profoundly (296), the relationship between maternal stress and fetal growth becomes paramount to understanding island syndrome.

Restricted fetal growth and island dwarfs

Reduction of available T_4 due to maternal stress from any cause will impact fetal growth. It has been shown that small offspring produced because of reduced fetal growth seldom recoup this loss: even if growth is not restricted afterward, premature and low-birth-weight individuals remain small throughout life, and these traits appear to be inherited (235, 468, 642). Reduced fetal growth resulting from maternal stress can thus have population-level impacts over time.

It has also been demonstrated that small individuals mature earlier on average than larger ones (574), and for late-maturing large animals, such as elephants, this disparity can amount to several years difference. For example, a small female elephant that matures at eleven years may be capable of producing several more offspring over her lifetime than a female that matures at fifteen years—and so will all of her small, early-maturing daughters.

An extreme example of this phenomenon is seen in the profoundly dwarfed straight-tusked elephant, *Elaphas falconeri,* that lived on Sicily during the Pleistocene (about one million years ago): these one-metre-tall elephants appear to have reached sexual maturity as early as three to four years of age. In smaller animals, especially those that first breed at one year of age or less, early maturation may be very subtle, as we saw in the experimental population of Siberian silver foxes discussed in Chapter 2: in those silver foxes, the earliest breeding females every season were ahead of the rest by only a few weeks but they were always stress tolerant ("less fearful") individuals (50, 755). Smallish, early-breeding animals may well be the most stress tolerant individuals within many large-bodied mammalian populations.

Deliberate colonization and protodomestication bring such early maturing individuals together within founding populations because only stress tolerant individuals participate in these events. And because the only growth programs available among these founders are the specific thyroid rhythm patterns that produce stress tolerant early breeders, most offspring will inherit similar stress tolerance (along with other behavioural and morphological traits associated with those specific growth programs). However, a few offspring of these original colonizers will invariably have somewhat lower stress tolerance than their founder parents, a phenomenon that may recur in subsequent generations. Under circumstances of protodomestication and mainland colonization, individuals with reduced stress tolerance are free to leave the stressful habitat and return to the original population; in island colonization, such freedom to emigrate doesn't exist.

Pregnant females of large animals born during the early generations of island colonization which have reduced stress tolerance are likely to produce smaller, earlier maturing offspring. This is a direct result of the unique stresses involved in living on an island in conjunction with the effects of maternal stress on fetal growth. In small founding populations especially, smaller, earlier maturing individuals will quickly outnumber large individuals simply because of their faster generation times, not because of food shortages. Large individuals don't disappear entirely—indeed, as already mentioned, one of the characteristics of island populations is a pronounced variation in size and morphology relative to mainland forms (101, 282); for example, the population of extinct mammoths on Santa Rosa Island off California has been estimated to have had one large mammoth for every ten dwarfed ones.

In other words, shortage of food due to rapid population growth (habitat saturation) is unlikely to be a driving force in the early stages of island colonization: on the contrary, unless an island is extremely small there should be enough resources for many generations before the equilibrium point between population size and available food is surpassed (613). However, this is not to say that food *cannot* be depleted past a critical point: once the equilibrium point between number of animals and available food is surpassed, different levels of nutritional stress can have a similar impact on a population as the initial stresses of colonization because animals must compete for food.

As discussed above, pregnant females under stress due to competition for food will produce smaller offspring which subsequently produce small offspring themselves. Thus, a trend toward small size may also take place among island populations if conditions of prolonged competition for food exists, as has been proposed for Spitsbergen reindeer (Table 7.1).

To summarize, then: in protodomestication, stress tolerant individuals of large mammals are interbreeding exclusively within a small founder population where escape—*emigration*—is possible, and these produce slightly dwarfed offspring due to the reduced variation in growth programs available among founders (emigration of any stress intolerant individuals born in early generations keep the effects from being extreme). In contrast, in island colonization, stress intolerant individuals born in the early generations produce much more profoundly dwarfed offspring because the environment is emigration-proof; if island founders include a few stress intolerant individuals from the very beginning, as may happen when once-connected islands populated during times of lower sea levels suddenly become cut off by rapidly rising seas, even more profound dwarfing can occur (Figure 7.2).

Box 7.1 Reduced food & fetal growth

A short aside is called for here to discuss the different effects on growth precipitated by food limitations and maternal stress, especially since food limitation is often suggested as a major factor precipitating island dwarfing (109, 613, 703).

In a pregnant female, food deprivation (fasting) primarily affects the condition of the dam, because reduced food intake reduces the amount of T_3 available for her cellular and tissue functions (via reduced conversion of T_4 to T_3, which causes a reduction in basal metabolic rate). Because the fetus can utilize unconverted T_4, it is buffered from the direct metabolic effects of fasting unless fasting becomes prolonged. However, even short-term psychological stress caused by competition for food, or panic induced by futile searches for food (especially for females with underlying stress intolerance), will result in profound reduction in T_4 secretion from the thyroid gland—which immediately affects the developing fetus.

Therefore, it is primarily the *stress* experienced by the dam that negatively affects the developing fetus, not the reduced maternal basal metabolic rate resulting from the food shortage itself. And while food shortages may indeed restrict the growth of juveniles directly, growth-restricted youngsters may eventually catch up on growth if conditions improve. The effects of maternal stress on fetal growth, however, are not only more profound but irreversible.

Both moderate and extreme dwarfing can occur on the same island at different times due to repeated colonization events by the same ancestral species (as happened, for example, on the island of Sicily during the Pleistocene: here the extremely dwarfed elephant *Elaphas falconeri* came first and the only moderately dwarfed form, *E. mnaidriensis,* came later—although both are now extinct). Remains of extremely dwarfed individuals have been recorded on islands within 4,000 to 6,000 years after initial colonization events in extinct large mammals (such as red deer on the Isle of Jersey in the English Channel and mammoths on Wrangel Island on the northeast coast of Siberia), although the process almost certainly begins soon after arrival and proceeds rapidly (455, 456, 747, 771).

Stegodons on Flores Island, Indonesia—one example

The generalities outlined in the previous section can be used to explain the fossil record of the primitive Southeast Asian elephants known as stegodons on Flores Island, one of the larger islands of the Lesser Sunda archipelago that lies east of Java (Figure 7.3). Flores was never connected to the islands on either side, but was isolated by deep channels that narrowed just enough (to about 25 km) during times of extremely low sea levels to allow deliberate colonization by *Stegodon trigonocephalus* from Java (via Bali and Lombok to the west).

Colonization of Flores by stegodons from Java occurred at least twice: once at about 1,500,000 years ago (ya) and again at about 800,000 ya. The first colonization resulted in the generation of *S. sondaari,* the smallest of all Indonesian stegodon species. This dwarf stegodon had an estimated weight of about 300 kg (vs. 1,000–1,700 kg for ancestral forms) and molar teeth with more complex cusp patterns than its ancestor (766). The most recent well-documented remains of this dwarfed form date to about 900,000 ya but the species appears to have become extinct shortly after, suggesting it survived on Flores for about half a million years (767).

At about 800,000 ya, a second *Stegodon* colonization of Flores occurred, associated with yet another major drop in sea level. This group of stegodon colonizers evolved into *Stegodon florensis,* which survived somewhat longer than the previous species. However, *Stegodon florensis* is estimated to have weighed about 850 kg and so was only moderately dwarfed compared to mainland forms (766). *Stegodon florensis* is described as the most common fossil of Pleistocene Flores, suggesting that a thriving population of this species existed before it became extinct sometime before 12,000 ya in the wake of a major volcanic eruption (767, 541, 542).

The critical question appears to be why the first wave of stegodons became extremely dwarfed but the second did not, given similar lengths of occupation of the same island. (Reports that the very late Pleistocene stegodons on Flores Island associated with *Homo floresiensis* remains were also extremely dwarfed, as described by Mike Morwood and colleagues (541), are apparently an error (643).

As I mentioned briefly above, most examples of extremely dwarfed large mammals are associated with islands that were once connected to other land masses, thus presenting an opportunity for a small colony of both stress tolerant and stress intolerant individuals (a hormonally heterogeneous mix) to become stranded together. This pattern suggests that if early Flores stegodons were extremely dwarfed, they should be descendants of a hormonally heterogeneous founding population. But how could a hormonally heterogeneous group of stegodons get to Flores, when it was never connected to other islands in the archipelago? Turned around this way, perhaps the really significant question is not why Late Pleistocene stegodons on Flores did not become extremely dwarfed but why the earlier population *did*. At least two possible scenarios can be suggested.

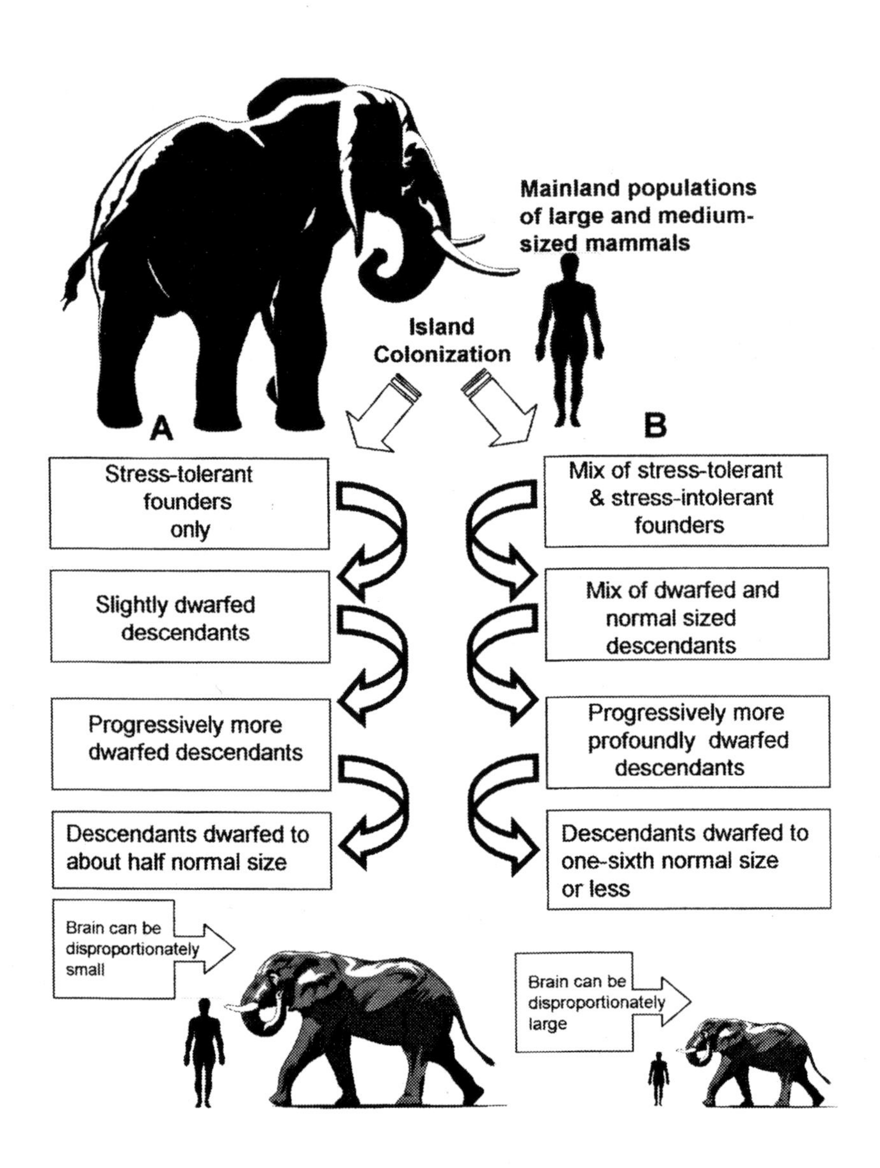

Figure 7.2 A model for insular dwarfing in large to medium-sized mammals

The first possibility is that the ancestral populations were different: while both early and middle Pleistocene ancestors technically belonged to the same species of *Stegodon*, they may not have been precisely equivalent. Colonizing founders from the two time periods may have differed enough in degree of stress tolerance to generate distinct descendants. This is precisely what has been suggested by researchers trying to assign appropriate scientific names to the many populations of elephants that inhabited islands of the Mediterranean during the Pleistocene, which all descended from a mainland straight-tusked elephant, *Elephas antiquus* (577).

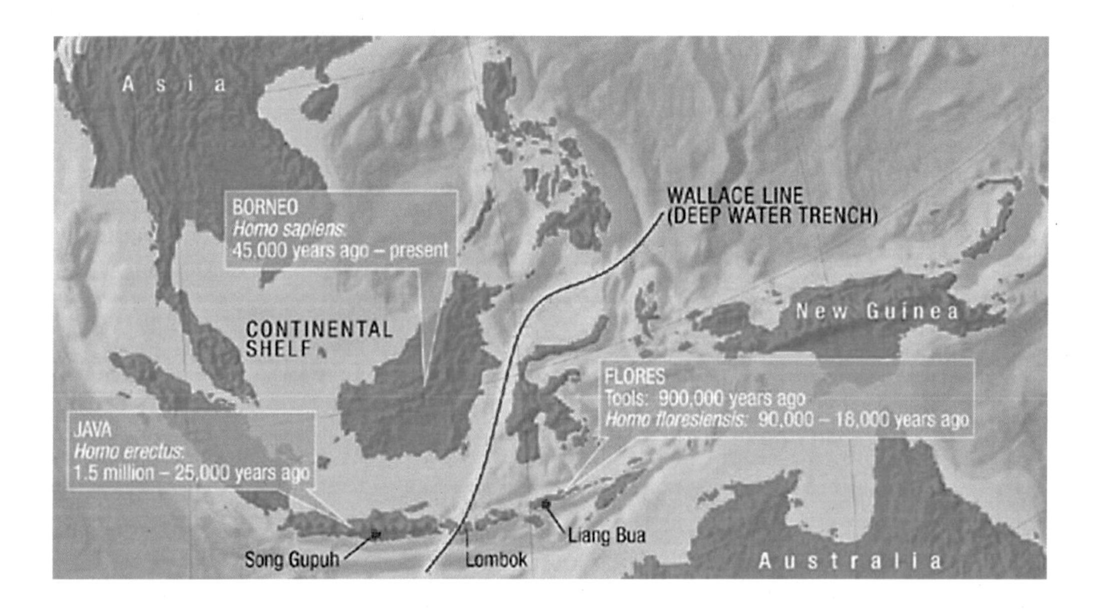

Figure 7.3. Map of Indonesia showing the location of Liang Bua, the cave on Flores Island where Homo floresiensis was discovered (from Nature, Vol. 434:432, R. Dalton, 2005. Copyright Nature Publishing Group)

Figure 7.4. Above, giant komodo dragon on Rinca Island, just east of Flores Island, Indonesia (photo courtesy Jesse McMillan). Right, Giant house mice left behind by ships live on the eggs and young chicks of sea birds (including the enormous young of various albatross species) that nest on the island of Gough in the South Atlantic (photo courtesy BBC News UK edition July 24, 2005, Jonathon Amos, http://news.bbc.co.uk)

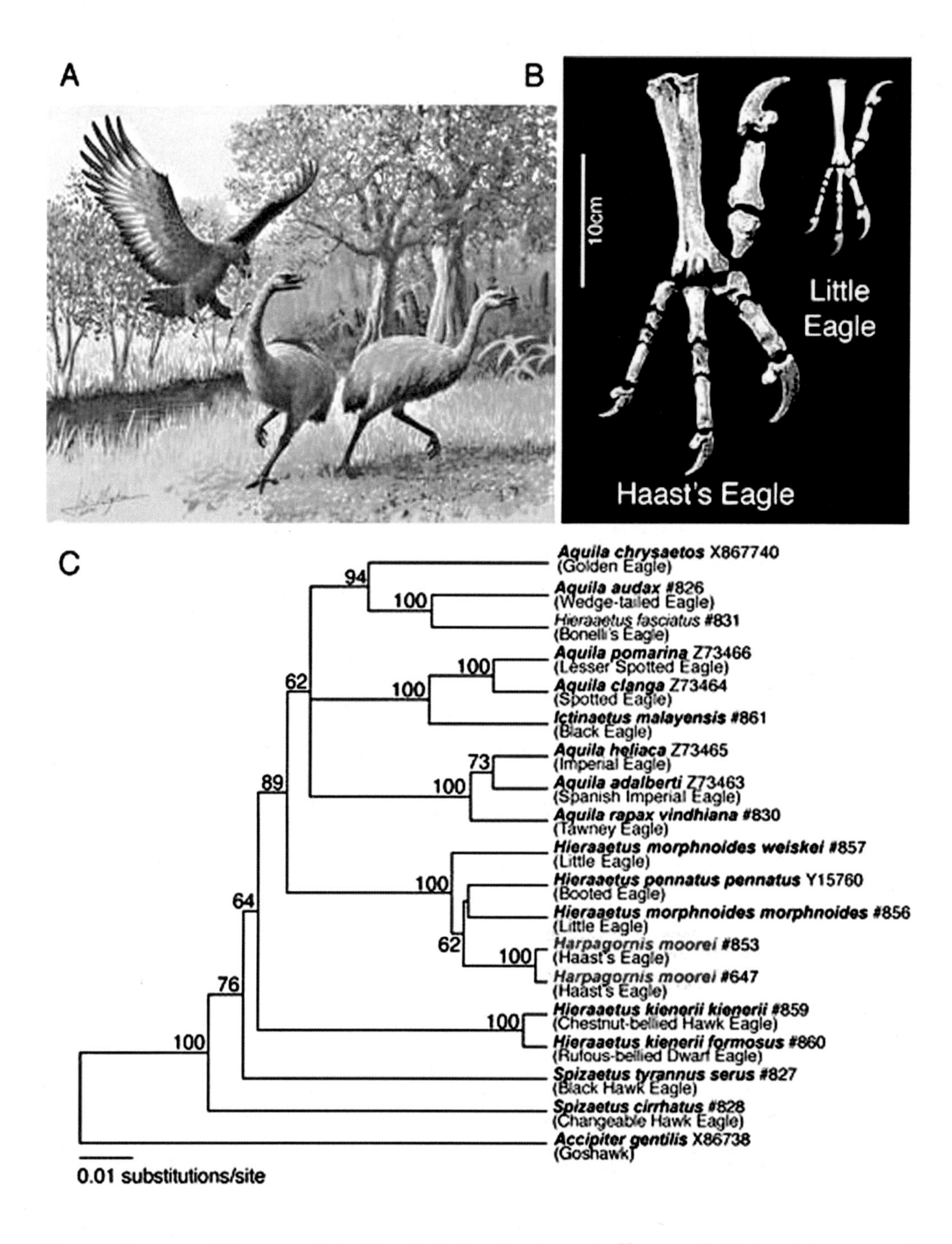

Figure 7.5. Images and phylogenetic analysis of New Zealand's extinct giant eagle, H. moorei *(A) An artist's impression of* H. moorei *attacking the extinct New Zealand moa (John Megahan). (B) Comparison of the huge claws of* H. moorei *with those of its close relative the* Hieraaetus morphnoides, *the "little" eagle. (C) Maximum-likelihood tree based on cyt* b *data (circa 1 kb), depicting phylogenetic relationships within the "booted eagle" group.* Harpagornis moorei *(red) groups exclusively with the small* Hieraaetus *eagles, and genetic distances suggest a recent common ancestor about 0.7–1.8 million years ago. Bunce et al., 2005 Public Library of Science*

There is another, perhaps equally likely explanation for extreme dwarfing in the earlier Flores population, and that is that the extended social bonds of elephant mothers and offspring encouraged a more physiologically and behaviourally heterogeneous mix of stegodons to swim offshore than occurs in other deliberately-colonizing large mammals. The prolonged mother/calf bond characteristic of elephants and their Pleistocene relatives especially, and perhaps to a lesser degree in Pleistocene hippos—as for modern *Hippopotamus amphibious* (301)—might induce stress intolerant offspring (even as adults) to accompany stress tolerant mothers on an offshore swim they would not undertake alone. This would explain why primitive elephants and their relatives, as well as hippos, often (but not always) became extremely dwarfed even when deliberate colonization events were involved.

The second wave of stegodon colonizers may have been only moderately dwarfed because on this occasion the population derived from a deliberate colonization event by a few stress tolerant individuals only, generating less extreme changes in growth programs. However, the possibility also exists that the second colonization event was different from the first because low sea levels during that period of the Pleistocene brought a predator to Flores: *Homo erectus*.

A viable population of near normal-sized stegodons may have survived longer than the previous extremely dwarfed population because hominid predation—hunting by *Homo erectus*—kept the animals relatively large and late-maturing. Hunting would also have kept the population small enough to prevent habitat saturation and subsequent dwarfing due to food shortages. As stated previously, a similar pattern of early colonizers becoming extremely dwarfed while later ones did not is also documented for the Pleistocene straight-tusked elephants (*Elephas antiquus*) that colonized the Mediterranean island of Sicily: remains of large carnivores are found on Sicily in association with fossils of medium-sized *E. mnaidriensis*, suggesting predation on small individuals of that elephant population may have occurred (101, 575, 576, 577). Evidence does indeed exist for late Pleistocene predation of stegodons on Flores: age profiles done on associated *Stegodon* remains suggest that juveniles were preferentially targeted by late Pleistocene hunters, although in earlier times very small adults (if they existed) may also have been taken (541, 542).

Dwarf Homo on Flores Island?

Why would adult stegodons have been avoided by late Pleistocene *Homo* on Flores, when adult mammoths were clearly hunted in other regions (226)? Enter the evidence that *Homo* on Flores was an island dwarf itself. Flores appears to have been colonized by *Homo erectus* from Java (via Lombok and Bali to the west), with finds of simple flaked stone tools dating their arrival at about 840,000 ya (766, 767, 817).

Skeletal remains of this small species, designated *Homo floresiensis*, suggest it stood only about one metre tall (426). With features strongly indicative of a *Homo erectus* ancestor, but with an unusually small brain, the remains nevertheless date to a period between 35,000 and 14,000 ya (average about 18,000 ya), perhaps extending back to about 74,000 ya (38, 96). While a continuous link between the earliest and latest finds has not yet been demonstrated (i.e., no skeletal remains of hominids, dwarf or otherwise, have been found in the oldest deposits, only tools), like other large mammals that colonized islands *Homo erectus* on Flores appears to have became a dwarf endemic species (159, 452, 824).

Stress tolerant *Homo erectus* individuals may have colonized Flores either by swimming or via constructed or naturally-occurring rafts. It's the deliberate nature of the journey, however, that makes this scenario equivalent to colonization by other large mammals. It's unlikely that *Homo erectus* was capable of constructing advanced watercraft as *Homo sapiens* later did, but eminently reasonable to assume it had discovered the floating characteristics of bamboo and mangrove logs. It's also unreasonable to assume *Homo erectus* was incapable of swimming moderate distances, especially in warm seas, given that such a journey could be assisted by a flotation device as simple as a bundle of bamboo sticks.

Note that *Homo floresiensis*, while definitely small, is more than one-half ancestral size (a range of 1.55–1.78 m for *H. erectus*, ref. 426), similar to the moderate dwarfing seen in *Stegodon* of the same time period on Flores. Although surprisingly petite, the small brain relative to body size reported for *Homo floresiensis* (about 380 cm^3 vs. 650–1,260 cm^3 for *H. erectus*, ref. 426) is not unique among medium-sized dwarfed mammals: Majorcan cave goats, for example, also had disproportionately small brains (405). In contrast, severely dwarfed elephants, such as *Elephas falconeri* from Sicily (reduced to one-sixth ancestral size or less), had relatively large brains for their body size (575, 576).

Debate continues regarding the status of this hominid and is not likely to end soon. After comparing the virtual brain *endocast* size and shape of *Homo floresiensis* with those from modern (*Homo sapiens*) microcephalic and pygmy individuals (plus those from *Homo erectus*), palaeoanthropologists have concluded that *Homo floresiensis* represents an endemic island dwarf descended from *Homo erectus* (or, less likely, that *Homo floresiensis* and *Homo erectus* both descended from an unknown small-brained and small-bodied ancestor (215), a possibility that is not well supported by any other evidence).

I see no reason (besides non-scientific anthropocentrism) why a colonizing *Homo erectus* would be exempt from the natural dwarfing processes that affected other island mammals. The presence of *Homo erectus* on the nearby island of Java is well-established (766). It's both plausible and biologically sound to consider *Homo floresiensis* an endemic island dwarf descended from a *Homo erectus* ancestor, via the same juvenilizing effects of deliberate island colonization noted in other medium- and large-sized mammals; whether it deserves a unique species name is another question entirely.

If this hominid is indeed an endemic island dwarf, we should anticipate that subsequent finds will show the marked size variation characteristic of all island dwarfs. We should also recognize that evolutionary changes undoubtedly continued in *Homo floresiensis*, perhaps along the same trajectory as the transformation of *Homo erectus* to *Homo sapiens,* and that these might produce some unique characteristics vaguely reminiscent of anatomically modern humans.

Another pertinent question is why more remains of dwarf island hominids—whether descendants of *Homo erectus* or *Homo sapiens*—haven't been found. Chance plays a huge role in this, of course, particularly the relatively slim chances of bone preservation in tropical locales and of modern discovery of fossil remains. In addition, *Homo erectus* may actually have been more susceptible to island dwarfing than later *Homo sapiens* because it had fewer and less sophisticated ways of manipulating its environment. *Homo sapiens* definitely used fire for cooking and had the ability to make shelter, clothing and boats, skills that put them at much less risk for the kinds of maternal stresses that produce dwarfing in other large mammals.

While thyroid rhythm theory does not prove that *Homo floresiensis* was an endemic dwarf of *Homo erectus,* it does provide a plausible and testable mechanism to explain such an outcome. However, the dwarfing characteristic of large herbivorous mammals is only one type of change associated with island colonization—what about the small animals that become giants (birds and reptiles as well as mammals) and the birds that lose the ability to fly?

Gigantic rodents and flightless birds

We cannot say that protodomestication and island colonization always produce dwarf individuals because in rodents (and some birds) the opposite appears to be true. Not surprisingly, early forms of domestic rodents (and birds) are larger than ancestral forms and the same often holds true for rodents that have colonized islands (6, 124, 723). Giant Indonesian rats and the Caribbean hutia (Table 7.1) are but two extinct examples. This reversal of the dwarfing phenomenon reflects basic differences in the effects of stress on growth and reproduction in these small animals: *stress tolerant rodents have been found to be larger, to mature earlier and to be more fecund (produce more offspring) than stress intolerant individuals* (523).

To understand the consequences of this difference between large and small mammals, you simply need to follow the reasoning outlined earlier for island colonization in large mammals. Colonization of islands by rodents usually occurs by chance, via rafting on storm wreckage or other kinds of random flotsam. Because these are not deliberately initiated events, founding populations of rodents are more likely to contain a mix of stress tolerant and stress intolerant individuals (even if it's only one of each). Larger-than-average individuals of small mammals are not only more stress tolerant but earlier to mature than smaller individuals. As a consequence, large individuals reproduce faster than small ones. Large individuals will soon come to outnumber smaller ones within a population, not because of plentiful resources or lack of predators, but because they reproduce faster. Offspring born to any stress intolerant rodent dams will be smaller and mature later, and as outlined above, will soon become outnumbered by large forms.

Birds that colonize islands and subsequently generate descendants with altered limb and beak proportions are also explainable by thyroid rhythm theory because in birds the thyroid hormone required for embryonic development is deposited in egg yolk by the hen (348, 644). Species-specific rhythms of maternal thyroid hormone in birds almost certainly generate a species-specific yolk characterized by layers with different thyroid hormone concentrations. Although species-specific thyroid layering in yolks has not yet been demonstrated experimentally, I contend that this (or something similar) must be present to create a concentration gradient of thyroid hormone in the yolk for the developing embryo to consume as it grows. In other words, *some mechanism must exist that allows the yolk to reflect maternal thyroid rhythms very closely*, because timing of delivery and amount of hormone delivered to thyroid-dependent genes in the developing embryo are critical (e.g., 86).

Changes in timely availability of thyroid hormone to developing chicks, caused by stress-induced disruption of maternal provisioning of thyroid hormone to yolk, are more likely to produce flightless descendants in *precocious* species (which, like chickens and ducks, are well developed at birth and feed independently within a few days) than in species with *altricial* young (which, like the starling and other small forest birds, hatch at a much less developed stage). Just as for large mammals, stress intolerant female offspring born in subsequent generations to an initial colonization by stress tolerant hens would tend to produce offspring with growth-rate related changes. Some body parts may reflect growth-rate reductions (such as underdeveloped wing bones) while others reflect growth-rate increases (such as overdeveloped legs and beaks), like those found in the dodo (see Figure 7.1 and refs. 77, 155, 459-461). Such disparate growth outcomes are probably due to under-provisioning of maternal thyroid hormone to some yolk layers in response to stress and perhaps compensatory over-provisioning to others, since limb and beak regions do not grow at the same time (535, 632, 746). Once these changes occur, however, they will be heritable over many generations, just as small size in mammals is (as discussed in Chapters 3 and 4, this is due in part to the fact that the growth and development of embryos—including the cellular architecture and nerve connections within the brain, and particularly the pineal and hypothalamus glands—is controlled by maternal thyroid hormones). The brain of the embryo must develop in such a way as to allow the offspring to grow after hatching in an appropriate, species-specific manner—alter the thyroid hormone pattern from maternal sources during embryonic development and not only do limb size and shape change but so does the cellular architecture of the hormone-controlling portions of the brain, perpetuating altered thyroid rhythms in subsequent generations.

When no predators are present, as is usual for remote oceanic islands, new traits such as reduced wings may not be particularly disadvantageous and so they can perpetuate. Consequently, completely flightless birds tend to survive over time *only* on remote islands (even if they are generated in other circumstances). Due to the rather strict requirements necessary for generating new flightless forms, these birds tend to descend from ancestral species that are strong-flying and migratory (behaviour required for getting them to remote islands in the first place) and from those that have relatively precocious development (extended egg time required to expose the developing chicks to the maximum impact of stress-induced disruption of maternal provisioning of thyroid hormone to yolk).

As a result, most of the flightless or near-flightless island species have descended from relatively few bird families: they derive almost exclusively from pigeons, rails, alcids, ducks or geese (109, 254, 459, 460). Some examples include the extinct giant dodo of Mauritius (below) and the large flightless solitaires of Rodrigues and Réunion, two small islands north of Mauritius (678)—both descended from pigeonlike species (note that the common rock dove and domestic pigeon, the best known members of this group, are anomalous in producing altricial young—most other members of the family produce precocious young and are strongly migratory). Rails, geese and ducks have many flightless or near-flightless members, including the nearly extinct flightless rail of Guam, *Rallus owstoni* (609), and a small duck, the endangered Auckland Island teal of New Zealand (461). Hawaii once sported a giant flightless goose that has never been named, recently shown by genetic analysis of skeletal remains to have descended from the common Canada goose, *Branta canadensis* (583). Ducks and geese are not only precocious (although slightly less so than gallinaceous birds like quail) but most are strongly migratory (103, 631, 675).

The now extinct great auk of several North Atlantic islands was a very large flightless member of a little-known bird family, the Alcidae. Alcids are common Northern Hemisphere sea birds; they are migratory and produce chicks as precocious as ducks and geese (108, 254, 459, 543, 705).

A number of flightless island species exist that don't at first glance appear to fit the above profile. However, in most cases, consideration of their particular life history traits explains their susceptibility to growth disturbances during island colonization and, in some cases, why they might colonize an island in the first place. One such example is the giant Galápagos cormorant: cormorants are large coastal sea birds, and although most are non-migratory and relatively altricial, they have large eggs and the chicks are quite large at hatching—which simply means that quite a lot of growth is going on inside the egg rather than after hatching (197, 705). A similar example is seen in the kakapo, a flightless parrot of New Zealand. Although most parrots are altricial, they nevertheless produce relatively large eggs and large chicks; most are also strongly migratory (776).

Another species that may seem at first glance an unlikely candidate to have become a flightless island species is the extinct giant owl of Cuba (609). However, many owls are strongly migratory, and like many birds of prey, most owls produce relatively large hatchlings compared to other altricial species (75, 108, 470).

Worth mentioning here are the extinct moas of New Zealand: although moas were also flightless birds, their origins (based on fossil and DNA evidence) predate the point at which New Zealand become an island (18, 133, 134, 770). As a result, moas and their kin are not normally considered examples of island syndrome.

Gigantism in reptiles and birds

Most reptile families tend to generate enlarged island endemics: think of giant island tortoises and the enormous Komodo dragon (a particular type of lizard known as a *varanid*) that still lives on a few islands in Indonesia (109, 609; see Figure 7.4, a colour photo in the "Stegodon on Flores" section). In reptiles, where growth does not stop with sexual maturation, rapid early growth can generate much larger sizes than usual, even if growth slows after maturation; in some species, relatively faster growth rates may also continue after sexual maturation.

Just as for small mammals, for reptiles and birds that colonize islands the critical feature is *individual variation in early growth rates*, because rapid early growth allows individuals to reach larger sizes before sexual maturity is attained. In contrast, dwarfing of large mammals is associated with marked *individual variation in timing of sexual maturation* because early maturation stops growth before maximum sizes are reached. Also, in reptiles, amphibians, fish and some birds, *fecundity* increases with increased size (59, 203): larger individuals produce more offspring. Over time in an isolated island population, larger animals of such egg-layers will become the most common form—small individuals don't disappear, but they can become very rare.

In some lineages of reptiles, such as tortoises, the natural response to stress is to slow early growth and delay maturation, which generates a larger body size at maturity. For example, in at least one tortoise, slow growth and delayed maturation resulting in larger body size have been shown to be a response to the stress of low temperatures (425). In other groups, such as pythons, more rapid early growth without a change in the age at maturity leads to larger sizes. While such growth increases in pythons are assumed to be simply the result of more abundant prey (486), another interpretation is also possible. For snakes that colonize islands (and perhaps in other isolated locations as well), dietary change may have a significant impact on descendant populations, if one takes age as well as size of prey into account.

Does the prey composition of snakes (and perhaps carnivorous birds) change much when they colonize an island, and if so, how and why might this contribute to size change? One good example of this is found in the giant tiger snakes that inhabit some islands off southern Australia. These big snakes eat mainly seasonally available shearwater ("muttonbird") chicks, because for much of the year there is little else available. Shearwater chicks are considerably larger than the prey consumed by small tiger snakes on the nearby coastal mainland, where no shearwater rookeries exist. As a consequence, the authors of a regional study on tiger snakes concluded that it's the absolute size of the prey item that controls snake size: large prey makes for large snakes (387).

However, I would point out that not all prey items are equal, gram for gram: a rapidly growing bird chick of any species has high levels of thyroid hormone coursing through its system, higher than levels found in adults (67, 348, 665, 814). Carnivorous bird and reptile species that colonize any new habitat (including islands) and consequently shift from eating adult prey animals (such as rodents) to young animals or even eggs, would almost certainly experience significant increases in their own early growth and in the early growth of their offspring because of the increased consumption of thyroid hormone. Increases in juvenile growth rates without a delay in age at maturity would lead to larger adult sizes. And because large adults produce more eggs, large individuals would soon dominate the population. Several other island populations of snakes that have markedly larger body sizes are known to eat primarily unfledged seabird chicks (70) and the same is true for some island rodents (such as the giant house mice that inhabit the island of Gough in the south Atlantic; see Figure 7.4, a colour photo in the "Stegodons on Flores" section).

Another example is equally illustrative, although it involves a carnivorous bird rather than a reptile. The giant Haast's eagle of New Zealand, now extinct, grew to tremendous size: it probably weighed more than 16 kg, larger than any eagle living today, see Figure 7.5. Although it has been assumed that the ancestor of this bird was a somewhat smaller eagle that colonized New Zealand from Australia, ancient DNA analysis of its bones by Mike Bunce and colleagues (82) have shown that its closest relative was very small indeed, an Indonesian eagle weighing only about 1 kg. How could such an incredible increase in size have occurred, in this case without a loss of flight? The authors of the genetic study imply that the giant eagle got big because it preyed on the adults of large flightless moas, a group of birds that's unique to New Zealand. In other words, they assume that, as for snakes, larger prey generates larger predators. I have another explanation.

I suggest the giant eagle got really large without losing its ability to fly because the ancestors that colonized New Zealand had a unique and very different prey item to subsist on: the newly-hatched chicks of flightless moas. Some of these moa species were very large (18, 81, 349) and would have produced very large chicks that would have been relatively easy to catch. As for the shearwater chicks mentioned in the tiger snake scenario, newly-hatched moa chicks would have had much higher levels of thyroid hormone than prey items—such as adult rodents or even adult small birds—that comprise the usual diet of many small eagles. Ingestion of prey items with higher than usual levels of thyroid hormone does not normally disrupt reproduction in carnivorous animals, but it could increase growth rates for developing embryos and newly hatched chicks.

Living eagle species are altricial: their chicks are born quite helpless but grow rapidly over a period of several months, totally dependent on their parents for food (75, 470). If fed thyroid hormone-laden foods during this time of rapid growth, size would almost certainly be enhanced; at maturity (given the same diet of moa chicks), these birds would tend to produce egg yolks with higher than ancestral levels of thyroid hormone, but without a disruption of thyroid rhythm patterns. Adult eagle females with the same diet of moa chicks would tend to produce egg yolks with thyroid hormone levels conducive to more rapid growth, leading to a relatively rapid attainment of gigantic size in the entire population without loss of flight ability.

Mike Bunce and colleagues include in their report an illustration of a Haast's giant eagle attacking a full grown moa of one of the largest species (reproduced as Figure 7.5, inset A, on the colour page in the "Stegodons on Flores" section), a drawing based on skeletal evidence of injuries on adult moas that would be consistent with such an attack. I suggest it's actually more likely that such conflicts arose as a result of a moa hen defending her victimized chicks, and not because this eagle actually depended on a steady diet of adult moas.

Selection on islands

As soon as variations in thyroid hormone rhythms generate distinct growth patterns and other life history traits among island colonizers, selective forces (e.g., differential predation or access to resources) can act on one segment of the population to shift the balance. For example, some researchers have suggested that dwarf *elephantoids* (true elephants, mastodons, mammoths and stegodons) may be capable of negotiating steeper slopes than normal-sized individuals, allowing them increased access to available resources (7, 703). The same may be true for other large mammals, such as hippopotamus and deer, that habitually colonize continental islands. Given such an advantage and without predation, very small individuals on mountainous islands might indeed be more reproductively successful than larger individuals.

If predators exist, however, small individuals of large taxa may be at risk of being picked off as fast as they can be produced. With selective predation on small individuals only, a founding population would increase much more slowly over time, and the average size would stay relatively close to that of the original founders or be only somewhat smaller.

Summary and discussion

Thyroid rhythm theory defines a biological mechanism capable of generating growth program changes in island colonizers fast enough to be effective in immediate *ecological time* (275), probably between one and twenty generations. If individual differences in thyroid rhythms alone account for virtually all evolutionarily significant differences in life history traits, as I maintain must be the case, the particular thyroid rhythms of individuals within small founding populations should modify life history traits of descendants in a few distinct ways. In other words, the particular thyroid rhythms possessed by founders profoundly affects the outcome of colonization events: this should be as true for island colonization as in mainland occurrences.

I contend that it is the particular combination of stresses unique to islands plus the thyroid rhythms possessed by the individuals who make up the founding populations that interact to shape island colonizers so consistently. Many of these island stresses are similar to those associated with protodomestication, where deliberate colonization leads to rapid but moderate changes in growth programs and thus also to changes in size at sexual maturity. Over time, the stresses unique to island environments (especially lack of emigration potential for stress intolerant individuals), drives further size reduction or enhancement. When chance or tightknit social bonds between individuals result in islands being colonized by a small group with variable stress tolerance levels, changes in growth programs can reach extremes.

Deliberate colonization by stress tolerant individuals is required to explain why island syndrome is moderate in some instances and extreme in others, although predation may explain a few cases of less extreme size reduction. Differences between mammalian orders in the physical form that's stress tolerant account for the disparity in direction of growth-rate change; in birds, differences in developmental stage and size at hatching explain disparities in shape change, which I've summarized at the end of this section. Thyroid rhythm theory provides a more comprehensive explanation for the patterns of dwarfing, gigantism and flightlessness documented for island mammals and birds than previously proposed hypotheses because it describes a testable biological mechanism to account for observed trends in behavioural and morphological change.

Of course, this application of thyroid rhythm theory still hasn't been tested. As stated earlier, thyroid rhythms are not easy to measure because they require automated, surgically-implanted blood sampling devices and assays capable of measuring minute quantities of hormone. However, these should not be insurmountable problems. The fact that both rodents and birds have long been used in laboratory experiments means they should also make suitable models for testing of thyroid rhythm theory as it applies to island syndrome, since island colonizing rodents are affected by a suite of physical and behavioural changes as predictable as those affecting larger mammals. In fact, some selection experiments have already been performed on birds that rival to some extent the Russian silver fox study discussed in Chapter 2. For example, a population of Japanese quail was used to select, over forty-seven generations, only the heaviest individuals at four weeks of age; at the end of the selection period, adult selected birds were twice the size of those from an unselected population and they were differently proportioned, with relatively larger pectoral muscles, reduced wings and smaller brains (631).

The following implications need to be tested in conjunction with the proposition that thyroid rhythm theory can be usefully applied to the phenomenon of island syndrome: 1) that thyroid rhythms are species-specific, with individual variations; 2) that individuals with similar thyroid rhythms should also possess similar stress-response behaviours, physical traits and reproductive characteristics; 3) that interbreeding small groups of individuals that possess only thyroid rhythms associated with stress tolerant behaviour should produce many offspring with thyroid rhythms similar to the parent population but a few with disparate rhythms; 4) that after several generations of breeding stress tolerant individuals together, the population should show evidence of changes in growth-rate programs that should be somewhat less pronounced if individuals with disparate rhythms are removed, and more pronounced if they are left to interbreed.

Bird models in particular are needed for testing two additional hypotheses: 5) that maternal provisioning of thyroid hormone to egg yolk results in species-specific layers of thyroid hormone concentration or some similar concentration gradient; 6) that subtle changes to the thyroid hormone concentration of these concentration gradients can be initiated by administering appropriately-timed stress to stress intolerant hens *and* that such disruption will generate non-lethal changes in size and/or body proportion in the offspring.

In conclusion, I maintain that life history traits are certainly important in bringing about the changes in morphology, behaviour and reproductive physiology associated with island syndrome, but recognizing that these factors affect populations over time does not address the issue of how such changes are initiated, coordinated or controlled. It is my contention that unique thyroid rhythms alone may account for all individual differences in life history traits and that none of these traits are actually controlled in any evolutionarily significant way by discrete, independently inherited genes. Because a shift in thyroid rhythms affects all thyroid hormone-controlled traits simultaneously, both rapid short-term adaptation of individuals and permanent transformation of populations over time are possible.

Thyroid rhythm theory offers a unique model for investigating island syndrome that can be applied to all animals, including *Homo erectus*. The argument for this is summarized here:

- Large mammals get smaller with protodomestication and island colonization (*paedomorphosis*, or juvenilization) because colonizers are usually stress tolerant phenotypes, with thyroid rhythms that typically generate early maturation at smaller sizes. Stress intolerant phenotypes born in any generation (or included by chance with the original founders) respond to stress by producing even smaller offspring (very extreme dwarfing). The greater the number of stress intolerant individuals present among founders (or born in early generations), the greater the degree of dwarfing that is likely to occur. Small individuals soon outnumber large ones because they have faster generation times (a process equally applicable to *Homo erectus* colonizers), even when plenty of food is available.
- Small mammals and precocious birds get larger with protodomestication and island colonization (*peramorphosis*, or gigantism) because stress tolerant individuals with their particular thyroid rhythms are relatively large, have early maturation and increased fecundity. Large individuals soon outnumber small ones because they have faster generation times, not because there is a lack of predators or especially plentiful food supplies.

- Maternal stress can affect fetal development unevenly in stress intolerant precocious birds, resulting in the stunting of some traits (*paedomorphic* change) and overdevelopment of others (*peramorphic* change), due to differences in the precise timing of stress-induced disruption of maternal provisioning of thyroid hormone to yolk combined with differential timing in development of various body parts.

Recommended reading

Anderson, A.J. 1989. *Prodigious Birds: Moas and Moa-hunting in Prehistoric New Zealand.* Cambridge University Press, Cambridge.

Balter, M. 2005. Small but smart? Flores hominid shows signs of advanced brain. *Science* 307:1386-1389.

Bunce, M., Worthy, T.H., Ford, T., Hoppitt, W., Willerslev, E., Drummond, A. and Cooper, A. 2003. Extreme reversed sexual size dimorphism in the extinct New Zealand moa Dinornis. *Nature* 425:172-175.

Bunce, M., Szulkin, M., Lerner, H.R.L., Barnes, I., Shapiro, B., Cooper, A. and Holdaway, R..N. 2005. Ancient DNA provides new insights into the evolutionary history of New Zealand's extinct giant eagle. *Public Library of Science* 3(1):e9 (3 pgs). [this journal is freely available online—no subscription needed]

Carlquist, S. 1965. *Island Life: A Natural History of the Islands of the World.* The Natural History Press, Garden City.

Gaskell, J. 2000. *Who Killed the Great Auk?* Oxford University Press, Oxford.

Geist, V. 1998. *Deer of the World: Their Evolution, Behavior, and Ecology.* Stackpole Books, Mechanicsburg.

Gough Island sea birds: http://news.bbc.co.uk/1/hi/sci/tech/4708899.stm and http://www.ezilon.com/information/article_7046.shtml

Williamson, M. 1981. *Island Populations.* Oxford University Press, Oxford.

Wong, K. 2005. The littlest human. *Scientific American* (February):56-65.

Quammen, D. 1996. *The Song of the Dodo: Island Biogeography in an Age of Extinctions.* Scribner, New York.

Vartanyan, S.L., Garutt, V.E. and Sher, A.V. 1993. Holocene dwarf mammoths from Wrangel Island in the Siberian Arctic. *Nature* 362:337-340.

Chapter 8

A Final Example: Explaining Human Evolution

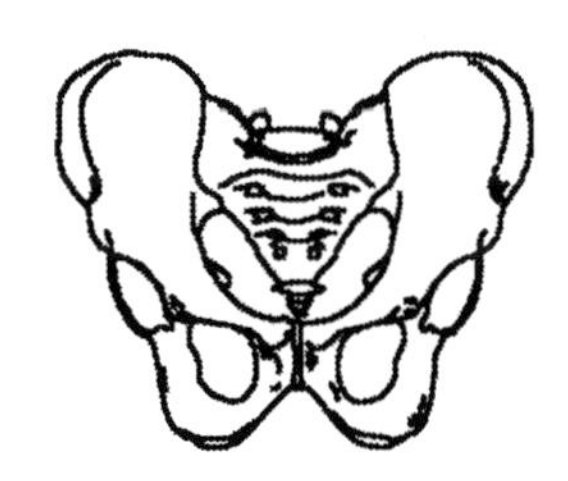

The conundrum of the first hominids

Several aspects of the traditional human evolution story have always bothered me and one of those regards the emergence of hominid hunters from stocks of vegetarian apes. The earliest ancestors that have been identified within our lineage were clearly *bipedal:* they had skeletons so modified from their ape ancestors that they were able to walk upright on two legs all or most of the time. These smallish hominids are assumed to have been rather strict vegetarians—eaters of leaves, fruits and seeds of fruits—based on patterns of wear on fossil teeth. The next bipedal species in line was a slightly larger creature that may not have been entirely carnivorous but does appear to have gotten most of its calories from animal products. Even though crude stone tools are found associated with these *Homo habilis* fossils, current opinion is that this second bipedal hominid species was probably a scavenger rather than a hunter. By waiting for larger four-legged hunters like lions to capture prey and consume what they wanted, these hominids were supposedly able to compete successfully with other scavengers, like hyenas and vultures, for possession of an abandoned carcass. The smashed bones of large prey animals found in association with *Homo habilis* remains suggest it was adept at breaking the long bones and skulls of these stolen carcasses to expose fat- and nutrient-rich marrow and brain tissue.

What bothered me about this proposed scenario was how an animal that ate primarily fruits and vegetation would become, in one speciation step, an animal that was almost exclusively carnivorous. This is an enormous change for any animal to undertake. It suggests that some of those first bipedal hominids, known as Australopithecines, made a decision to move—from a forested area full of fruit and seeds—onto the open savannah to eat marrow from scavenged carcasses. How would they know marrow existed if they were primarily fruit eaters?

When I decided the time had come to apply my thyroid rhythm theory to human evolution, the first challenge was to explain how apes came to walk upright. And the key to answering that question, it seemed to me, lay in explaining why and how apes became carnivores.

Whence came the hunter?

Australopithecines are assumed to be fruit and nut eaters *only* because of wear patterns found on their teeth. The many pits and scratches found on fossil teeth appear to match well with the wear patterns found on the teeth of modern fructivorous primates. Although this evidence seemed very slim to me, it's rarely questioned. However, a fairly new technique known as *isotope analysis* (which measures the kinds of carbon atoms, the chemical "backbones" of all plant material, that are left in bone and other tissues) has recently advanced to the point where it can be used on fossil bone (that is, bone that's been replaced by rock). Isotope analysis methods can distinguish between the kinds of carbon atoms that are found primarily in grasses and those found in trees and shrubs. As a consequence, isotope analysis can tell if an animal ate primarily grasses (like a gazelle would) or shrubs (like a giraffe would).

Isotope analysis performed on several fossils of early hominids as well as on some other animals that lived alongside them yielded surprising results. The isotope signatures of the first hominids (Australopithecines) were not like those of plant-eaters of any sort but were similar to carnivorous hyenas of their day (708, 709). This means that the initial dietary change from fruit and leaves to animal prey took place before or concurrently with the changes that made those first hominids bipedal. This confirmed what I had already surmised about thyroid hormone and speciation events: that the skeletal and muscular rearrangements that allowed Australopithecines to walk upright had to have been associated with a major change in both habitat *and* diet.

Such a major evolutionary event is what biologists call *macroevolution.* In macroevolution, changes occur that are of such magnitude that they prompt us to classify the descendants into a new genus or even a new family. Such profound changes clearly exceed what we see in what is called *microevolution*—the speciation events that generate animals with small, yet distinct differences. And yet, the extreme changes associated with macroevolutionary events appear as suddenly, and in as well-coordinated a fashion, as in microevolution (590, 611). The question is, do the same evolutionary processes control both—or are these events fundamentally different in some way? I suggest that if we use thyroid rhythm theory to explain what's going on, it's clear that both are part and parcel of the same process. Macroevolutionary change simply involves more profound changes to an animal's lifestyle than simple speciation does: the habitat change is more extreme, and profound changes in diet almost certainly occur. As I explain in this chapter, the story of human evolution is a perfect example of how the environmental variables of habitat and diet contribute to the stability and the flexibility of animal life over long stretches of geological time.

The first hominid to walk

The earliest hominids are the Australopithecines, bipedal ancestors that appear in the fossil record at ca. 4.4 million years ago (mya). The pelvic, vertebral and femoral shape changes that allowed bipedal locomotion in Australopithecines preceded other morphological characteristics that make later hominids unique (229), see Figure 8.1. But what initiated such particular morphological changes in the first place? Various suggestions have been advanced, most of which assume that bipedal morphology arose because it conferred some survival or reproductive advantage in and of itself (such as for carrying infants or food, or for helping animals keep cool in hot tropical climates) (354, 446, 803, 804).

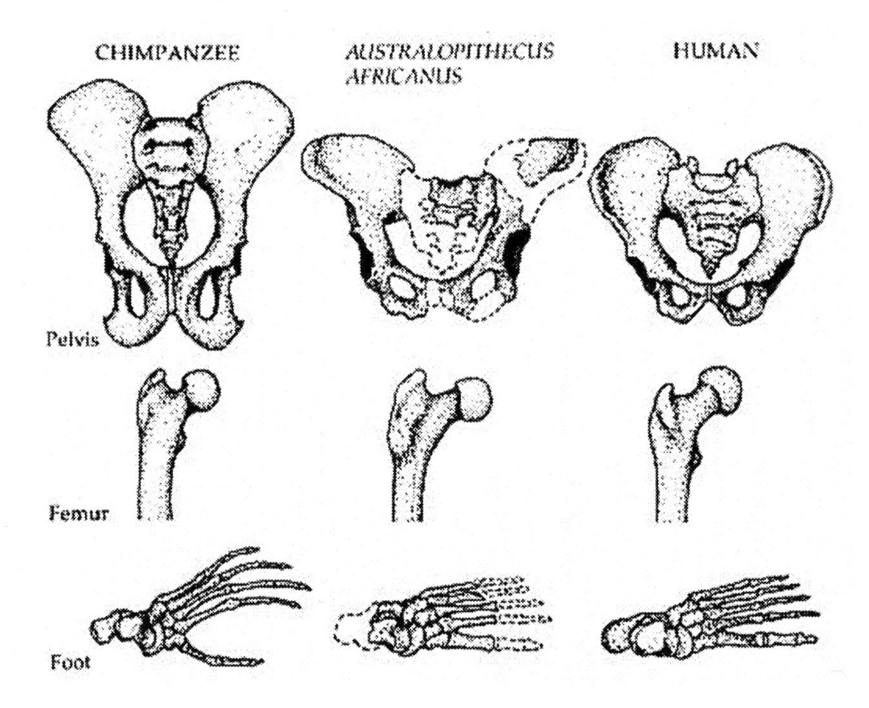

Figure 8.1 Some skeletal changes that accompanied the transition to bipedal locomotion (from Fleagle 1999)

However, you don't get bipedal morphology (or a bigger brain, or a shorter gut) because they would be advantageous to survival. *Evolution is not a mail-order catalogue*: natural selection can act only on traits that are already present within a population, and those traits must convey either distinct survival advantages or disadvantages (508, 511). This means that some members of the first generation of hominids had to have had bipedal morphology while others did not. Recent analysis of Australopithecine skeletal anatomy confirms that their bipedal traits were caused by changes in developmental rates or shifts to the timing of growth spurts (58). This suggests strongly that a speciation event associated with changes in growth rates precipitated the changes associated with bipedalism. And because the changes are fairly profound, we can assume that this speciation was very likely associated with colonization of a distinctly different habitat that also involved a change in diet.

Early Australopithecines fossils are found with fossils of other animals that typically live in closed woodland forests or forests mixed with bush (181, 618). I suggest that Australopithecines evolved with a novel bipedal morphology because climatic and subsequent environmental changes pressured some of their ancestors to colonize a woodland habitat in which the prevalent foods were not fruits but small animals: insects and grubs, bird eggs and fledgling birds, small mammals, reptiles and amphibians. The dietary change associated with this move out of fruit-bearing forests would have been profound because it involved the consumption of vastly increased amounts of dietary thyroid hormone.

Small prey items such as rodents, reptiles, amphibians and young birds are generally eaten whole, which means that their thyroid glands and livers (which contain especially high concentrations of thyroid hormone) are consumed as well. As discussed in Chapter 3, not only do egg yolks of all vertebrates also contain thyroid hormone, but young animals in their fast-growing newborn stage contain much more thyroid hormone (gram for gram) than do adults of the same species (348). This thyroid hormone can be absorbed unaltered through the digestive tract and utilized by the body as if it were self-produced.

Animals that have always been carnivores appear to handle massive influxes of dietary thyroid hormone without problems (379). However, herbivorous or fructivorous animals, even if they occasionally ate small animals, would have possessed a thyroid hormone metabolism unprepared for a real excess of thyroid hormone. Consumption of large quantities of thyroid hormone-laden foods (rather than occasional small amounts), day after day and month after month, would have had a major impact on populations of early Australopithecines ancestors, especially on females of reproductive age.

I suggest that only those individual Australopithecine ancestors who were relatively tolerant of high stress situations would have chosen to colonize a radically new environment in the first place. Experimental domestication, as discussed in Chapter 5, suggests this that stress-resistant component could comprise as much as 20% of an existing population (50). Even stress tolerant colonizing individuals, however, would have varied somewhat in their ability to accommodate a dramatic change in food resources without a major disruption of their reproductive potential.

In experimental animals such as rats, there is abundant evidence that thyroid hormone from external sources crosses the placenta. High doses of thyroid hormone given to pregnant rats result in offspring with birth defects (149, 150). Significant changes in normal thyroid hormone levels during pregnancy (either too much or too little) certainly have profound effects on the developing embryos of modern humans as well (24, 157, 598, 651, 797). There is no reason to expect that incipient Australopithecines would have responded much differently than rats or modern humans to consumption of dietary thyroid hormone that far exceeded their normal intake.

It's very probable that the major shift in diet I propose for early Australopithecines, made necessary by the change in habitat, would initially have resulted in some instances of reduced fertility (failure to ovulate or conceive, plus repeated miscarriages and stillbirths) and a relatively high incidence of birth anomalies of various kinds. Offspring afflicted with profound anomalies probably died young. However, survival rates of infants with relatively minor anomalies, such as the changes in skeletal shape and musculature that allowed them to stand upright with ease, may have been quite high. Such changes need not have been advantageous to be perpetuated. As long as bipedal characteristics did not *negatively* impact the survival of afflicted individuals, the animals would have had a reasonable chance of living to sexual maturity and passing their genes on to the next generation. If we view bipedalism as something akin to an unavoidable birth anomaly that could be adapted to by adjusting behaviour, it gives quite a new perspective to hominid evolution.

A possible modern analogue of this scenario exists in another primate, although the underlying causes are somewhat different. The minor birth anomalies long known to exist in some populations of Japanese macaque, *Macaca fuscata* (and possibly caused by chemical contaminants in provisioned food), are clearly mitigated by the determination of some mothers to raise afflicted offspring regardless (760). Some afflicted offspring with truly crippling anomalies not only survive but are reproductively successful. One well-known female named "Mosu" lived for twenty-six years and raised five offspring despite severe defects in both hands and feet, in part because *her* mother didn't abandon her as an infant. This ongoing natural experiment suggests that primates can indeed adapt their maternal behaviour enough to successfully raise offspring with relatively minor birth anomalies.

I'm suggesting that early Australopithecine offspring with viable new skeletal anatomy were generated because their mothers possessed the genes for a particular thyroid rhythm that was not totally disrupted by a diet high in thyroid hormone. This "resilient" thyroid rhythm allowed almost-normal fetal growth and development to proceed in the face of a major dietary shift. The offspring of these first hominids would not only have had some minor physical anomalies, but probably some behavioural and reproductive function differences as well (based on what we know about the wide-ranging effects of thyroid hormone). Colonization of a radical new habitat, and the associated dietary switch it necessitated, precipitated the expression of several new skeletal and behavioural variations, but natural selection was responsible for the fact that bipedalism was the option that survived over time.

Offspring with bipedal changes seem to have been favoured because these changes were advantageous, or at least not disadvantageous, to survival. Animals with such anomalies would have had a good chance of inheriting a thyroid rhythm similar to that of their mother, making it more likely that they would produce bipedal infants themselves. Over the next few generations, the specific growth programs that produced a bipedal body type would have become the norm for the whole founding population.

What about the behavioural and physiological effects of disruptively high levels of dietary thyroid hormone? Altered newborn growth programs were probably also associated with changed adult behaviours and reproductive physiology, through slight but significant shifts in the placement of fetal SCN clock genes that control thyroid rhythms. We know that timing and absolute amounts of thyroid hormone are critical to the development of embryonic brain cellular architecture (252, 437), and so excessive amounts of dietary thyroid hormone could have disrupted maternal thyroid rhythms in such a way as to alter the normal migration, proliferation and maturation of SCN oscillating cells in their offspring. In other words, slight differences in the physical relationship of these neuron cells to each other might have altered their combined output just enough to produce thyroid rhythm differences capable of affecting postnatal growth and behaviour in the offspring.

It's possible that excessive levels of dietary thyroid hormone impacted founding populations of early bipedal hominids by both providing strong selection on resilient maternal thyroid rhythms *and*, through disrupted maternal rhythms, generating slight but permanent changes in fetal SCN architecture (and if you think about it, this might also explain why so-called "cloned" animals, which are genetically identical to each other but often spend their fetal development time in different mothers, are far from "identical," as explained in Box 8.1; for more information, see the recommended readings at the end of this chapter).

Support for the association of excess dietary thyroid hormone with this speciation event comes, as I suggested earlier, from carbon isotope analysis of *Australopithecus africanus* fossils from South Africa. The studies indicate that these first hominids consumed animal prey that ate grasses (208, 209). As mentioned, their isotope signatures are actually more similar to those of carnivorous hyenas than to any other herbivorous grazers or browsers of their time. Fossils of other members of the same family (genus *Paranthropus*) recovered from deposits dated shortly after this time have very similar isotope signatures. (Some palaeoanthropologists insist *Paranthropus* is not different enough to be placed in a new genus, that it is simply another species of *Australopithecus*: alas, such taxonomic debates are only too common, making it awfully confusing to sort out fact from opinion in the human evolution literature.)

Ironically, while the authors of one of these carbon isotope studies conclude "we must seriously consider the possibility that these hominids [*Australopithecines*] were ^{13}C-

enriched [contained the type of carbon found in grasses rather than shrubs] because they consumed animal foods" (ref. 708:369), they offer only insects (termites) or the young of grazing bovids as possible choices. What about snakes that consumed grass-eating rodents, or the rodents themselves? Have we become so fixated on later hominid big-game hunting that we can consider no other prey items for their ancestors? It seems far more likely that small animals, fledgling birds, and eggs of all kinds would serve as an easy and natural transition from a life of herbivory to carnivory.

My diet-plus-habitat change explanation for the rapid generation of bipedal characteristics raises the question of whether bipedalism is exclusive to the hominid lineage. Claims of bipedal morphology (either based on direct evidence, from limb bone shape or inferred from skull shape) have been made for a number of other species, including:

- *Ardipethecus ramadens* (4.4 mya, refs. 229, 805)
- *Australopithecus anamensis* from east Africa (4 mya, refs. 785, 440)
- *Orrorin tugenensis* (5–6 mya, ref. 451)
- *Sahelanthropus tchadensis* (6–7 mya, ref. 825)
- *Oreopithecus bambolii*, the "swamp ape" from Sardinia (7–9 mya, refs. 404, 645)

Box. 8.1 Thyroid rhythms & non-identical "clones"

Maternally-mediated effects on SCN architecture represent a possible mechanism to explain an intriguing phenomenon disclosed by researchers working on the new technological process of "nuclear transfer" (often colloquially, but erroneously, called "cloning"). While using genetically identical cells implanted into a surrogate mother, this technique has so far failed to produce truly identical offspring. For example, Carol Ezzel's editorial (213) cites work by Ted Friend and Greg Archer (Texas A & M University) that demonstrates this effect on pigs: among two litters of pigs (one consisting of four, and the other of five, offspring) produced by implantation of genetically identical embryos into two different surrogate sows, there was as much individual variation in physical and behavioural traits (e.g., in coat "bristliness," number of teeth, food preferences and temperament related to being handled) as within two natural pig litters. Similarly, the 24 healthy cows generated so far by implantation of genetically-identical embryos into different surrogate mothers show entirely normal differences in temperament (432). In fact, behavioural differences are so pronounced in these genetically-identical cows that the social dominance hierarchies typical of naturally-produced herds have become established in the "cloned" herd as well; and the ages at which cloned cows attained sexual maturity varied as much as the range exhibited by naturally-produced offspring (10 to 12 months). See also the reports on a cloned dog (441), rat (848), mule (827), horse (248) and cat (683), and a review of the typical genetic explanation for such effects (616).

If some or all of these species prove to hold up as bipedal, we may not only have to revise the earliest date for hominid origins (721), we may have to acknowledge that bipedalism is not exclusive to this lineage. The pertinent question is: which of these genera are ancestral to the line that produced later *Homo sapiens* and which specimens represent cases of *convergent evolution* (similar or identical changes evident in unrelated lineages)? If convergent evolution seems the best interpretation of these fossils, can we explain why?

At least two periods of climatic change near the end of the Miocene (one at about 6.2 mya and an earlier one at about 7.8 mya) resulted in more arid regions across East Africa (181). The primary result of these changes was a marked reduction in the variety and abundance of fruit-bearing trees, as zoologist Jonathon Kingdon has pointed out in his recent book (394). This suggests that more than one ape lineage may have been forced to abandon fruit in favour of a diet heavily dominated by small animals. Kingdon himself comes to precisely this conclusion, but without realizing what the developmental consequences of such a dietary change would have been for vegetarian primates.

Consumption of high levels of dietary thyroid hormone would have had similar effects on any ape population as that described for Australopithecines: not identical, but similar enough to cause us confusion in sorting out the scanty fossil remains of closely-related lineages. Just because *Orrorin* was bipedal doesn't automatically make it a direct human ancestor (although the same could also be said for *Australopithecus*). Sorting them all out will take many more years of work and more quality specimens, but thyroid rhythm theory makes possible a number of new options in the interpretation of early hominid fossils.

To reiterate, I maintain that the ancestors of *Australopithecus* had to have been much more omnivorous that we have so far assumed, with small animals a minor but consistent part of their diet. Early Australopithecines later became bipedal apes with slightly larger brains as a direct result of a significant increase in their consumption of whole small animals—made necessary by a major shift in habitat from one vegetation type to another (the vegetation changes themselves being driven by changes in rainfall and temperature). Population growth of this reproductively successful lineage would have led to further subdivisions (additional species) as the group expanded into adjacent territories and adopted local dietary specializations.

For example, fossils of several species of the late robust Australopithecine genus, *Paranthropus,* are found associated with animals typical of wet grasslands (618). Such a habitat suggests that *Paranthropus* may have had access to a greater range of small animal species, perhaps including ground-nesting water-fowl, small aquatic mammals, turtles, and shallow-dwelling fish, crayfish and shellfish. These items, especially eaten whole (including the shells and bones), would likely have been as tough to chew and as abrasive as hard nuts and seeds, which could account for wear patterns found on their fossil teeth.

The dietary shift proposed here for *Paranthropus*, from consumption of quite small animals to larger and more varied ones (including things with shells and spines), seems a likely stimulus for the invention of simple cutting or crushing tools. Thus we might expect to see some evidence of simple tool use at this stage and, indeed, this seems to be the case. The skeletal anatomy of hand and foot remains of *Paranthropus robustus* suggests that they possessed the manipulative ability to use tools, although direct evidence for tool use is still lacking.

The first hominid tool user

Homo habilis retains the title of first hominid tool user. *Homo habilis* appears to have diverged from a late Australopithecine (*Paranthropus boisei, Australopithecus afarensis*, or one of their close relations) at the beginning of the Pleistocene, about 2.3 mya. Fossils of this species are associated with somewhat drier, more open habitats where bovid species like gazelles were plentiful (618). Although these first members of the genus *Homo* likely scavenged carcasses of large species rather than actively hunting them, to do so meant early *Homo habilis* had to forage in exposed habitats where carcasses and predators abounded. In other words, they must have had thyroid rhythms that rendered them tolerant of stresses inherent to much more open habitats than their ancestors experienced. The scenario for speciation via colonization of a new habitat is the same here as in protodomestication: a few stress tolerant individuals leave the old habitat for a new one, where interbreeding amongst the few thyroid rhythms represented in the group of colonizers results in distinct growth patterns, physical attributes and behaviour. After only a few generations, the colonizers are different enough in behaviour and reproductive physiology to discourage interbreeding with individuals from the ancestral population. From this point onward, they represent a distinct species that has a unique evolutionary future.

Since the dietary changes faced by early *Homo* colonizers were not nearly as large as those faced by early Australopithecines, we would not expect the physical changes associated with these speciation events to be as dramatic, and indeed they are not. *Homo habilis* was larger than any Australopithecine, with a slightly larger brain. However, other changes are so minor that some palaeoanthropologists argue for placing *Homo habilis* in the genus *Australopithecus* (where they would also put *Paranthropus*).

The slightly different growth and maturation schedules possessed by *Homo habilis* seem to have affected brain development in such a way as to give them slightly more dexterity (for tool manufacture and use), and perhaps also better decision-making skills. Such neural rearrangements almost certainly gave this species distinct survival advantages in its new exposed habitat. Although *Homo habilis* probably still ate some small animals and vegetation, their diet appears to have included a substantial amount of marrow and brains from scavenged carcasses (229). Such a diet would have been proportionally higher in the essential fatty acids necessary for brain development and function—particularly arachidonic acid (AA) and docosahexaenoic acid (DHA) (340, 341, 343).

While brains are also a good source of dietary thyroid hormone, they aren't as rich in the hormone as liver, kidney or thyroid glands themselves. Unprotected organ tissues are almost always consumed by the primary hunters of any prey and are therefore not usually available to scavengers. Consequently, *Homo habilis* (and similar species of that era, such as *Homo rudolfensis*) must have consumed smaller quantities of dietary thyroid hormone than their Australopithecine ancestors did, even if a modest proportion of their diet still consisted of small whole animals. Selection would thus have favoured individuals whose thyroid glands could produce adequate amounts of thyroid hormone in the absence of high dietary quantities.

If thyroid hormone was at times available in less than optimal amounts for early *Homo habilis,* especially for the brain development of offspring and the mental function of adults, a proportional excess of essential fatty acids such as AA and DHA may have compensated, optimizing what thyroid hormone was available. We now know that DHA in particular is required for the production of transthyretin (397), the hormone transport molecule that's particularly important for moving thyroid hormone around the brain. This suggests that abundant fatty acids in the diet of *Homo habilis* may be of evolutionary significance: when this hominid was faced with reduced amounts of dietary thyroid hormone, abundant fatty acids in its diet may have optimized available thyroid hormone and buffered the selection processes that might otherwise have precipitated more profound and dramatic changes.

Big bodies and big brains on the move—*Homo erectus*

The next major change in the hominid lineage, especially for physical attributes, is seen in *Homo erectus. Homo erectus* had a significantly larger body and brain than its predecessors, with smaller teeth and a more gracile mandible (229). Tools and associated animal bones unequivocally establish this hominid as a hunter of a wide range of medium- and large-sized terrestrial mammals that inhabited open savannah-type environments. Although some of the larger prey species, such as elephants, may still have been scavenged, smaller species were undoubtedly hunted. As primary predators of such prey, *Homo erectus* would have had ready access to thyroid glands, and other unprotected organ tissues would also have been readily available. As a consequence, total dietary thyroid hormone consumed by *Homo erectus* would have increased over levels consumed by *Homo habilis.* Just as for early Australopithecines, individuals that had particular thyroid rhythms would have retained reproductive function better than others under circumstances of increased consumption of dietary thyroid hormone.

As thyroid hormone so strongly controls growth and maturation programs, shifts in thyroid rhythms are probably responsible for the morphological changes to brain size and body proportions seen in *Homo erectus.* The increased brain function and manual dexterity that must have accompanied these developmental timing changes (476, 579) would have conferred distinct survival advantages. Once present in the founder population, traits such as larger body size and a more complex brain could increase in frequency via selection for the growth and maturation programs that underlay them (and controlled them, I maintain, by distinct thyroid rhythms).

Homo erectus survived as a species for well over one million years and was the first hominid to move beyond Africa. Fossils have been found in both southeast Asia (Indonesia) and western Asia (Georgia) that date to about the same time as these fossil forms are found in Africa (about 1.8 mya). Lack of a continued presence in western Asia suggests this area may have been a transition zone for a single out-of-Africa emigration by *Homo erectus*, although little palaeontological work has been completed. There is no doubt, however, that *Homo erectus* lived as successfully in Asian habitats as in African ones. The early presence of *Homo erectus* in Indonesia, as well as small but distinct skeletal differences and other evidence of a continuous isolated existence in Asia, are the reasons some palaeoanthropologists give for referring to the African form as *Homo ergaster* and the Asian one as *Homo erectus* (the Asian fossils, discovered first, retain the original name if these are considered separate species).

In biological terms, however, this situation almost certainly qualifies for the application of *subspecies* rather than species-level distinctions. Subspecies are used to indicate geographically isolated populations of the same species (which would make the Asian form *Homo erectus erectus* and the African form *Homo erectus ergaster*). I therefore refer to both in general discussion as *Homo erectus*. The very recent report of fossils with the skeletal distinctions of the Asian form (*Homo erectus erectus*) in Ethiopia dated at about 1 mya—a form not found in Africa before—seems to confirm this single-species interpretation (34). Nevertheless, what is really critical to understand is that a hominid of essentially identical form that originated in Africa came to live in both Asia and Africa. That hominid species appears to be the direct and undisputed ancestor of the *Homo sapiens* lineage (224). See Figure 8.2 at the end of this chapter for a summary.

Life on the cold front—*Homo neanderthalensis*

The speciation event that occurred next in the hominid lineage was the emergence of *Homo heidelbergensis*. A large-bodied, large-brained hominid, this species appears in the fossil record of Africa, Europe and Asia just after a major global climate change that occurred about one million years ago (721). There is little doubt that this speciation event occurred as a divergence from *Homo erectus*, but what this species should be called is still hotly debated (527). While many researchers regard the species that diverged from *Homo erectus* (tagged with the scientific moniker *Homo heidelbergensis*) as distinct from our own, others consider this the first appearance of *Homo sapiens*, although an "archaic" form. Proponents of the term "archaic *Homo sapiens*" (an unorthodox category not used for other mammalian taxa) also lump the relatively distinctive Neandertal and an earlier, possibly ancestral form into this group. I will treat them separately here for reasons that I hope will become apparent.

Early *Homo heidelbergensis* fossils are relatively abundant. They are known from both Africa (Ethiopia at about 600,000 ya; Zambia at about 400,000 ya or so) and Europe

(Germany and England at about 320,000–500,000 ya; Spain, France and Hungary at about 250,000–300,000 ya). *Homo heidelbergensis* is considered by many respected palaeoanthropologists to be the common ancestor of *Homo neanderthalensis* in Eurasia, and *Homo sapiens* in Africa. (The significance of so-called *Homo antecessor*, a fossil hominid with a peculiar mix of both modern and primitive features found in deposits about 780,000 years old in northern Spain, is still heavily disputed, so I'll leave them out for now.)

The emergence of *Homo heidelbergensis* coincides with the onset of cooler and drier Pleistocene environments, at a time when many habitats around the world experienced a major turnover of animal communities. Such changes seem to have required hominids to intensify their hunting activities, perhaps in response to increased competition from other carnivorous scavengers for available carcasses—less scavenging, more hunting, more dietary thyroid hormone.

Again, an increase in dietary thyroid hormone would have favoured individuals with the particular thyroid rhythms capable of handling higher levels of thyroid hormone without disruption of reproductive function. At the same time, much cooler temperatures would have effectively eliminated those with especially cold-sensitive thyroid rhythms from the population (since cold-sensitive animals would likely, at the very least, either fail to breed or be unsuccessful at raising healthy offspring).

As the Pleistocene epoch intensified, further adaptations became necessary in the northern extremes of *Homo heidelbergensis* territory. Early Neandertals were *Homo heidelbergensis* colonizers of arctic steppe and tundra. These habitats expanded and contracted in response to the pronounced glacial and interglacial periods that dominated the climate of Eurasia at this time. Northern colonizers faced even colder temperatures than *Homo heidelbergensis* had before them. In addition, there were seasonal fluctuations in quantity and variability of prey, both natural consequences of the extreme conditions.

The oldest fossils classified as *Homo neanderthalensis* were found in Germany and date to about 225,000 ya. The limited geographic area where fossil remains of this species are found suggest that the total range of this species never expanded beyond Europe and western Asia. In other words, *Homo neanderthalensis* represent successful adaptation of early hominids to a particular set of conditions that were geographically limited and of relatively short duration (in evolutionary terms).

Neandertal ancestors who chose to colonize the harsh and ever-shifting Pleistocene steppe environments must have consisted of a small group (or groups) of individuals who were physiologically tolerant of the relatively severe climatic conditions: they would have possessed one of several naturally-occurring physiological variants that existed naturally within the ancestral *Homo heidelbergensis* population. Since thyroid hormone metabolism is the body's mechanism for adjusting to cold, individually-unique thyroid rhythms within the ancestral *Homo heidelbergensis* population would have given some individuals a higher tolerance for the physiological stress of reduced temperatures than others. Temperatures need not have been frigidly arctic, just significantly colder than the habitats of their ancestors. Non-adaptive individuals didn't necessarily die, they just produced few or no surviving offspring.

The reduced variation of thyroid rhythms present within this small population of early Neandertals would have established a distinct hormonal pattern in their descendants that differed significantly from the common ancestral pattern. Compounding the selection pressure of cold temperatures would have been the extremely high proportion of the diet necessarily composed of raw meat and organ tissue, simply because far fewer plant products would have been available. Raw meat still appears to have been the dominant dietary component for all hominids at this time, with little compelling evidence for the use of fire for cooking rather than for heat and light (229). *Homo heidelbergensis* in more temperate climates may have consumed moderate quantities of plant material (either year round or seasonally), but it is doubtful that option was available to Neandertals. Neandertals were in many ways simply a more cold-tolerant, more carnivorous form of *Homo heidelbergensis.*

Many studies in the literature compare Neandertals to modern humans, with the often unstated inference that since Neandertals are our closest hominid relatives (in terms of geological time), they should be very similar to us in most ways—in fact, some consider them little more than a geographically-distinct subspecies. I suggest, however, that in contrast to the situation for *Homo erectus* (see discussion in the previous section), these two hominids deserve distinct species status.

The most obvious difference between Neandertals and modern humans is diet. Analysis of mineral levels in *Homo neanderthalensis* bones attest to the fact that a very high proportion of the Neandertal diet was composed of red meat (38), as could be predicted from the diet of their ancestors and the habitat in which they chose to live (see earlier discussions in this chapter). Such a diet would have provided significantly more dietary thyroid hormone than exists in diets of modern *Homo sapiens,* especially if thyroid glands and livers were eaten. As a consequence, Neandertals must have had a turn over rate for thyroid hormone closer to that seen in carnivorous modern dogs and cats (about twelve to thirteen hours) than the rate of about seven days recorded for modern humans (379). A faster turn over rate for thyroid hormone implies that a distinctly different thyroid rhythm must have existed for *Homo neanderthalensis.* And as I've argued in previous chapters, distinct thyroid rhythm patterns are one way of distinguishing discrete species.

However, more direct evidence that Neandertals possessed distinctive thyroid rhythms comes from new chronological aging techniques that measure incremental growth lines in tooth enamel (*perikymata*). Analysis of growth lines in the tooth enamel of selected fossils suggests that *Homo neanderthalensis* had faster postnatal growth rates than do modern humans (614). In addition, analysis of the changes in skull shape that occur during early childhood growth periods reveals that craniofacial growth patterns are distinct for each (527).

Even more convincing perhaps is a study that compared modern human and Neandertal craniofacial growth patterns to another closely related species pair—the common chimpanzee, *Pan troglodytes,* and the pygmy chimpanzee or Bonobo, *Pan panisus.* This study confirmed a faster childhood growth rate for Neandertals relative to modern humans, from soon after birth onwards. More significantly, these differences were shown to be at least as great, or greater, than those manifested by the two distinct species of *Pan* (Williams et al. 2002, cited in ref. 150). Such evidence that different post-natal growth rates existed for *Homo neanderthalensis* is the most reliable indicator that this hominid must have possessed a distinct thyroid rhythm.

The growth-rate evidence that Neandertals possessed a distinctive thyroid rhythm supports the view that they represent a distinct species of *Homo* rather than a sub-species or "archaic" form of *Homo sapiens.* This position is now supported by genetic evidence. Analysis of Neandertal mtDNA sequences suggests strongly that these hominids were a genetically distinct lineage that had been reproductively isolated for a considerable length of time (150, 151). While some critics are still not convinced, the combined evidence presents a very strong case for making a taxonomic distinction between Neandertals and modern humans (721).

Continued reluctance to classify Neandertals as a distinct species is due in part to a misguided belief that this would necessarily rule out interbreeding between the two when declining numbers of *Homo neanderthalensis* populations met increasing numbers of early *Homo sapiens* at the end of the Pleistocene. However, recent molecular data confirm that hybridization between closely related species is much more common than previously thought. Hybridization has been shown to be neither rare nor "unnatural" in closely-related fish (such as char and trout, and various Pacific salmon species), birds (such as European and domestic Japanese quail) and mammals (such as red-backed and bank voles, fin and blue whales, Dall's and harbor porpoises, various dolphin species, coyotes and wolves, bison and cattle) (30, 49, 61, 64, 147, 185, 284, 315, 442, 447, 673, 737, 794, 807). Even some not-so-closely-related animals are now known to hybridize, as documented by viable crosses between species of dolphins, porpoises, seals and quail from different genera (365, 412, 812, 853).

Early *Homo* species were not likely an exception to this phenomenon. A few Neandertals could very easily have been assimilated into populations of early *Homo sapiens.* Evidence of just such a scenario is provided by a putative modern-Neandertal hybrid specimen found recently in Portugal, although not all researchers accept this interpretation of the remains (150).

An additional argument against classifying some *Homo neanderthalensis* fossils as representatives of a distinct species lies in their skeletal similarity to some early *Homo sapiens* fossils. However, I suggest that the thyroid hormone rhythm and associated growth programs that generated a cold-tolerant Neandertal didn't cease to exist in *Homo heidelbergensis* ancestral populations simply because *some* individuals with those rhythms chose to leave the group—some cold-tolerant individuals would have stayed, contributing the rhythms for this physical form to future generations. Early *Homo sapiens,* as bona fide descendants of the same ancestral species (*Homo heidelbergensis*), very likely contained at least a few individuals with thyroid rhythm patterns that were relatively cold-tolerant. In other words, early *Homo sapiens* seem to have possessed a fairly broad range of thyroid rhythm and growth program variation—which may be what allowed them to become a successful, globally distributed species.

Anatomically modern humans—*Homo sapiens*

The end of the line for the *Homo* linage, as we now perceive it, is the emergence of so-called "anatomically modern humans": the species to which we all belong, *Homo sapiens.* A few fossils from Africa attributed to this species are estimated to be almost 150,000 years old, while some found in Israel are closer to 100,000 years old (721). However, abundant specimens are not found until about 36,000 years ago in Europe and 30,000 years ago in Asia (150), with a few more from 40,000–50,000 years ago.

The biological changes in this species are slight: the teeth are smaller, heavy brow ridges are essentially gone, and the skeleton overall is less robust than that of its ancestor. An almost vertical forehead in *Homo sapiens* gives the face a flatter profile with a protruding chin. Culturally, there are the beginnings of truly "modern" human behaviour: more complex and varied tool types, highly decorated art objects, musical instruments and painting, and most important to this story, the regular use of fire for cooking (ref. 229:537).

Definitive evidence for the frequent use of fire for cooking is available for only the most recent finds (from about 30,000 years ago) and is perhaps not coincidentally associated with the most dramatic of the cultural artifacts, such as Aurignacian tools and elaborate cave paintings. I suggest that whenever (and wherever) cultural stimuli and technical skills led to the regular practice of cooking meat, it precipitated the physiological and physical changes in descendant populations that we recognize as *Homo sapiens.* This would be a natural consequence of cooking, which reduces the amount of thyroid hormone in meat.

Although data on the response of thyroid hormone to heat of various temperatures is slim, it is known that the heat required for drying thyroid tissue (about 100–120°C.) is insufficient to affect either T_3 or T_4 (296). Much higher temperatures, especially those required for thorough cooking by simple direct methods like spit-roasting (perhaps yielding temperatures of 400–600°C.), probably cause either complete or partial degradation of T_3 (since it is a less stable molecule) and at least modest degradation of T_4. The regular practice of cooking meat rather than eating it raw would have decreased the amount of dietary thyroid hormone (but especially T_3) intake of many widespread populations of incipient *Homo sapiens.* And as in previous declines of thyroid hormones (such as for *Homo habilus*), biological consequences ensued.

A profound decrease in dietary T_3 seems to have precipitated selection pressure on the thyroid glands of incipient *Homo sapiens* to provide virtually all of the T_3 required for normal physiological function. Individuals whose thyroid metabolism could not function optimally without some influx of dietary T_3 would gradually have become minor constituents of the population because of correlated reduced fertility. I suggest that only those individuals whose thyroid glands could readily supply all of the T_3 needed would have produced abundant offspring.

Regardless of when and where the meat-cooking innovation occurred, rapid and widespread changes would have followed. The fact that many physical traits are linked to brain development through thyroid hormone-controlled embryonic and postnatal growth programs means that the survival advantages conveyed by a more complex brain to any early *Homo sapiens* could easily have generated descendants with smaller teeth, gracile musculature and a shorter gut as an indirect consequence—as opposed to being selected for as discrete traits, as proposed by anthropologist Leslie Aiello (10). This means that the changes in skull features documented in the archaeological record are real and significant manifestations of selection for increased brain complexity. Brain and bone change together, along with other body tissues. Neurological pathways were added that had not been present before (476, 580), circuits that made it possible for complex language and culture to develop.

While other reasons have been offered, I suggest there really is no *plausible and testable* evolutionary explanation, other than the regular cooking of meat, for the sudden explosion in cultural behaviour accompanied by marked biological changes in early *Homo sapiens*—no other major changes in climate, diet or culture have been documented. The appearance of a few anatomically modern humans a bit earlier may reflect a local reliance on fish, remains of which have been found associated with some of these older sites (160). Fish diets may have mimicked a cooked meat one because while they would have been rich in essential fatty acids, they might not have been especially rich in thyroid hormone (especially fresh-water species).

Such local dietary adaptations could have generated anatomically modern *features* (of both the skull bones and underlying brain) within relatively small and isolated populations but perhaps couldn't precipitate the dramatic burst of cultural innovations we see later, simply because there were not as many individuals involved. Alternatively, the innovation of using fire for cooking could also have arisen independently and sporadically at first, with local morphological effects as described above. The full range of cultural and behavioural effects may have required the involvement of many individuals, a situation that did not arise until almost 40,000 years ago. If so, as soon as enough incipient *Homo sapiens* were using fire to cook their meat, the behavioural effects would have become widespread. Eventually, enough brain-enhanced human minds were interacting that it fueled an associated explosion of cultural innovations.

If the regular use of fire for cooking was indeed the stimulus that precipitated the biological and cultural changes associated with fully modern *Homo sapiens* in Africa, it is entirely possible that *Homo sapiens* could have arisen independently in Asia from local populations of *Homo heidelbergensis*. Indeed, Asian forms of *Homo heidelbergensis* could themselves have descended from local Asian *Homo erectus* populations just as easily as they could have migrated in from Africa (if not more so). Geographically-isolated subspecies of *Homo erectus* lived in both continents and would have responded in similar fashion to similar environmental pressures. Whether that actually occurred would be very difficult to tell: I suggest that the skeletal form of both populations could have been so similar that we would be hard pressed to tell their fossils apart.

Genetic support for the controversial palaeontological evidence that *Homo sapiens* may have descended from more than one ancestral population (721) comes from one way of interpreting the results of a recent study of mtDNA extracted from ancient Australians (150). However, criticisms of these data have been so diverse that it's virtually impossible to determine if indeed the study provides evidence, independent of skeletal criteria, that multiple origins are involved (see ref. 150). More ancient genetic data and well-dated fossils are still needed to resolve the issue. In any case, it is also true that even if *Homo heidelbergensis* and *Homo sapiens* did arise independently in Asia, their populations could still have been culturally and genetically swamped by later-arriving representatives of African origin.

Box 8.2 Modern human big brains: the result of genes or thyroid rhythms?

How did modern humans end up with such disproportionately large brains? A research team from the University of Chicago recently announced the discovery of particular mutations in two genes (both involved in determining brain size), whose worldwide distribution indicate they were favoured by natural selection. The authors suggest that individuals who possessed these particular alleles had a reproductive advantage over those who did not because they had somewhat larger brains (209, 515). Speculating further, they correlate the approximate time when these "advantageous" mutations arose with two significant events in human history: the explosion of artistic representations at about 40,000 ya and the emergence of domestic plants and animals at about 10,000 ya.

However, other geneticists (37) point out two facts: 1) both of these genes (*microcephalin* and *ASPM*) act on other tissues as well as on the brain; 2) all humans are capable of symbolic expression regardless of which allele (mutation) of the gene they possess. They also suggest that if these particular alleles conveyed an adaptive advantage, it was either very small or *directed at some other function of the gene (rather than its impact on brain size)*. There are other studies that take a similar genocentric approach (190, 493, 494, 701, 813).

It appears to me very likely that these two genes are thyroid hormone dependent (as many genes controlling brain development are) and that the selected alleles were advantageous because they worked somewhat better *in conjunction with a particular thyroid rhythm*. As I've argued in this chapter, significant shifts in thyroid hormone metabolism of modern humans must have occurred at these times (about 40,000 and 10,000 ya) as a result of dietary changes that were uniquely independent of speciation events. While brain size and complexity increased at both times as a consequence, selection favoured the thyroid rhythm that worked best with the new diet:: big brains were simply a fortuitous bonus that could be fine-tuned by selection later.

Recent dietary adaptations

Evolution is a continuous process, of course, and small additional changes in human developmental programs (as reflected in skeletal conformation and physiology) are still under way. Some of these changes began only after what is known as the *agricultural revolution* (when deliberate cultivation of cereal grains began). Generally speaking, before the agricultural revolution, most human populations were hunters and gatherers who depended on meat and fat sources for most of their calories, supplemented by whatever edible plants, nuts and seeds were available. Only when cereal grains and starchy root crops could be produced in dependable quantities and stored for future use did carbohydrates come to represent a dominant staple of the human diet.

Although local reliance on collected wild ancestors of grain preceded organized cultivation by as much as 10,000 years, grain crop domestication occurred quite late in our long evolutionary history (not until about 8,000–9,000 ya in the Middle East and China, and much later, about 5,000 ya, in Mesoamerica (307). The relevance of the agricultural revolution to my argument is that an abundance of staple cereal or starchy root crops in the diet reduced further still whatever component of T_4 remained. Of course, some cultures did not adopt agriculture at all (as on the Northwest Coast of North America or in the Arctic). However, in regions where agriculture has its longest history, such as the middle East and Asia, diets came to be increasingly dominated by carbohydrates rather than meat as early as 20,000 ya (150). Because crop cultivation meant a dependable carbohydrate source, both culture and social organization changed dramatically in every population that later mastered the knack of controlling plant production (187, 188).

As the relative proportion of meat consumption in modern humans dropped, thyroid hormone function would have had to shift to accommodate the loss of virtually all dietary T_4. Due to the tight association of thyroid hormone with reproductive function, individuals whose thyroid rhythms generated optimal amounts of thyroid hormone would have been more reproductively successful than those individuals who needed significant supplemental T_4 from dietary sources.

Although the physical and skeletal changes associated with these adaptations would have been small in comparison to previous evolutionary shifts, such changes are nevertheless evident. The skeletal changes that occurred are clearly the result of growth rate or timing shifts. In her study of the skulls of early versus late modern humans, for example, paleoanthropologist Nancy Minugh-Purvis (ref. 527:496) found that features in the back of the skull (the *posterior neurocranium*) found in recent humans appear only *after* the emergence of *Homo sapiens*—in fact, these features are apparent only in individuals who lived after the agricultural revolution (in other words, samples from about 30,000 years ago differed from those less than 6,000 years old). "Early Upper Paleolithic Europeans," she says, "usually regarded as early modern humans based on both behavioral and morphological criteria, were not fully modern in their growth patterning."

I contend that the tight correlation between growth at all stages and reproductive function—controlled as these are by thyroid hormone—had to have been instrumental in continuing to adapt late modern humans to dietary changes that were largely independent of changing environmental conditions. This disconnection between adaptation and environmental change in *Homo sapiens* appears to be unique to modern humans: no other animal has changed so markedly independent of environmentally controlled triggers.

Summary and discussion

I suggest that thyroid rhythm theory explains how selection for particular, naturally-occurring variants of thyroid rhythms in hominid ancestors, when isolated within small interbreeding populations of colonizers, could quickly generate new hominid species with distinctive skeletal, physiological and behavioural characteristics. Dietary changes associated with colonization of new habitats, or adaptation to changes in existing habitats, involved shifts back and forth in relative amounts of dietary T_3 and T_4. These habitat shifts appear to have driven all but the most recent biological modifications, which were entirely diet-driven. Although many authors have tried to explain the acquisition of bipedalism as well as other hominid-exclusive traits (10, 342, 394, 439, 465, 599), without a biological mechanism to explain *how* such characteristics could be generated by natural selection, all become "just-so" stories. In contrast, I have been able to use thyroid rhythm theory to suggest, in explicit biological terms, how particular speciation events in the hominid lineage could have generated particular changes in traits that are repercussions of alterations in growth and development, including the physically unique Neandertals.

The particular physical characteristics that emerged over the course of hominid evolution (including increased height, increased brain size and complexity, changes to the pelvis and femur, changes to facial structure, shortening of the gut, delayed tooth eruption) are all indicative of altered growth programs. As a consequence, all of these traits could have been produced by the effects of shifts in thyroid rhythms during fetal and postnatal development.

Because thyroid hormone ultimately controls many physical characteristics, dietary shifts that impacted developmental programs even slightly would have affected many seemingly unrelated traits simultaneously, even skin colour—in contrast to being selected directly as a discrete trait, as proposed by some researchers (355, 356). Indeed, two distinctive features of modern humans that have been especially hard to explain as selected traits are our unique pattern of fat accumulation/production of fat newborns (see discussions by David Horrobin, ref. 341, Christopher Kuzawa, ref. 423, and Stephen Cunnane, ref. 160), and our relative hairlessness. I suggest these traits may also have been inevitable consequences of the rapid changes in growth programs that took place in the hominid lineage, perhaps quite early on, and may never have been actively selected traits themselves—in contrast, for example, to suggestions that hairlessness provided an advantage related to thermoregulation (803, 804).

In addition, maternally-produced and -consumed thyroid hormone, through its effects on fetal growth, would have been a significant non-genetic modifier of an individual's ultimate physical and physiological form. Consequently, maternally-controlled effects of thyroid hormone on embryonic growth were almost certainly somewhat different in nature than later effects on postnatal growth and brain development governed by the individual's own genes.

Fluctuating diets may have been instrumental in maintaining a large pool of individual thyroid rhythm variation in the hominid lineage and may have been essential to the success of our species. More than once over the course of hominid evolution, when the amounts of dietary thyroid hormone was low, the consumption of foods rich in essential fatty acids, or iodine-rich foods, such as marine fish, shellfish and kelp (199), may have buffered the natural selective forces that might otherwise have eliminated some variants.

The innovation of using fire to cook meat appears to be the stimulus that finally pushed our ancestors over an important evolutionary line. Changing the thyroid hormone content of an established diet simply by preparing it differently— in contrast to making it more tender or digestible, as suggested by some researchers (445, 829)—although not considered before now, is something only hominids have done. Therefore, the use of fire for cooking (rather than for light or heat) may have been responsible for making modern humans the truly unique species we are (although certainly the ability to make and use tools in the first place gave our ancestors distinct choices in food and life history strategies that helped set the path of their biological evolution).

Thyroid rhythm theory provides the first truly plausible biological explanation for a number of aspects of hominid history that have confounded palaeoanthropologists for so long. The debates surrounding these issues have often been particularly acrimonious. However, my theory offers testable hypotheses to explain the following features of hominid evolution:

1) How and why convergent evolution of bipedal morphology occurs in non-hominid apes
2) Why there are such coordinated changes in skeletal conformation, brain function and gut length over evolutionary time
3) Why Neandertals qualify as a distinct species of hominid
4) The very real possibility of an independent evolution of *Homo heidelbergensis* and *Homo sapiens* from genetically distinct geographic subspecies of *Homo erectus*
5) Why the explosion of creative innovations unique to anatomically modern humans came so late in our evolutionary history
6) Why individual modern humans react so differently to distinct stresses and challenges, despite our close genetic similarities, making the diagnosis and treatment of many chronic health problems particularly difficult (the topic of the next chapter)

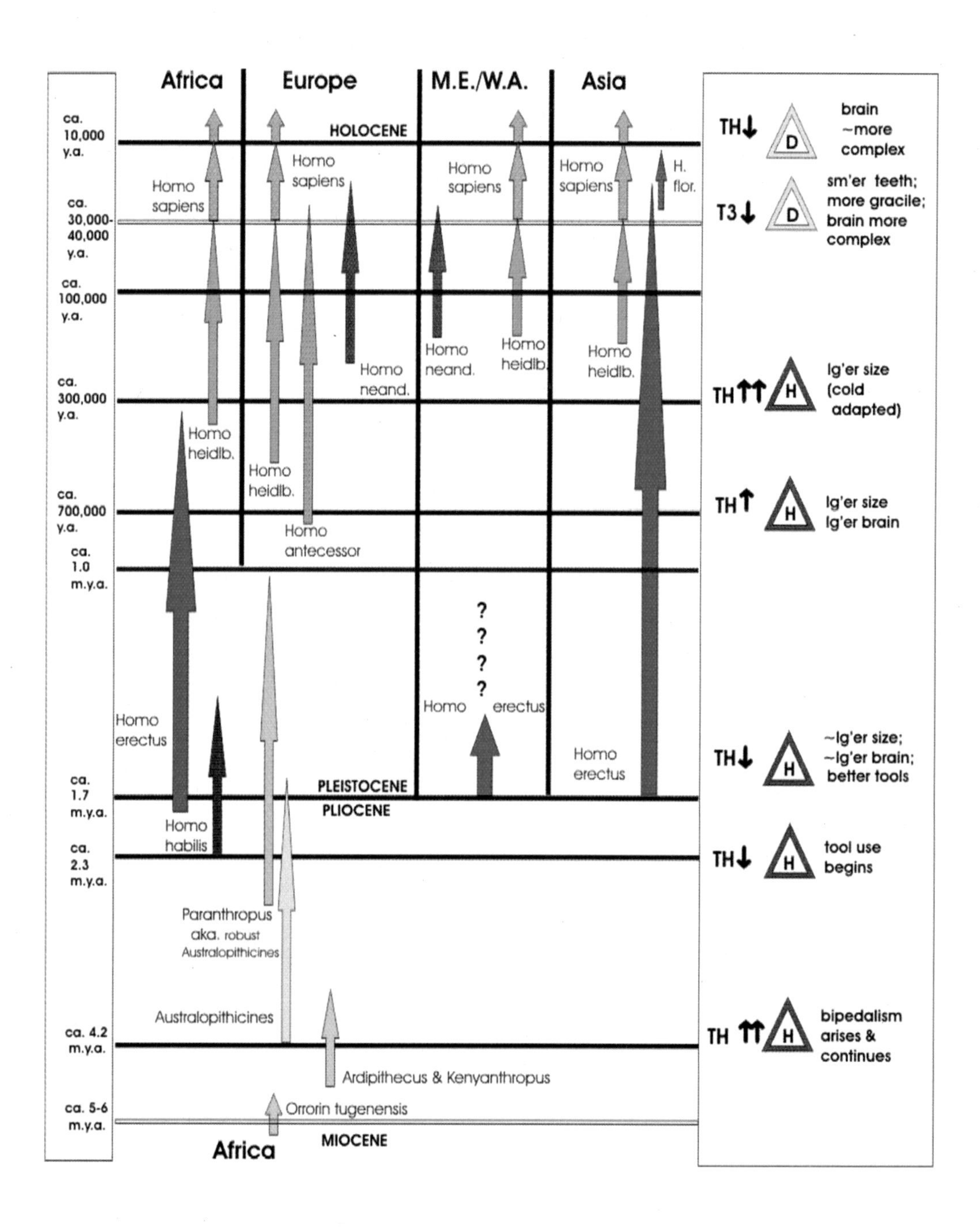

Figure 8.2 Summary of the most widely accepted chronology of hominid evolution, by approximate date and major geographic region (M.E./W.A. is the Middle East/West. Asia); "H.flor." in top right corner is Homo floresiensis of Indonesia; TH ↑ and TH↑↑ indicate slight and profound increases in dietary thyroid hormone, respectively; TH↓ and T3↓ indicate decreases in thyroid hormone and T3, respectively; **D** *in yellow triangles indicates* **dietary change only***;* **H** *in red triangles indicates* **habitat change plus diet change.**

Although we will never know for sure if the particular scenarios proposed here actually produced past hominid changes in the ways I've suggested, we can test the underlying biological assumptions using animal model systems. This is a viable approach because the concept should work similarly for all vertebrates. There are compelling reasons, however, apart from understanding our history, for needing to know if this theory presents a useful model for evolutionary change, and these relate to point number 6 listed above: if my theory is correct, there are enormous implications for human health.

The truth is that we are in desperate need of research that documents the range of normal human thyroid rhythms and demonstrates how disruption of these rhythms (in the face of psychological stress, disease, age and biochemical pollutants) affect day-to-day health. My interpretation of our evolution suggests that individual variation in thyroid rhythms were essential to the ability of hominids to adapt to changing conditions of life: this appears to be as true today for modern humans as it was five million years ago for our ancestors.

In the next chapter, I'll address the manifestations and implications of these individually unique thyroid rhythms on modern health.

This is where the story of evolution gets personal, for all of us.

Recommended reading

Culotta, E. 2005. New "hobbits" bolster species, but origins still a mystery. *Science* 310:208-209.

Ezzel, C. 2003. Ma's eyes, not her ways. *Scientific American* (April):30.

Finlayson, C. 2005. Biogeography and evolution of the genus *Homo. TRENDS in Ecology and Evolution.*20(8):457-463.

Galli, C., Lagutina, I., Crotti, G., Colleoni, S., Turini, R., Ponderato, N., Duchi, R. and Lazzari, G. 2003. A cloned horse born to its dam twin. *Nature* 424:635.

Horrobin, D.F. 2001. *The Madness of Adam and Eve: How Schizophrenia Shaped Humanity.* Bantam Press, London.

Jablonski, N.G. and Chaplin, G. 2002. Skin deep. *Scientific American* (October):172-179.

Kingdon, J. 2003. *Lowly Origin: Where, When and Why Our Ancestors First Stood Up.* Princeton University Press, Princeton.

Lanza, R.P., Dresser, B.L. and Damiani, P. 2000. Cloning Noah's ark. *Scientific American* (November):84-89.

Leakey, M. and Walker, A. 2003. Early hominid fossils from Africa. *Scientific American* (December):14-19.

Lee, B.C., Kim, M.K., Jang, G., Oh, H.J., Yuda, F., Kim, H.J., Shamim, M.H., Kim, J.J., Kang, S.K., Schatten, G. and Hwang, W.S. 2005. Dogs cloned from adult somatic cells. *Nature* 436:641.

Leonard, W. R. 2002. Food for thought. *Scientific American* (December):262-271.

Lieberman, D.E. 2005. Further fossil finds from Flores. *Nature* 437:957-958.

Minugh-Purvis, N. and McNamara, K. (eds.) 2002. *Human Evolution Through Developmental Change*. Johns Hopkins University Press, Baltimore.

Riddle R.D. and Tabin, C.J. 1999. How limbs develop. *Scientific American* (February):74-70.

Schwartz, J.H., 1999. *Sudden Origins: Fossils, Genes, and the Emergence of Species*. J. Wiley and Sons, New York.

Shin, T., Kraemer, D., Pryor, J., Liu, L., Rugila, J., Howe, L., Buck, S., Murphy, K., Lyons, L. and Westhusin, M. 2002. A cat cloned by nuclear transplantation. *Nature* 415:859.

Wills, C. 1998. *Children of Promethius: The Accelerating Pace of Human Evolution*. Persius Books, Reading.

Chapter 9

Health Implications of Thyroid Rhythms

An evolutionary context for thyroid dysfunction

I honestly never considered that my thyroid rhythm theory would have implications for human health until I hit the early stages of menopause myself at the age of forty-five: night sweats, lethargy, mental fog, weight gain. I was having a hard time writing because I couldn't concentrate. I wasn't sleeping well and just couldn't find the energy to exercise. When it finally dawned on me that I was old enough for impending menopause (otherwise known as *perimenopause*) to be an explanation, I knew that declining estrogen levels were considered the prime culprit. Estrogen replacement therapy (or its herbal analogue) is the most common treatment prescribed for this common side effect of female aging (677, 524, 592).

But hang on, I thought, estrogen doesn't regulate body temperature—that's what thyroid hormone does (296, 348, 684). What if most of this hormonal upheaval isn't about estrogen alone but is a result of declining thyroid hormone levels leading to declines in estrogen and other hormones (699, 818, 819)? When I looked more closely at the literature, I found out that the root cause of menopausal hot flashes and night sweats is not truly known. While fingers are still pointed primarily at declining estrogen levels, changes in levels of a variety of hormones and neurotransmitters have also been implicated (243, 516, 592, 592).

I wondered if perimenopausal symptoms could be caused by a predictable disruption of thyroid rhythms with age and its associated repercussions in women (including me). What if, at a certain age, the morning spike of thyroid hormone that we normally experience on awakening no longer hits a critical peak—a certain level that other hormones depend on to set them up for the next twenty-four hours? Correct or not, a daily dose of two raw egg yolks (a chicken's source of thyroid hormone; see Chapter 3) every morning relieved my symptoms, including the mental fog. I was off again, this time delving into the scientific *and* medical literature on the fine points of hormonal interrelationships and thyroid imbalances.

Thinking about my own health gave me quite a new perspective on what it means to be human. Modern humans, with their very slow turn over rate for T_4, *appear* to be well adapted to a diet lacking any appreciable amounts of thyroid hormone. However, our tolerance to fluctuations in thyroid hormone levels is very low indeed, and either too much (*hyperthyroidism*) or too little (*hypothyroidism*) makes us decidedly ill (91). Hypothyroidism is by far the most common imbalance: too much thyroid hormone is comparatively rare (24).

Although we don't have data from 10,000 years ago to know if the prevalence of hypothyroidism is a new phenomenon for humans, we do know that millions of people worldwide are currently afflicted. The list of indicators of hypothyroidism is long and diverse—not surprisingly, considering how many functions depend on thyroid hormone. The symptoms of hypothyroidism, which can be present in a diverse and often idiosyncratic manner, with some (but not necessarily all of those listed below) occurring together, include:

1) **Brain function:** depression; poor memory; inability to concentrate; mood swings; sleep disturbances; anxiety
2) **Skin/hair quality:** dry skin; puffy eyes; hair loss; coarse or brittle hair
3) **Reproductive functions:** irregular menstruation; heavy menstruation; infertility; miscarriage; birth defects; erectile dysfunction, delayed ejaculation
4) **Circulation/respiration functions:** high cholesterol; high blood pressure; congestive heart failure; sleep-associated breathing problems
5) **Muscle/joint function:** muscle cramps; muscle weakness; muscle and joint pain; deep voice; hoarse voice; "pins and needles" in hands/feet
6) **Metabolic functions:** heat/cold intolerance; weight gain; constipation; general tiredness

Hypothyroidism has also been strongly linked to hyperactivity disorders in children (310-312, 777), to postpartum depression (418, 570) and to some kinds of post-traumatic stress disorder (particularly where the situation involves a "sit and endure" type of stress rather than a "fight and run for your life" type). People do not often die of these disorders but they are very often rendered under-productive or even non-productive (24, 157, 105-107, 636). See Box 9.1 for a referenced list of body malfunctions known or suspected of being associated with thyroid hormone.

How many people are affected by thyroid hormone imbalances? All we have are estimates. For example, a comprehensive study undertaken a few years ago (hereafter, the *Hollowell* study, ref. 338) tested over 17,000 Americans for serum TSH, T_4 and thyroid antibody levels. Extrapolation of their survey results predicted that within the

United States alone more than eight million people would probably test positive for hypothyroidism and an additional 700,000 people unknowingly have hyperthyroidism. In addition, the survey suggested that a significant number of people (more than 30%) who are already taking thyroid supplements for hypothyroidism are not getting enough (another study on an even larger population of almost 26,000 participants (105) got similar results). As high as these numbers seem, they are probably still far below the real figures because both studies used a very wide range of serum TSH concentration readings (0.5 to 5.0 mU/liter) as indicators of normal thyroid function (values exceeding *high end values* for "normal" TSH—in this case, over 5.0—indicate hypothyroidism).

Many doctors now consider that TSH readings above 2.0mU/liter (especially when accompanied by symptoms) are valid indicators of hypothyroidism or at least the early stages of it (24, 25, 302, 567, 797). In a press release issued January 2003, the American Association of Clinical Endocrinologists (see www.aace.com\publications) stated that the number of Americans affected by thyroid imbalance (including undiagnosed cases) now exceeds the number with diabetes and cancer combined; it also drew attention to clinical guidelines published for physicians a few months earlier, in November 2002. Those new guidelines recommended narrowing the range indicating normal TSH function down to 0.3–3.04mU/liter, while admitting that 95% of normal individuals actually have values below 2.5mU/liter) (182). Other researchers not only concur but suggest this upper limit might warrant even further reduction (786).

Why the change of heart? In part the answer lies in the nature and history of the TSH test itself. Remember from Chapter 3 that TSH is the hormone released by the pituitary gland in response to hormonal stimulation by the hypothalamus, when thyroid hormone levels in the blood fall or rise. In principle, a small drop in thyroid hormone concentration should generate a very large rise in TSH values (about 100X more, which is one of the reasons TSH, rather than T_4, was chosen as an indicator of thyroid function—all other hormones are measured directly). Developed in 1976, the laboratory test for TSH used a reference population of only twenty-six individuals to establish what levels are "normal" for humans (211, 716). At that time, the value separating a diagnosis of hypothyroidism from normal function was set at about 10.0UmU/liter, a figure determined in large part by technical limitations—the test was just not very sensitive when it came to detecting small amounts of TSH in a sample (ref. 182:34). Over the last two decades, the sensitivity of TSH tests improved enough to drop this upper limit to about 5.0mU/liter (726).

However, the Hollowell survey indicated that the range of values describing normal TSH function was probably still too large. The usual explanation offered for the large range is that people with undiagnosed hypothyroidism were inadvertently included in the original laboratory reference population of healthy individuals (769, 797). But my question is this: is narrowing the range of values for the TSH diagnostic test going to solve the problem of undiagnosed hypothyroidism in the human population identified by the Hollowell survey? Is it going to give physicians the information they need to diagnose and treat hypothyroidism effectively?

I suggest not. In part, that's because, despite having a more sensitive thyroid function test, we still do not see any suggestion for testing an *individual's* TSH while they're healthy and strong to get an idea of what's normal for them—even though experimental evidence suggests individual variation can be quite pronounced (711). For example, a recent Danish study on sixteen healthy males tested thyroid function repeatedly over one year and found a wide range of variation *between* individuals for all values (serum T_4, T_3, free T_4 and TSH) but a very narrow range of variation within them (17]. In other words, each of these men felt well at one particular level of thyroid function that was fairly consistent for him over the space of a year—the test values were not bouncing all over the place from one month to the next but were individually stable. Similarly, a later study that compared thyroid function values (free T_3 and T_4, and serum TSH) among and between sets of identical and non-identical twins generated results that prompted the authors to report that "each individual may have a genetically determined thyroid function set-point" (303:1181)—interestingly, non-Hispanic African-Americans consistently show a much lower incidence of hypothyroidism, suggesting they have a much lower "set-point" for thyroid function than Caucasians or Hispanics (338, 636, 786). Although none of these experiments were measuring hormone rhythms (testing many times per hour), the results are nevertheless consistent with my contention that thyroid hormone function should show distinct individual variation.

However, along with the issue of individual variation in thyroid function, there is also a problem with the reliance on the TSH test as the primary diagnostic tool for determining when thyroid hormone levels have gone awry. Its use is defended on the grounds that TSH and thyroid hormone levels in the blood are intimately tied (182:35): "the pituitary monitors the level of thyroid hormone in the blood and increases or decreases the amount of TSH released, which then regulates [adjusts appropriately] the amount of thyroid hormone in the blood." On the other hand, the new clinical guidelines also state that there is a considerable *lag effect* between changing thyroid hormone levels and TSH readings: "at least 6 weeks is needed before retesting TSH following a change in dose of synthetic thyroid hormone." This is why replacement doses for patients with hypothyroidism are usually given in 25 microgram increments every 6–8 weeks, with retesting for TSH at those intervals (usually until the level falls below 2.0 mU/L). I don't call a 6–8 week time lag an immediate and sensitive indicator of thyroid status.

Box 9.1 References for demonstrated or suspected effects of thyroid hormone (TH), direct or indirect, on aspects of human health

Effect and actions of TH	References
general metabolism (genomic & non-genomic)	Franklyn 2000; Hadley 2000; Hulbert 2000; Yoshida et al. 1997; Ahima & Flier 2000; Kershaw & Flier 2004; Armario et at. 1987
nutrition & obesity	Douyon & Schteingart 2002; Knudsen et al. 2005; Skibola et al. 2005; Mantzoros et al. 2001; Blüher & Mantzoros 2004; Bodosi et al. 2004; Licinio et al. 1998; Knudsen et al. 2005
aging	Greenspan et al. 1991; Smith et al. 2005; Velduis et al. 2005; van Coevorden et al. 1991; Woller et al. 2002; van den Beld, in press
disruption of embryonic and postnatal growth	Bassett & Williams 2003; Piosik et al. 1997; Kilby et al. 1998; Hulbert 2000; Rovet 2004; Chan & Rovet 2003; LaFranchi et al. 2005
disruption of development & function of the CNS (including schizophrenia & autism)	Chan & Kilby 2000; Bernal 2002; Yen 2001; Köhrle 2000; Brosvic et al. 2002; Larsson et al. 2005; Jones et al. 2005; Farwell & Leonard 2005; Martinez & Gomes 2005
disruption of development &/or function of teeth, hearing structures, and skeletal muscle	Pirinen 1995; Brucker-Davis et al. 1996; Hulbert 2000; Jones et al. 2005; Clément et al. 2002
disruption of skin and hair pigment production	Hadley 2000; Burchill et al. 1993
behaviour disorders, including depression, postpartum depression, juvenile hyperactivity	Othman et al. 1990; Kuijpens et al. 2001; Henley & Koehnle 1997; Hauser et al. 1997; Hauser & Rovet 1998; Vermiglio et el. 2004; Eravci et al. 2000; Kalsbeek et al. 2005
direct or indirect effects on sleep and/or circadian rhythm disruption	Buckley & Schatzberg 2005; Peterson et al. 2005; Turek et al. 2005; Richardson & Tate 2000; Bodosi et al. 2004; Behrends et al. 1998; Kalsbeek et al. 2005; Everson & Nowak 2002; Goichot et al. 1994; Mantzoros et al. 2001; Everson & Nowak 2002; Baumgartner et al. 1993; Brabant et al. 1990; van Coevorden et al. 1991
cancer	Hercbergs 1999; Hercbergs et al. 2002, 2003; Cristofanilli et al. 2005
stress response (including general illness, surgery)	O'Connor et al. 2000; Peeters et al., in press (x2); O'Malley et al. 1984; Juma et al. 1991; Chowdhury et al. 2001; Custro et al. 1994; Yehuda et al. 2005; Safer et al. 2005; Baumgartner et al. 1998
disruption of development and function of the reproductive organs	Yoshimura et al. 2003; Chan et al. 2001; Hayssen 1998; Velduis et al. 2005; Kennaway 2005; Anderson et al. 2002

disruption of heart development, function and circulation	Sussman 2001; Miller et al. 2004; Kahaly & Dillman 2005; Fazio et al. 2004; Chowdhury et al. 2001; Michalopoulou et al. 1998; Staub 1998; Chubb et al. 2005; Forhead & Fowden 2002
disruption by environmental contaminants	Hauser et al. 1998; Porterfield 2001; Hawaleshka 2005; Soloman & Schettler 2000; Zoeller 2003; Zoeller et al. 2005; Thayer & Houlihan 2002
effects on daily & seasonal hormone rhythms (all considered)	Greenspan et al. 1986, 1991; Campos-Barros et al. 1997; Lucke et al. 1977; Dardente et al. 2004; Morton et al. 2005; Kronfol et al. 1997; Richardson & Tate 2000; Anderson et al. 2002; Goichot et al. 1998; Baumgartner et al. 1998; Brabant et al. 1990

In addition, the thyroid gland is now known to have a direct nerve connection (via *sympathetic nerves*) to the hypothalamus and the rhythm control centre of the brain (the SCN); this means that *thyroid hormone secretion can be stimulated or depressed independently of changing TSH levels* (377, 378). In fact, it appears that the response of the thyroid gland to TSH varies depending on the activity of these nerves (843). All together, mounting evidence suggests that steadfast reliance on the TSH test as the best indicator of optimal thyroid function, rather than measuring thyroid hormone itself (as is done for all other hormones), may be causing physicians undue problems in diagnosis and treatment.

The same may be true in veterinary medicine. The TSH test is also the primary diagnostic tool used by veterinarians for measuring hypothyroidism in animals, where its use has been similarly disputed (287). In dogs, the normal range for TSH concentrations established by laboratories came from testing a few dogs from many breeds to generate a general "dog" value. It's now acknowledged that each dog breed probably has a different range of values for "normal" (220). However, the possibility that there might be *individual* variation in values for normal thyroid function are not taken into account here either.

Most importantly, the rhythmic nature of both thyroid hormone and TSH secretion is still rarely acknowledged in either human or veterinary clinical medicine. All hormones, including TSH, are now known to be secreted in rhythmic fashion, and some of these hormonal rhythms (such as leptin and growth hormone) are known to be different between men and women (383, 450, 490, 638). Most significantly, experiments on several hormones (including insulin, growth hormone and testosterone) have shown that when these rhythms are disrupted, symptoms of clinical malfunction appear. However, a similar approach in terms of analyzing daily thyroid rhythms has not been taken in the study of thyroid disorders.

As a consequence, for both human and canine cases, when single-sample lab tests are used as a guide, the diagnosis of hypothyroidism is haphazard and the treatment prescribed is often ineffective in resolving all symptoms (24, 105, 106). In the old days, before the TSH test was developed, physicians used a combination of reported symptoms and cholesterol levels to diagnose and treat hypothyroidism. This is because one of the most immediate effects of suboptimal thyroid hormone is that of elevated cholesterol levels, a detail which often seems to get left out of general discussions of the dangers of high cholesterol. However, research has shown that patients with TSH readings between 2.0 and 4.0 mU/L (still considered "normal," even within the new TSH guidelines) can have high cholesterol levels that drop significantly with T4 supplementation (121, 573, 521). With the modern reliance on TSH test results alone when diagnosing and treating hypothyroidism, many patients complain that lingering symptoms are not taken seriously.

Physicians are very likely to dismiss symptoms of hypothyroidism, especially for women, who are acknowledged to be five to eight times more likely to be afflicted than men. That many physicians don't take indicators of hypothyroidism seriously in women is probably not surprising, considering that two of its leading symptoms are weight gain and depression. The first (weight gain) is still easily written off as character weakness ("lack of self-control") or vanity, while the second (depression) has traditionally been dismissed as psychological rather than physical in origin.

Thyroid hormone and depression

Today, however, the pendulum has swung in the other direction and depression is now seen as *primarily* physical—or rather, physiological—in nature. Depression is currently viewed as the predictable result of particular brain chemical imbalances that can be corrected by specific drug treatment. Antidepressant drugs such as Prozac or Zoloft (*selective serotonin reuptake inhibitors* or SSRIs) are now widely used in treating depression (800). However, these drugs take several weeks to provide improvement and are associated with significant side effects; more importantly—and more often than drug companies and physicians would lead you to believe—they often don't do the job (326, 362).

The premise behind the use of SSRIs in the treatment of depression is that they restore compromised neurotransmission between synapses in the brain. However, antidepressants cannot fix depression caused or exacerbated by hypothyroidism. Thyroid hormone, specifically T_3, is now known to be essential for adult brain function as well as for fetal and early childhood brain development. T_3 is critical to the communication that goes on in the adult brain, in part because it has a fundamental role in controlling brain-specific genes, but also because it's required for brain synapses to function properly (326, 372, 377). Rarely is insufficient thyroid hormone considered as a cause of or complicating factor in depression, even though depression is not only the most common symptom for hypothyroidism, but for chronic stress, congestive heart failure, chronic pain and menopause as well—see the AACE website press release for 1999: "the Missing T in HRT" (also 68, 222).

At the very least, it's almost certain that undiagnosed hypothyroidism in many patients underlies the enormous individual variation when it comes to success rates in treating depression with drugs (24, 326). In fact, the horror stories of patients (especially women) seeking relief from depression abound, many of which detail a litany of treatment with one drug after another over a period of *years*. Such stories are found increasingly often in print, like the one described in *Maclean's* magazine a few years ago (193). However, you only need to raise the topic of depression with friends and neighbours to hear your own local version. I expect you'll hear what I did: cases of flagrant clinical cases of hypothyroidism (not marginal "subclinical" cases) that should have been easily caught by a reasonably astute physician—debilitating depression of many years duration relieved *within days* by thyroid hormone supplementation.

Thyroid hormone and obesity

Thyroid hormone is required at several stages for the conversion of food into metabolic energy, whether that energy is utilized immediately or stored as fat: thyroid hormone is the ultimate controller of *basic metabolic rate* (BMR). As a consequence, there is also a close relationship between thyroid hormone function and body weight. As I mentioned at the beginning of this chapter, one of the most common symptoms of thyroid hormone deficiency, as characterized by elevated values of TSH and low T_4, is weight gain. Even relatively minor hypothyroidism has been found to be correlated with weight gain—a greater *body mass index* (403).

Ideally, high leptin levels tell the body to stop eating and start moving: in experimental rats, high leptin levels stimulate physical activity and turn off the production of the hormone that stimulates appetite (*ghrelin*, which also increases fat production). This weight-normalizing relationship between ghrelin and leptin is currently under investigation, particularly in regard to abnormal interactions or imbalances between them that might contribute to certain eating disorders, such as binge eating and anorexia.

Recent increases in the prevalence of obesity and its associated health problems in many Western countries has prompted a proportional increase in research directed at understanding the role of hormones in surplus fat production (192, 344, 390). Some of the most heavily studied topics in this context concern the functions and control of *leptin*, a relatively simple hormone produced primarily by fat cells. Leptin levels appear to be what keeps the hypothalamus informed about the status of energy stored as fat: the more leptin released, the more fat reserves are available. Leptin, as is true for virtually all hormones, is secreted in a rhythmic manner. Each of the large number of fat cells present in obese patients releases a burst of hormone, pumping considerably more leptin into the bloodstream than is present in lean patients (see Box 9.2 for expert-level details and references).

But what has largely been left out of the equation in discussions about leptin and ghrelin is their relationship with thyroid hormone. In fact, studies have shown that without T_3 to stimulate the gene that makes it, fat cells may not produce leptin in adequate quantities. As we've seen before, in the action of other genes (see Figures 3.6 and 4.2), this relationship between leptin and T_3 presents two possible scenarios that may be hard to distinguish experimentally:

1) Those with a mutation in the *obese* gene get fat because their fat cells can only make leptin in small amounts (or not at all), due to an *impaired gene.*
2) Those without enough T_3 to stimulate their *obese* genes get fat because their fat cells can only make leptin in small amounts, due to an *impaired gene switch.*

Option two is certainly consistent with the evidence that people and animals who aren't generating enough thyroid hormone tend to put on weight. But what's also interesting is that, just as for the brain, what fat cells need to keep them happy is not T_4, but T_3 (more on the significance of this point shortly).

Box 9.2 Hormonal control of fat

A recent study demonstrated that even relatively minor thyroid hormone deficiency (as measured by both TSH and T_4 levels) is correlated with an increased level of obesity (403). A pair of studies found synchronized rhythms of TSH and the hormone *leptin* in both sexes of normal weight and in obese women (408, 490): in obese patients, TSH and leptin rose together. This relationship is significant because leptin has an important role in fat metabolism.

Leptin is the product of a gene called *obese* (*ob* for short), a gene that operates only in white fat cells (adipose tissue). Leptin appears to be the hormone responsible for sending signals to the hypothalamus regarding the status of energy stored as fat: the more leptin released, the more fat reserves are available. Ideally, high leptin levels stimulate physical activity and turn off the production of the hormone that stimulates appetite (*ghrelin*): animals stop feeding and start moving. Ghrelin is produced by cells in various organs (such as the stomach and kidney) as well as by the hypothalamus—it not only stimulates appetite but increases fat production and body growth. Although constantly shifting levels of leptin and ghrelin are normally able to regulate both feeding and fat storage, there has been much interest over the last few years regarding how and why abnormal interactions between them might contribute to eating disorders, such as binge eating and anorexia (69, 408, 503).

But what about the relationship between leptin, ghrelin and thyroid hormone? Recent experimental evidence has shown that T_3 (*not* T_4) increases the production of leptin quite dramatically (841). As a consequence, people or animals with a mutation in the *obese* gene get really fat (because their fat cells can't make leptin at all), while those without enough T_3 to stimulate their *obese* genes make more fat than they really need (because their fat cells can't make as much leptin as they should). This is consistent with the evidence that patients with hypothyroidism gain weight and that, as for the brain, fat cells need T_3. Several studies have shown that leptin secretion is not only distinctly rhythmic, but that leptin rhythms rise and fall in synch with TSH rhythms (408). Some authors have suggested that leptin may actually *regulate* the rhythm of TSH to a large degree (490).

However, leptin has other functions besides balancing fat stores. Leptin also stimulates inflammation, brain growth, and perhaps sleep. One study showed that obese patients were fatter than those who got a good night's sleep (see Box 9.3). It isn't clear from this study whether lack of sleep makes you fat or being fat makes you sleep less. But since sleep is known to be very strongly tied to thyroid hormone rhythms, it's probable that disruption of hormone rhythms causes you to sleep badly, which then contributes to weight gain.

Box 9.3 Hormonal control of sleep

Leptin is not only involved in balancing fat stores, it also stimulates inflammation and brain growth, and interacts with many other hormones (69, 503, 606). As a consequence, there is some evidence that leptin's influence correlates with both sleep *and* fat: a recent study showed that people who slept less than 8 hours a night were fatter than those who got a good night's sleep; they also had lower blood levels of leptin and higher levels of ghrelin (606).

So, does lack of sleep make you fat or does being fat make you sleep less? Since sleep is known to be very strongly tied to rhythms of hormones secreted by the hypothalamus, as shown by the many experiments on sleep disorders (44, 78, 89, 540), it is more likely that disruption of thyroid rhythms causes you to sleep badly, which then contributes to weight gain. So far, it has been determined that: 1) blood levels of thyroid hormone decline dramatically in profoundly sleep-deprived rats, without the usual increase in TSH— suggesting that the hypothalamus is getting a direct incoming signal to turn off secretion of the hormone TRH that would normally result in TSH release and subsequent thyroid hormone production (212); 2) alteration of normal sleep patterns is often noted in patients with thyroid dysfunction; 3) rhythms of TSH (the hormone that triggers the release of thyroid hormone) become disrupted in sleep-deprived animal and human patients (44, 88, 270, 271).

Although I've found no published studies that correlate disruption of thyroid rhythms with sleep disturbances directly, there is circumstantial evidence that an intimate relationship between the two will eventually be documented. So far, we know that: 1) profoundly sleep-deprived rats become hypothyroid; 2) patients with thyroid dysfunction report sleep disturbances; 3) rhythms of many hormones become disrupted in both sleep-deprived animals and humans.

Thyroid hormone and birth defects

It has long been known that acute hypothyroidism (especially common in women who are severely iodine-deficient, a condition that is still common in some parts of the world) can cause the peculiar combination of mental retardation and dwarfism known as *cretinism* (258, 296). However, it's becoming increasingly clear that even mild hypothyroidism can cause detectable birth defects. More pronounced hypothyroidism appears to be associated with increasingly profound birth defects and an elevated risk of premature delivery (116, 424, 710). Premature delivery itself is dangerous to infant development, as shown by a recent study indicating that at least 80% of preterm babies born at 22–25 weeks show either neuromotor or mental disabilities, or both, by the age of six (495, 781).

Because mild to moderate maternal thyroid hormone deficiency primarily affects fetal brain development, repercussions not only involve modest mental retardation, but deafness and colour vision defects as well (97, 437, 651). Thyroid hormone deficiency has been implicated in autism spectrum disorder because of the abnormalities in brain development now shown to be associated with the condition (436, 808, 823). (For a unique perspective on autism, see a new book by high-functioning autism sufferer Temple Grandin, listed at the end of this chapter.) The realization that thyroid hormone deficits can have significant effects on fetal development, coupled with an awareness that many women may have untreated hypothyroidism, has prompted a suggestion that perhaps women of childbearing age should have their thyroid function tested routinely (182), although many physician consider this to be an unnecessary waste of time and resources.

Diagnosis and treatment of hypothyroidism

Even when a diagnosis of hypothyroidism has been made, treatment is not always successful at relieving all of a patient's symptoms. Some of this failure may be due to a reliance on TSH values alone to indicate when normal levels of T_4 have been restored. As we've seen, the value for what's considered normal thyroid function has not only changed enormously over the years, but it's also clear that this approach may not supply enough T_4 for all individuals. For example, some people are almost certainly relatively hyperthyroid by nature (high energy types), and if their thyroid function gets out of whack, true restoration might require aiming for a TSH level that's actually *below* normal to get them feeling "their old selves." Individuality may also explain why some people who are clinically hypothyroid (TSH values above the normal range) don't complain of hypothyroid symptoms (106, 636).

Another factor complicating the task of maintaining optimum thyroid hormone levels for people who are thyroid hormone deficient involves tissue-level variation in the efficiency of T_4 to T_3 conversion. Supplementation with synthetic T_4 has been the standard treatment for human hypothyroidism for at least thirty years (24), a strategy that assumes the body's natural conversion mechanism of T_4 will yield all of the T_3 required. However, for some patients, treatment with a combination of T_3 plus T_4 has been shown to resolve symptoms that persist when T_4 alone is given (24, 99). Such results suggest that the T_4 to T_3 conversion mechanism can be impaired in some patients.

It has also been shown that some tissues, such as the brain and fat cells (as discussed above), have a higher requirement for T_3 than others (15, 326). This suggests that hypothyroid symptoms that reflect brain function (such as depression, irritability and impaired concentration) may persist in the face of a supplementation regime that resolves all others. Clearly, the mechanism responsible for regulating optimal T_3 levels in various tissues (especially the brain) is more complex and variable than previously thought.

Disruptors of thyroid function

Untreated maternal hypothyroidism, as discussed previously, can cause significant birth defects in infants, particularly in brain growth and development. Not surprisingly, disruption of maternal thyroid hormone production is suspected of being the mechanism underlying the physical and brain development anomalies known as *fetal alcohol syndrome.* Prompted by the realization that the birth defects associated with fetal alcohol syndrome are very similar to the defects seen in infants born to mothers with thyroid hormone deficiency, recent research seems to support this interpretation. Subjecting sheep in their third trimester of pregnancy to levels of alcohol equivalent to human "binge drinking" has been shown to suppress thyroid hormone levels in both mother and fetus (157). However, more work is clearly required to sort out exactly what is going on (156).

Some common pharmacological agents (including ASA, general anesthesia, Phenobarbital, etc.) are also known to suppress or disrupt thyroid function (221, 302), as are some naturally occurring biochemical compounds. Such *goiterogens* are present in various amounts in certain vegetables, for example, in all members of the cabbage family (including broccoli), sorghum, sweet potatoes, maize, almonds, and cassava. Goiterogens are often very similar in molecular structure to thyroid hormone, at least in the portion of the molecule that interacts with others. Goiterogens can attach themselves to thyroid hormone receptors and, in large enough quantities, can produce a hypothyroid state by what is known as *competitive exclusion* (296). For example, the extremely high incidence of goitre (enlarged thyroid gland accompanied by hypothyroidism) caused by insufficient dietary iodine in some African regions is exacerbated by a reliance on the *cassava* tuber as the primary source of calories (258).

Synthetic chemicals, such as dioxins, polychlorinated biphenyls (PCBs), perfluorinated chemicals (PFCs, substances used in the manufacture of such brandname products as Teflon, Gore-tex, Scotchgard, Stainmaster, etc.), and bisphenol-A (BPA, used in the production of many plastics and epoxy resins, including food and beverage containers, baby bottles, toys) may similarly affect normal fetal and early postnatal brain development because they either reduce or mimic thyroid hormone actions (310, 313, 598, 702, 741, 851, 852). As might be expected given the importance of thyroid hormone to normal body function, humans and animals contaminated by such synthetic chemicals often display many seemingly unrelated symptoms that range from skin problems to reproductive failure.

As much as we fear such biochemical contaminants, however, there are two inescapable factors known to inhibit thyroid hormone production: stress and age (296). Stress can be emotional, physical or physiological, and may have temporary or permanent effects on human health. Illness is one kind of stress where thyroid hormone declines are well documented (161, 218, 374). Sleep deprivation and acute trauma are other forms of stress whose associated symptoms of thyroid hormone decline are generating increasing scientific attention (212). Stress-induced reductions in thyroid hormone production can precipitate coordinated changes in growth hormone, cortisol and gonadal hormones, resulting in an often confusing assortment of physical and psychological symptoms (24, 25). Thyroid hormone is not usually considered to be a classic stress hormone (that award usually goes to *cortisol,* an adrenal hormone). However, stress has been shown to have immediate and profound effects on relative amounts of T_4 and T_3 in specific areas of the brain, including the hypothalamus itself, and the same may also be true for other tissues (45, 208).

Surprisingly, perhaps, exposure to cold does not invoke a typical stress response, but rather stimulates a rapid *increase* in secretion of both thyroid hormone and adrenaline (565, 684).

However, given that stress-associated declines of thyroid hormone in pregnant women may be capable of compromising normal brain development in their babies, I have to wonder what effects various potential stressors (including medical testing and hospitalization) may have on thyroid gland function and thyroid-dependent brain growth of late-term fetuses and newborn babies themselves. Could various fetal and neonatal stresses, if they coincide with particular stages of brain development, be the cause of the wide range of behavioural traits we call autism spectrum disorder? Some forms of autism have now been shown to be associated with particular brain anomalies that must reflect disrupted brain development. The prospect that fetal and/or neonatal stress may cause or contribute to autism is currently one of many possibilities being explored.

What about age? In addition to reductions associated with stress, thyroid hormone levels also decline with age (636). By age eighty-five, basic metabolic rate has dropped to 52% of what it was at age three—largely the result of declining thyroid function (24). Note that many age-related disorders (such as poor temperature regulation, insomnia, decreased mental activity, dry skin, constipation, depression, muscle weakness or stiffness) are the same as those for hypothyroidism: it is not inconceivable that such symptoms of aging may ultimately be caused by declining thyroid hormone levels or by disruption of thyroid rhythms.

At least a few clinicians have suggested that because we expect ourselves and others to remain active and useful well into old age, finding effective diagnosis and treatment regimes for thyroid disorders will become increasingly urgent (24, 198). I contend that one critical component of future studies in this area must focus on thyroid rhythms and what happens to those thyroid rhythms with chemical contamination, age and the unavoidable stresses that come with living in our increasingly complex society.

The future—personalized medical treatment?

It should be clear from this discussion that doctors do not currently have a very precise understanding of what constitutes normal thyroid hormone function or its range of variability. Nor do many physicians have a good understanding of the full range of consequences when disruption of the thyroid system occurs. Therefore, it's not surprising to find that precious few of the papers or books published on thyroid hormone disorders in human or veterinary medicine even acknowledge the rhythmic nature of thyroid hormone production.

The few studies that have been done, however, document that individual humans do indeed have individually distinct rhythms of both thyroid hormone and TSH—precisely what I've determined must be so for all animals (48, 161, 408, 466, 718). In these studies, it appears that because high and low values (the peaks and valleys of the daily rhythms) do not exceed either end of the accepted range of variation that's considered normal, the rhythmic nature of the secretion is assumed to have no biological impact.

As a consequence, few studies have measured the magnitude and extent of thyroid hormone rhythm changes under potentially disrupting conditions. Experiments on other hormones, however, suggest that in individuals with diagnosed medical conditions, one of two things happen: 1) rhythms become increasingly irregular; 2) critical threshold peaks of hormone concentration are not attained at appropriate times. Rhythm irregularities in secretion profiles of insulin have been particularly well documented in diabetes (100). Disruption of rhythmic secretion coincident with specific health problems has been documented in a number of other hormones, including growth hormone, testosterone, cortisol and leptin (71, 88, 450, 490, 502, 663, 764, 772-775, 663). Clearly, the nature of rhythmic thyroid hormone secretion and consequences of rhythm disruption need more study in both humans and animals.

In addition, research into individual differences that exist at the receptor level (the receiving end of thyroid function as opposed to the production end) are also needed. There is increasing evidence that some people can develop a resistance to thyroid hormone at the tissue level, as also occurs with insulin (102, 496, 620). In most cases, however, clinical researchers view thyroid hormone resistance as a rare disorder that is only valid if it is associated with mutations in thyroid hormone receptor genes.

Similarly, the reigning opinion of many researchers and physicians involved in both human and veterinary clinical medicine seems to be that virtually all thyroid function abnormalities are the result of antibodies attacking various components of the thyroid metabolism mechanism—what is called *autoimmunity* (179, 636). However, I contend that such an interpretation is not consistent with an evolutionary perspective of thyroid hormone function and its variation across all life forms; this is not to suggest that thyroid hormone antibodies don't exist, but only that they may not be totally or directly responsible for the apparently associated effects to which they are currently attributed (777).

I suggest that once we have a better understanding of thyroid rhythms—including how they are generated, how they interact with other hormonal rhythms and why rhythm disruption occurs—a completely novel level of personalized medical care becomes a possibility for the future. But the kind of personalized care I envision is very different from what I'm reading about in the scientific literature and news reports. Right now, some geneticists suggest that what they call "personalized medical care" is possible and imminent, based on their ability to generate a personal dossier of key disease-controlling genes. Knowing your genetic constitution, they claim, will allow physicians to identify and treat certain medical conditions more effectively, perhaps even before they cause trouble (335, 549).

However, as we've seen, bad genes aren't necessarily the culprit in a good number of debilitating medical conditions. Even if someone has a particular variant of a gene, seldom is a 100% cause-and-effect relationship between it and an illness established (102, 342). In reality, while a particular genetic variant (or even a number of bad genes together) may be shown to increase a person's likelihood of developing an illness, it does not guarantee it (e.g., ref. 725, on the genetics of schizophrenia; ref. 3, on the genetics of asthma). In many cases, I suggest, hormone rhythm disruption is more likely to be a better predictor or indicator of many chronic medical conditions.

It appears to me that repeated short-term monitoring of hormone rhythms offers a more effective and precise method of delivering truly personalized medical care than does genetic profiling. If individual thyroid rhythms can be monitored regularly as we move from childhood into adulthood, the changes indicative of an impending system malfunction—or the start of normal aging processes—should be recognizable early.

And if disruptions in hormone rhythms rather than absolute declines in hormone levels are what sidetrack us as we mature and age, as appears to be the case, we may be able to help stabilize all hormonal rhythms by stabilizing the pacemaker—the governing thyroid rhythm.

Recommended reading

Arem, R. 1999. *The Thyroid Solution.* Ballantine Books, New York.

De Groot, L.J. and Hennemann, G. (eds.) 2005. www.thyroidmanager.org Endocrine Education Inc.

Demers, L.M. and Spencer, C.A. (eds.) 2002. Laboratory Medicine Practice Guidelines: Laboratory Support for the Diagnosis and Monitoring of Thyroid Disease. Monograph published by the National Academy of Clinical Biochemistry, Washington, DC, see www.nacb.org

Doyle Driedger, S. 2001. Overcoming depression: one woman's terrifying odyssey through a nightmare of despair. *McLean's* 114(46):34-38.

Grandin, T. and Johnson, C. 2005. *Animals in Transition: Using the Mysteries of Autism to Decode Animal Behavior.* Scribner/Bloomsbury.

Hawaleshka, D. 2005. Teflon trouble. *Maclean's* 118(21):70-72.

Horrobin, D.F., 2001. *The Madness of Adam and Eve: How Schizophrenia Shaped Humanity.* Bantam Press, London.

Jenkins, K., 2001. Not tonight, dear – I'm feeling better: the drugs that relieve depression also sap the libido. *Maclean's* 114(46):40-42.

Prinz, P. 2004. Sleep, appetite and obesity: what is the link? *Public Library of Science Medicine (*www.plosmedicine.org*)* 1(3):e61. [online, available to all]

Sullivan, P.F. 2005. The genetics of schizophrenia. *Public Library of Science Medicine (*www.plosmedicine.org*)* 2(7):e212. [online, available to all]

Thayer, K. and Houlihan, J. 2002. Perfluorinated chemicals: justification for inclusion of this chemical class in the National Report on Human Exposure to Environmental Chemicals. *Nomination Report by the Environmental Working Group* (Washington, DC) to the US Center for Disease Control (Atlanta, GA).

Wickelgren, W. 2005. Autistic brains out of synch? *Science* 308:1856-1858.

Wise, P.M. 1998. Menopause and the brain. *Scientific American* Women's Health: A Lifelong Guide (Special Issue):78-81.

Wong, K. 2001. The search for autism's roots. *Nature* 411:882-884.

http://www.aace.com/publications.htm

http://www.helpguide.org/mental/depression_signs_types_diagnosis_treatment.htm

http://helpguide.org/mental/depression/

http://www.thyroidmanager.org.htm

http://www.nacb.org.htm

Chapter 10

Evolution Made Personal

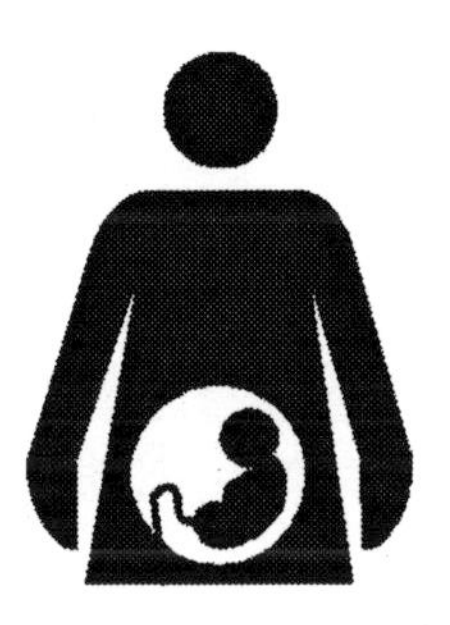

Summary

Charles Darwin spent a good deal of time early in his career trying to figure out how breed development in domestic animals worked. The practical answers he was able to tease out of his studies on animal breeding helped enormously in his big-picture philosophical endeavour to explain how and why evolution occurs. I've taken this thought process one step further by focusing on the process that generates a domestic animal from a wild one. I firmly believe that the biological mechanism responsible for the initial transformation of wild animals into domestic ones is the key to unlocking the mysteries of both evolutionary change and day-to-day health. I contend that the process of protodomestication (made possible by individual variations in species-specific thyroid rhythms) is an especially good model for describing the truly dynamic relationship that exists between populations and the environment. The actively responsive system of interconnected hormones described by this model, governed as it is by thyroid rhythms, not only permits individuals to adjust to local conditions that change on a daily and seasonal basis within their lifetimes (what biologists call *plasticity*), but allows populations of animals (species) to adapt to changes of greater magnitude that occur over evolutionary time scales.

I also contend that individual variants of species-specific thyroid rhythms hold more promise as a mechanism of evolutionary change than do randomly variable genes encoding discrete traits. Individual variation in species-specific rhythms of thyroid hormone secretion links individual variation in a wide variety of traits that may be subject to natural selection, including physical, skeletal, physiological, reproductive and behavioural characteristics. As a result, one unique thyroid rhythm has the power to coordinate an entire suite of evolutionarily significant individual characteristics. Selection for any characteristic controlled by an individual's unique thyroid rhythm selects all other thyroid-controlled traits, providing a mechanism for whole suites of traits to change in coordinated fashion in response to changing environmental conditions.

I propose that during colonization events, differences between individuals in their response or tolerance to stress (controlled by individually-unique thyroid rhythms) encourages small groups of stress tolerant individuals to become isolated in founder populations. Low levels of variation in thyroid rhythms (and their associated growth programs) among such small groups of founders inevitably precipitate shifts in founder population growth rates. Such shifts in growth rates cause a suite of associated physical and physiological repercussions among future descendants, including changes in the age of sexual maturity, size, coat colour and behaviour, which are evident, for example, in the transition from wolf to domestic dog, and in the transformation from brown bear to polar bear.

Colonization events are not the only evolutionary context where this concept can be applied. For example, the death of a particular portion of a population (*selective mortality*), which occurs when certain thyroid rhythm-controlled physiological traits are possessed by only one portion of a population, can also lead to evolutionarily significant changes among descendants of founders, as proposed to explain the transformation of wild to domestic rainbow trout. Adaptation of entire populations to changing environmental conditions can be achieved over both short and long time periods via selection for *any* thyroid rhythm-linked trait, due to shifts in the relative proportions of different thyroid rhythm profiles in the group, as proposed to explain recent fluctuating adaptation of beak shapes in Darwin's finches.

Adaptations of this kind can be temporary (adjustments to short-term climate change) or permanent (resulting in speciation), depending on the circumstances. Reduced thyroid rhythm variation in small founder or adapting populations can be increased via hybridization with a closely-related species or via a natural accumulation of thyroid rhythm variation over time (due to mutations in clock genes or as a consequence of slight changes in the physical relationships of individual oscillating clock neurons to each other during brain development, or both).

Using this thyroid rhythm model to explain evolutionary change provides such a novel perspective on human health that it makes understanding evolution truly personal. For one, it gives us a whole new perspective on individual variation in human form and function, especially for brain function. Brains must be as individually unique as fingerprints or the pattern of spots on a Dalmation: subtly different in "hard-wiring," subtly different in biochemistry. Being different doesn't mean any one pattern is better than another. Different is just different and is a characteristic of being human. Indeed, such differences have been essential to our evolutionary success and are key to our continued survival.

However, there are disturbing indications that new stresses in many societies and prolonged life-spans are presenting challenges that our species has not experienced until quite recently. Disruption of individually-unique thyroid rhythms by psychological stress, illness, biochemical contaminants and aging may account for a significant proportion of known health problems, including depression, heart failure, obesity, sleep disruption, learning disabilities and brain development disorders. If so, the innovative approach I propose presents the possibility of a genuine revolution in medicine: personalized health care that takes individual differences in thyroid hormone physiology into account at all stages of life, from conception to death.

Certainly, this concept of mine needs testing to see if it holds up. However, a strong body of circumstantial evidence supports the few assumptions I've had to make in developing this theoretical model—the rest is based on solid experimental fact. These assumptions are simple but pivotal:

1) Thyroid hormone rhythms are species-specific.
2) Species-specific thyroid rhythms are individually variable.
3) Individually variable thyroid rhythms are the pacemaker that controls and co-ordinates all other hormones, effectively linking a huge number of physical and behavioural traits and physiological functions.
4) Factors which modify maternally-controlled thyroid hormone delivery to the fetus can alter brain development of offspring in such a way that the thyroid rhythm of the offspring is also changed, introducing slight but heritable individual variation to future generations that is independent of identifiable genetic mutation.
5) Thyroid rhythm disruption may significantly impact human and animal health in a heritable manner that is independent of identifiable genetic mutation.

The facts, in summary, include these:

1) Rhythmic secretion is an essential feature of all hormones, generating hormone concentrations ("doses") that vary with time.
2) All hormone rhythms tested so far show individual variation.
3) There is strong evidence that an overriding *pacemaker* for non-thyroid hormone rhythms exists.
4) This hormonal pacemaker has not yet been identified, although thyroid rhythms have not been considered.

5) All hormone rhythms change in response to season and reproductive stage, both factors known to be controlled by thyroid hormone.
6) Thyroid hormone exerts a strong influence over the production and regulation of other hormones in a time- and dose-dependent manner.
7) Thyroid hormone is required for proper function of a huge number of genes involved in embryonic, fetal and postnatal growth, as well as those which govern day-to-day functions of cells, membranes and mitochondria in adult organisms (via effects on regulatory genes and DNA transcription, and basic cellular processes (including calcium transfer, oxygen consumption and the sodium pump).
8) All thyroid-dependent genes respond to thyroid hormone in a time- and dose-dependent manner.
9) Too much or too little thyroid hormone during pregnancy is known to have significant effects on the newborn, with the brain being particularly sensitive.
10) The genes responsible for generating hormone rhythms ("clock and rhythm genes") are located in brain cells ("clock and rhythm cells").
11) The physical relationship of clock and rhythm cells to each other appears to be as important to their function as their genetic makeup (their DNA).
13) Clock and rhythm cells are particularly responsive to signals from other cell in their complex as well as other parts of the brain (which relay signals received from receptors throughout the body of an individual).
12) There is a direct nerve connection between the retina and the hormone rhythm control area of the brain (the SCN), and between the SCN and the thyroid gland itself—this direct connection explains how thyroid hormone secretion can be adjusted quickly in response to changing environmental or psychological states.
14) Disruption of hormone rhythms is known to be associated with stress, illness and aging.

Over all, there is very strong support for my interpretation that, given the time- and dose-dependent relationship that exists between thyroid hormone and other hormones (and given that all hormones are known to be secreted in rhythmic fashion), thyroid rhythms may be the critical hormonal pacemaker that generates selectable individual variation over evolutionary time.

Although it has been known for years that daily variations in thyroid hormone concentration in humans result from rhythmic thyroid hormone secretion, such variations have been assumed to have little or no impact on normal biological function. However, since all thyroid-dependent genes respond in a time- and dose-dependent manner, the fluctuations of thyroid hormone concentration in blood and tissues that result from rhythmic secretion must be more critical to brain and body health than was previously thought. This time- and dose-dependency of genes on thyroid hormone is nowhere so critical as in brain growth and development, suggesting that very minor variations of

T_4 delivery to the fetal or newborn brain may alter the relationship of brain cells to one another, which may alter later brain function.

Current "genocentric" wisdom holds that an embryo draws only essential nutrients from its mother (through the placenta or egg yolk)—that a fetus maintains almost complete control over its own growth and development through expression of its own genes (i.e., that the complement of genes inherited by the fetus from its parents determines its final form and function). However, if what I have surmised here is true—that growth and development during embryonic life is influenced less by an individual's own genes than by the precise rhythm of maternally-supplied thyroid hormone—the power of fetal genes to determine individual traits of an embryo in early development is usurped (the fetus *does* direct its own destiny eventually, but not until its own thyroid rhythm becomes established near the end of fetal life).

The possibility that disruptively high levels of dietary thyroid hormone in mothers (ingested by them as a consequence of dietary changes associated with colonization of new habitats) may be capable of permanently altering the thyroid rhythms of their offspring is a truly revolutionary concept. It provides new and significant insight into the sudden appearance in the fossil record of profound physical changes that are often referred to as *evolutionary novelties*. Maternal thyroid rhythms permanently altered through dietary change could explain the sudden appearance of many such evolutionary novelties—especially those that are developmental in nature, such as bipedal morphology in early hominids and the profound changes required for whales to re-invade the seas. A mechanism such as this (alteration of maternal thyroid rhythms via dietary change), capable of bypassing the gradual accumulation of mutations in a number of independent genes, has long been suspected to exist.

As a consequence, thyroid rhythm theory offers the first truly testable hypothesis that explains how a single biological mechanism can generate distinct species-specific growth generation after generation, but provide an avenue for change when necessary. This is the only system proposed so far with the potential to control the kinds of morphological, reproductive and behavioural traits known to change in coordinated fashion over evolutionary time.

What's more, the concept addresses the long-standing debate on whether rapid evolutionary change is even possible, let alone common. Evidence from deliberate and experimental domestication in silver foxes and salmon, as well as data derived from analysis of adaptation in Darwin's finches over that last thirty-year period, gives us the fastest and most concrete time frame yet for rapid speciation and adaptation events: for some animals, significant changes can be seen within a single generation, while in others it may take up to twenty. This definition of "fast" gives us a whole new perspective on the ability of populations to respond quickly to changing environmental conditions. While Ernst Mayr once suggested, back in 1988, that "*several thousand years*" was the fastest rate he could conceive of for rapid speciation events, future generations of evolutionary biologists are likely to be investigating change over several *years*.

Lastly, the possibility that thyroid rhythm theory might apply to organisms other than vertebrates, at least in general terms, is truly intriguing. The molecular character of thyroid hormone is not only stable across all life forms but appears to be essential to the function of all cells (especially mitochondria) in virtually all animal species, whether they have backbones or not. The widespread presence of thyroid hormone in both plants and invertebrate animals, as well as the revelation that regulatory hormones in these organisms (such as *juvenile hormone* in insects) bear a striking functional similarity to thyroid hormone, both suggest that thyroid hormone, or molecules analogous to it, coordinate the expression of the genes that control development and reproductive functions in all species. Since we have firm evidence that rhythmic hormonal secretion occurs in plants and invertebrates as well as in vertebrates, this suggests the possibility that my theory, taken in its broadest sense, might drive adaptation and speciation in all multi-cellular organisms, although the pivotal hormone may vary between these major groups. Ironically, perhaps, I am already receiving very strong support for my theory from a number of invertebrate evolutionary biologists—even though the details of the concept as presented for vertebrates are not directly applicable to the organisms they study. The fact of the matter is, however, that the general model *is* relevant to all types of organisms, a concept that seems not to have been lost on starfish and insect researchers.

At the end of Chapter 1, I presented a list that summarized what I considered some of the most serious shortcomings of the story of evolution as it is currently presented. Here is that list restated, but revised to reflect the aspects of my concept that *improve* the story of evolution. Thyroid rhythm theory represents a significant and novel revision of the traditional population genetics-based explanation for speciation and adaptation changes because:

1) It provides a demonstrable link between genes, individuals and the environment.
2) It describes a heritable biological mechanism that allows populations to transform in response to changing environments over evolutionary time, a mechanism that is capable of bypassing the gradual accumulation of genetic mutations.
3) It provides an explanation for the fact that there are not enough genes available to account for the number of traits that actually change during evolution, and explains why whole suites of traits always change simultaneously, even between closely related species.
4) The biological mechanism utilized allows species to transform rapidly enough, in response to changing environmental conditions, for adaptation to be effective over the short term ("in ecological time")—in fact, the mechanism is so relevant to individuals in their own time that it makes evolution *personal.*
5) It provides an explicit explanation for why some species have more evolutionary flexibility ("adaptability") than others.

Final thoughts and questions

Once we have adequate technology for doing so, I predict that analysis of individual thyroid rhythms, especially for pregnant females and their developing young, will help unravel some important but puzzling aspects of human health and animal conservation. As I bring this volume to a close, I thought it might be useful to mention a few of the many questions that haunt me—questions inspired by intriguing bits of information (from references cited in this book, reports of studies in progress, preliminary experimental results and discussions with active researchers) coupled with what I've come to know or suspect about thyroid hormone actions.

First, consider the role of individual variation in the response to (or tolerance of) stress. As we gain a better handle on thyroid hormonal repercussions of both short- and long-term stress on mental health in adults, it seems clear to me that we must also apply this knowledge to fetal and newborn effects. For example, we know that babies (and especially their brains) don't grow continuously, but in short intense bursts every 7–10 days. This pattern has been documented by, among other measures, periodic increases in head circumference (429, 528). How does this pattern of brain growth relate to thyroid rhythms? If periodic infant growth is, at least in part, a reflection of periodic age-specific shifts in thyroid rhythm intensity, would disruption of thyroid rhythms have different effects during periods of active infant growth than it would during the lulls in between?

Does a late trimester fetus, which has begun to establish its own thyroid rhythm, grow in a similar episodic manner as a newborn—and if so, is the development of fetal brains more vulnerable to thyroid rhythm disruption during these short periods of intense growth, regardless of whether that disruption is due to stress experienced by the mother or by the fetus itself? Could stress of any kind that's capable of depressing thyroid rhythms during periods of active growth (whether physical or psychological, especially in stress intolerant newborns) account for the fact that infants hospitalized for medical problems in the first few months of life are far more prone to develop autism than those who are not? Could the periodic growth pattern, coupled with individual variation in tolerance to stress and in thyroid rhythms, explain why some infants exposed to such early stresses have impaired brain growth while others do not? Some of these stresses may be unavoidable but others might be easily prevented: while we cannot eliminate all such brain function anomalies, we might be able to reduce their incidence significantly.

Another issue is that of individual variation in thyroid function and female infertility. Infertility, in susceptible women, could very well result from thyroid rhythm disruption or "mild" hypothyroidism rather than being a consequence of profound thyroid hormone insufficiency. If so, what will be the long-term consequences of continually bypassing the effects of thyroid deficiency and rhythm disruption in such women by providing in vitro fertilization? There are strong indications that many infertile women have sub-optimal thyroid function, but most who have trouble becoming pregnant aren't aware of this as a possible cause (nor are their doctors). Indications are that women infertile due to sub-optimal thyroid function/rhythm disruption may pass this trait on to their offspring, even if the implanted egg is not their own. Such offspring also face the possibility of subtle brain abnormalities if underlying maternal thyroid dysfunction is not addressed during pregnancy. In a worst case scenario, in vitro fertilization as an increasingly common solution for infertile women with undiagnosed thyroid deficiency could have significant evolutionary consequences that won't be apparent for many generations.

Other questions of mine relate to conservation issues. For example, what are we really doing to endangered species kept in zoos and game parks when we control their breeding in ways that are not meant to change them (as in breed improvement) but to keep them the same? Is providing them with their natural food and with mates of the right kind enough to preserve a species? Does the stress of captivity that keeps most individuals of some species infertile allow only stress tolerant individuals to become breeders? If so, what will be the taxonomic status of the few offspring that are produced? Such populations appear to be experiencing conditions equivalent to enforced protodomestication; if this is the case, will their offspring eventually come to be domesticated versions of the wild species? Can we predict how descendant populations will change? In some animals with long generation times, change may not be apparent for many years to come; in others, change may be muddied or mitigated by practices such as in vitro fertilization.

Nevertheless, evidence is mounting that small populations of species "conserved" over long periods of time can change in surprising ways, like, for example, the last remaining population of South African mountain zebras, *Equus zebra*. "Rescued" from their alpine habitat almost one hundred years ago, and maintained in sea-level game parks, some mountain zebras within this small remnant population have been reported as showing subtle changes to the pattern of stripes typical of their species (more detailed research on this is currently underway). Since patterning of animal coats is set during fetal development (as described in Chapter 3), and fetal development is known to be dependent on thyroid hormone, it's possible that the small remnant population of mountain zebras isolated in this reserve either don't possess quite enough individual variation (or not quite the right mix of variation) in the individually-variable thyroid rhythms that give this species its unique striping pattern. Alternatively, because one of the main roles of thyroid function is to respond to seasonal and daily changes in temperature (which vary with altitude), this population may be adapting to being kept at sea level in a way that's having subtle effects on fetal growth patterns and thus, on the striping patterns of some offspring. See Stephen Gould's essay "How the zebra gets its stripes" or Sean Carroll's new book *Endless Forms*

Most Beautiful for more insight on zebra striping (112, 276).

What about fish confined and bred in fish farms? *Nature* magazine editorial writer Kendall Powell (600) recently discussed plans being considered by international fish farmers to switch carnivorous farmed fish (such as salmon, trout and cod) from a diet of ground dried fish meal to one that's vegetarian-based. From two to five tons of small fish (smelt, herring, anchovies, etc.) must be processed to produce a single ton of farm-raised carnivorous fish. Not only is this expensive, but it's depleting the oceans of small fish. However, knowing what we now know about the importance of thyroid hormone in reproduction and embryonic development (especially for carnivores), what impact might there be on salmon if they're switched to a vegetarian diet lacking thyroid hormone? Will they still be "salmon" after twenty years on such a diet?

The thyroid hormone content of animal diets is never taken into account, as far as I've been able to determine, whether in wild or captive situations: only protein, carbohydrates, fats, vitamins and minerals are considered. This nutritional blind spot to the importance of dietary thyroid hormone could have serious consequences for captive carnivores of all kinds, not just for salmon, including the carnivores trapped inside our own bodies (current recommendations to reduce the amount of red meat and eggs in Western-style diets, and to cook all such foods thoroughly to eliminate *Salmonella enteritidis* bacteria, could already be impacting individuals with marginal or unstable thyroid function, especially if cooking indeed reduces the majority of biologically active thyroid hormone from foods, as I suggested in Chapter 8).

Another conservation issue to be considered involves the use of a surrogate mother of one species (such as a domestic cat) to gestate fertilized eggs of a closely-related endangered species (such as an African wildcat)—a process technically called *interspecies nuclear transfer*—as described in Chapter 8, Box 8.1 (see also ref. 433). As of the year 2000, the following clones of rare or endangered species have resulted in live births, although no detailed descriptions of the offspring have been published: *guar* (in a domestic cow); *African wildcat and Indian desert cat* (both in a domestic cat); *bongo antelope* (in an eland); *mouflon sheep* (in a domestic sheep); *European red deer* (in a white-tailed deer). Knowing what we know now about how maternal thyroid hormone can impact embryonic and fetal growth, what is the taxonomic status of these offspring: how do we assign a scientific name? Are they true representatives of the species which generated the egg? Are they hybrids, or something else entirely? This dilemma raises practical and ethical issues that are becoming increasingly common in our twenty-first century world, topics that require serious consideration. If interspecies nuclear transfer becomes common, perhaps we won't end up saving endangered species at all, but rather creating totally new ones that would never have existed without our interference.

There are many more questions to be asked, of course, ones that may not be answered for decades. But the important point is that several of these are entirely new questions that could not been asked before now: some might say they are long overdue. A comment made by palaeoanthropologist Jeffrey Schwartz in regards to the current status of evolutionary theory (including human evolutionary theory) is worth repeating:

> *[F]ar from the expectations of the Synthesis—that we know enough of the basic outline of how evolution works that we can concentrate on the minutae of details—we are only now beginning to understand the broad picture. (Schwartz 1999, ref. 670:379)*

Thyroid rhythm theory is based on the premise that a simple biological mechanism exists that allows species (populations of individuals) to adapt and transform over evolutionary time in response to changing environmental conditions—a process that merely expands in scope the existing system individual animals are known to use in adapting to daily and seasonal changes in their lifetimes. If my assertion—that individually-unique variants of species-specific thyroid rhythms control an entire suite of unique physiological, physical and behavioural characteristics—is upheld, then the concept describes an avenue for natural selection to tap into an extremely critical reservoir of variation within a population that is not exclusively genetic in nature.

Such a statement represents a revolutionary step in our understanding of evolution, but what does it really mean? Let's return to the question of what makes us human: what differences exist that separate us from our closest living relatives, the chimpanzees? Traditional evolutionary rhetoric says these differences must lie in the genes. Accordingly, a draft sequence of the total nuclear genome of the common chimpanzee (*Pan troglodytes*) has just been published for comparison to the previously published human genome sequence (both are consequences of enormous investments of time and money).

The publication of the chimp genome has already spawned a number of scientific papers, and a deluge of others are expected to follow. A recent editorial in *Science*, the weekly scientific journal published by the American Association for the Advancement of Science, succinctly contrasts the trumped-up expectations of scientists in this endeavour with the little-publicized reality of the situation:

> *Can we now provide a DNA-based answer to the fascinating and fundamental question, "What makes us human?" Not at all! Comparison of the human and chimpanzee genomes has not yet offered any major insights into the genetic elements that underlie bipedal locomotion, a big brain, linguistic abilities, elaborated abstract thought, or any other unique aspect of the human phenome. This state of affairs may seem disappointing, but it is merely the latest example of a generalization that genomics research has already established – interpretation of DNA sequences requires functional information from the organism that cannot be deduced from sequence alone. (McConkey and Varki 2005, ref. 471)*

This last sentence is not apt to be challenged by even the most devoted molecular geneticist (158, 549). Since we know it to be true, we need a new evolutionary model, a new paradigm, that's capable of taking the complex interactions it implies into account.

My thyroid rhythm theory is just the kind of model we need: one that comes with a testable hypothesis rather than more theoretical rhetoric (e.g., 35, 300 278, 510, 511, 647, 802). Thyroid rhythm theory is the first credible new paradigm we've been offered so far.

Many biologists realize that increasing our knowledge of embryonic and postnatal growth will be critical for revamping currently accepted evolutionary theory. But I doubt anyone would have anticipated that a totally new theoretical model might be a necessary component of such modifications—what scientists call a *paradigm shift*. That's probably why we haven't seen one until now. Even still, who would have guessed that a paradigm shift in evolutionary theory could have the power to totally revolutionize human medicine, that unravelling the mysteries of dog domestication might be the key to making evolution truly personal? It's amazing, even to me, how much this new perspective changes the big picture.

Recommended reading

Carroll, S.B. 2005. *Endless Forms Most Beautiful: The New Science of Evo-Dev and the Making of the Animal Kingdom.* W.W. Norton and Co., New York.

Culotta, E. 2005a. Chimp genome catalogs differences with humans. *Science* 309:1468-1469.

Gould, S.J. 1983. How the zebra gets its stripes. *Hen's Teeth and Horses Toes,* 366-375. W.W. Norton and Co., New York.

Powell, K. 2003. Eat your veg. *Nature* 426:378-379.

Lanza, R.P., Dresser, B.L. and Damiani, P. 2000. Cloning Noah's ark. *Scientific American* (November):84-89.

McConkey, E.H. and Varki, A. 2005. Thoughts on the future of great ape research. *Science* 309:1499-1501.

Nielsen, R. 2005. Disclosure of variation. *Nature* 434:288-289.

Appendix

Timeline for human evolution and historical events

Years Before Present	Time	Millenium
2,000		21 st century
1950	"0" for radiocarbon years	
1,000		AD
	late	1st
	early	
		Christian era
753	start of Roman period	1st
1,000		BC
	late	2nd
	early	
2,000		
		3rd
3,000		
		4th
4,000		
	1st chickens ?	5th
5,000	1st horses & donkeys	
		6th
6,000	1st llamas & alpacas	
		7th
7,000		
		8th
8,000	1st pigs & cattle	
		9th
9,000	1st sheep & goats	
		10th
10,000		**End of the Ice Age**
12,000	1st dogs in Israel	
12,500	**start of Pre-Pottery Neolithic period**	
14,000	1st dogs in N. Eurasia	
40,000		Regular appearance of culture and modern form in *Homo sapiens*
100,000		Scattered skeletal evidence of *Homo sapiens*
400,000		First evidence of *Homo heidelbergensis*
1.8 million		First evidence of *Homo erectus*

Glossary of terms

achondroplasia. A specific genetic mutation leading to disproportionately stunted growth alleles.

adaptive radiation. Evolutionary divergence of members of a single species (or lineage) into different ecological niches, resulting in a number of new distinct species.

agouti. Alternating bands of colour in individual hairs that give wild animals a grizzled appearance.

allele. One of several possible forms of a given gene.

allopatric speciation. see **speciation, allopatric.**

altricial. A mode of vertebrate development characterized by large litters, rapid growth and short gestation, associated with the birth of relatively undeveloped, helpless young.

anadromous. Migratration up rivers from the sea to breed, moving from salt- to fresh-water.

anthropogenic environment. A term used to describe a localized set of environmental conditions created by the impact of human populations (e.g., conditions resulting from the formation of permanent settlements). An anthropogenic environment is also a habitat dominated by the continuous presence or proximity of people.

archaeozoology. The identification and interpretation of animal bones recovered from archaeological sites.

bipedal. Having two feet; the ablility to walk upright on two legs.

breed. A group of animals that has been selected by humans to possess a uniform appearance that is heritable and that distinguishes it from other groups of animals within the same species.

circadian rhythm. A biological rhythm having a periodicity of about one day (24 hours); a diurnal rhythm.

clade. A group within a species or other taxon whose members share a closer common ancestry with one another than with members of any other clade.

classic domestication. The gradual processes of conscious and unconscious human selection (working in concert with natural selection) that can modify any captive or commensal population, whether those animals are products of prior protodomestication or individuals deliberately removed from the wild.

clock genes. The genes within so-called "clock cells," most of which reside in the suprachiasmatic nucleus portion of the hypothalamus (a portion of the brain).

competitive exclusion. Refers to chemical compounds that have a similar molecular structure in the particular region of the molecule that reacts with other molecules; the molecule that arrives first occupies the reaction site.

congenital hypothyroidism. Hypothyroidism that exists at birth, originating in the fetus.

convergence. Phenotypic similarity of two taxa that is independently acquired and is not produced by a genotype inherited from a common ancestor.

convergent evolution. Similar or identical changes evident in unrelated lineages.

deiodinization. The chemical process that removes an iodine atom from a molecule.

dew claw. An extra digit in some dog breeds.

diestrous. Breeding twice a year.

dispersal. The movement of individuals from their birthplace; more broadly, the spread of individuals of a species beyond the current species range.

dogma. An idea adopted without question or accepted on faith.

dwarfing. Becoming smaller than normal; having stunted or restricted growth.

ecological time. Designating a time period that experiences short- or long-term ecological changes.

endemism/endemic. Native to, and restricted to, a particular geographic region.

epigenetic. Traditionally refers to all processes relating to the expression and interaction of genes (including instances where the first few cells of a specific embryonic tissue require the physical presence of an adjacent tissue (or group of cells) for the genes in its cells to be expressed, allowing the tissue to develop properly). However, in the last decade or so the term has been coopted by a group of geneticists to refer to the heritable changes in gene expression that are not mediated by DNA sequence alterations (such as degree of saturation of the amino acid sequence with methyl groups, see **methylation**). Note this new useage of the established definition when searching the literature for information.

evolvability. Flexibility for future change; having more evolutionary options (coined by Stephen Jay Gould).

founder population. A population beyond the previous species range, founded by a single female or a small number of conspecifics.

fossil. Bone, shell or soft tissue that has been replaced by rock.

fossil record. The sequence of fossils spanning a particular portion of geological time.

genetic drift. The occurrence of changes in gene frequency brought about not by selection but by chance. It occurs especially in small populations.

genocentric. Referring to a belief that all changes of evolutionary significance are controlled by genes.

genotype. A set of organisms with the same genetic constitution; the genetic constitution of an individual

goitre. A swelling of the neck caused by an overworked and enlarged thyroid gland.

half-life. The time taken for a measured amount of a substance to become reduced to half its original value.

haplotype. A particular sequence of DNA.

heterochrony. The evolutionary process where changes occur in rates of growth or in the timing of particular developmental events (embryonic or postnatal) of one lineage relative to its ancestral taxon.

hominid. Referring to any member of the ape/human lineage.

hominin. Referring strictly to members of the human portion of the ape/human lineage.

hypothyroidism. A condition that results from insufficient thyroid hormone.

IGF-I (insulin-like growth factor-I). A liver protein stimulated by GH, it is responsible for most of the essential functions usually associated with growth.

individual ontogenies. Reflects the idea that growth and development may be individually distinct for each animal over its lifetime.

introgression. Movement of genes from one taxon into another.

island syndrome. The phenomenon of dramatic change in size, shape and behaviour characteristic of isolated island animals as compared to their mainland ancestors.

isotope analysis. Measures the kinds of carbon atoms found in bone and other tissues; isotope analysis methods can distinguish between the kind of carbon atoms (the chemical "backbones" of all plant material) that are found primarily in grasses and those found in trees and shrubs.

juvenile hormone. The substance that controls growth in insects.

life-history trait. A selected set of adaptations to local environments, involving such quantitative aspects of life history as fecundity, the timing of maturation, and the frequency of reproduction.

lineage. An ancestor–descendant sequence of populations.

macroevolution. Evolution above the species level (of genera, families, etc.) and the production of evolutionary novelties, such as new structures; some researchers consider macroevolution to include the generation of new species as well as levels above (genera, families etc).

methylation. The chemical process that adds a methyl group (a sub-chain of one carbon and three hydrogen atoms) to a given molecule, used recently in reference to DNA, where it has been proposed that the extent of methylation of a gene affects its expression (see **epigenetics**).

microevolution. Evolution at or below the species level; some researchers consider microevolution to include *only* changes that occur within a species (e.g. local adaptation).

mitochondria. The **organelles** found within cells (each similar to a primitive unicellular organism) that produces the enzymes needed for a cell to produce the energy to live and divide; the cell's "energy powerhouse" (singular: mitochondrion).

monoestrous. Breeding once a year.

morphotype. The physical form of an individual or group of individuals that distinguishes them from others of the same species.

mtDNA/mitochondrial DNA. The genetic material found within each mitochondrion of an animal's cells.

mule. The offspring of a male donkey and a female horse.

nDNA/nuclear DNA. The genetic material found in the nucleus of a cell.

neoteny. Refers to examples of **paedomorphosis** where a stage of ancestral development may be truncated in the descendant but shape does not change (much less common than paedomorphosis).

neural crest tissue. The embryonic cells that migrate from a fold of tissue and give rise to many different adult organs and structures.

ontogeny. The growth and development of an individual, both embryonic and post-natal.

organelles. The small spherical bodies with double membranes found within all cells.

paedomorphosis. The process that generates individuals of smaller size at sexual maturity than their ancestors as a result of altered growth rates or shifts in the timing of particular developmental milestones. Also known as *juvenilization.*

pangenesis. Charles Darwin's explanation for how traits are transmitted from one generation to the next.

peramorphosis. The process that generates individuals that are larger at sexual maturity than their ancestors as a result of altered growth rates or shifts in the timing of particular developmental milestones. Can result in *gigantism.*

peripatric speciation. see **speciation, peripatric.**

phenotype. The total of all observable features of a developing or developed individual (including its anatomical, physiological, biochemical, and behavioural characteristics).

phylogeny. The evolutionary history of an organism.

polyphyletic origins. A taxon derived from two or more different ancestral sources, also called *sibling species* (e.g. sheep, pigs, horses and goats have *polyphyletic origins*: they are considered members of the same species, but clearly resulted from more than one protodomestication/speciation event).

precocial/precocious. A mode of vertebrate development characterized by small litters, slow growth (often extended gestation), and the birth of relatively well-developed, capable young.

plasticity. The capacity of an organism to vary morphologically, physiologically or behaviourally as a result of environmental fluctuations.

protodomestication. A natural speciation process whereby certain wild ancestors generate descendants with modified biological features; the initial event that transforms an animal from a wild to a domestic state.

punctuated equilibrium. A term used to describe short bursts of rapid changes in form followed by long periods of geological or evolutionary time without change.

regulatory gene. A gene that controls the expression or activity of other genes involved in developmental processes of an individual.

reproductive advantage. The notion that individuals with the combination of genes that best fit a new habitat leave the most surviving offspring.

scientific nomenclature. A hierarchical method of naming organisms according to their presumed evolutionary relationship; e.g. Class, Order, Family, Genus, Species.

selective breeding. The process where humans choose the mate of an individual; control over reproduction.

sexual dimorphism. Refers to the differences in size and/or shape that are often found between the males and females of a species.

speciation event. The event which precipitates the splitting of a species into two or more descendant lineages.

speciation, allopatric. The origin of a new species involving a geographically isolated portion of a parental species (this is the "divided by a newly formed river or advancing glacier" model).

speciation, peripatric. The origin of a new species via colonization by a small isolated founder population which colonizes a new niche in the newly-invaded habitat; can sometimes result in the transformation of a distinct new genera or family (also known as *budding*).

speciation, sympatric. Speciation without geographic isolation; usually considered rare but may explain the presence of several similar but non-interbreeding species that co-exist within one geographic area that provides several distinct ecological zones.

steroidogenesis. The synthesis of steroid hormones from a cholesterol substrate, which takes place in mitochondrial inner membranes and is required for manufacture of all glucocorticoids, catecholamines, testosterone and estrogen.

stress intolerant. Unable to tolerate or adjust to particular stressful events or circumstances.

stress tolerant. Able to tolerate or adjust to particular stressful events or circumstances.

subspecies. A distinctive geographical segment of a species; a group of wild animals that is geographically and morphologically separate from other such groups within a single species.

sweepstakes dispersal. Accidental colonization, often associated with islands.

sympatric speciation. see **speciation, sympatric.**

tandem repeats. Repetitive sequences of DNA bases that exist in multiple copies within a certain genetic region.

taxon. A taxonomic group, at any categorical level (e.g., a subspecies or a species).

thyroglobulin. A molecule derived from the essential amino acid *phenylalanine* with four atoms of idodine attached.

thyroid hormone. A proteinlike molecule with iodine atoms attached; a colloquial term that refers to either *thyroxine* (four iodine atoms, T_4) or *triiodothyronine* (three iodine atoms, T_3), or both.

transthyretin. The hormone transport molecule that is particularly important for moving thyroid hormone around the brain in humans.

tyrosine iodination. The process of attaching iodine atoms to a tyrosine molecule.

variable alleles. Slight mutational changes that occur randomly and continuously in genes.

Bibliography

001 Aarseth, J.J., Van't Hof, T.J. and Stokkan, K-A. 2003. Melatonin is rhythmic in newborn seals exposed to continuous light. *Journal of Comparative Physiology B* 173:37-42.

002 Abzhanov, A., Protas, M., Grant, B.R., Grant, P.R. and Tabin, C.J. 2004. *Bmp4* and morphological variation of beaks in Darwin's finches. *Science* 305:1462-1465.

003 Ackerman, K.G., Huang, H., Grasemann, H., Puma, C., Singer, J.B., et al. 2005. Interacting genetic loci cause airway hyperresponsiveness. *Physiological Genomics* 21:105-111.

004 Ádám, G., Perrimon, N. and Noselli, S. 2003. The retinoic-like juvenile hormone controls the looping of left-right asymmetric organs in *Drosophila. Development* 130:2397-2406.

005 Adcock, C.J., Ogily-Stuart, A.L., Robinson, I.C., Lewin, J.E., Holly, J.M., et al. 1997. The use of an automated microsampling system for the characterization of growth hormone pulsatility in newborn babies. *Pediatric Research* 42(1):66-71.

006 Adler, G.H. and Levins, R. 1994. The island syndrome in rodent populations. *Quarterly Review of Biology* 69:473-490.

007 Agenbroad, L.D. 2001. Channel Islands (USA) pygmy mammoths (*Mammuthus exilis*) compared and contrasted with *M. columbi,* their continental ancestral stock. In G. Cavarretta, P. Gioia, P. Mussi, and M. Palombo (eds.), *The World of Elephants: Proceedings of the First International Congress,* pg. 473-475. CNR, Rome.

008 Aguilar, A., Roemer, G., Debenham, S., Binns, M. Garcelon, D., et al. 2004. High MHC diversity maintained by balancing selection in an otherwise genetically monomorphic mammal. *Proceedings of the National Academy of Sciences USA* 101(10):3490-3494.

009 Ahima, R.S. and Flier, J.S. 2000. Adipose tissue as an endocrine organ. *TRENDS in Endocrinology and Metabolism* 11(8):327-332.

010 Aiello, L.C. and Wheeler, P. 1995. The expensive tissue hypothesis: the brain and digestive system in human and primate evolution. *Current Anthropology* 36:199-221.

011 Alberch, P. 1985a. Problems with the interpretation of developmental sequences. *Systematic Zoology* 34:46-58.

012 Alberch, P. 1985b. Developmental constraints: why St. Bernards often have an extra digit and poodles never do. *American Naturalist* 126:430-433.

013 Alberch, P. 1991. From genes to phenotype: dynamical systems evolvability. *Genetica* 84:5-11.

014 Alderson, L. 1978. *The Chance to Survive: Rare Breeds in a Changing World.* Cameron and Tayleur Books Ltd., London.

015 Amma, L.L., Campos-Barros, A., Wang, Z., Vennström, B. and Forrest, D. 2001. Distinct tissue-specific roles for thyroid hormone receptors β and $\alpha 1$ in regulation of type 1 deiodinase expression. *Molecular Endocrinology* 15(3):467-475.

016 Amos, W. and Balmford, A. 2001. When does conservation genetics matter? *Heredity* 87:257-265.

017 Andersen, S., Pedersen, K.M., Bruun, N.H. and Laurberg, P. 2002. Narrow individual variations in serum T_4 and T_3 in normal subjects: a clue to the understanding of subclinical thyroid disease. *Journal of Clinical Endocrinology & Metabolism* 87:1068-1072.

018 Anderson, A.J. 1989. *Prodigious Birds: Moas and Moa-hunting in Prehistoric New Zealand.* Cambridge University Press, Cambridge.

019 Anderson, G.M., Connors, J.M., Hardy, S.L., Valent, M. and Goodman, R.L. 2002. Thyroid hormones mediate steroid-independent seasonal changes in luteinizing hormone pulsatility in the ewe. *Biology of Reproduction* 66:701-706.

020 Anderson, G.W., Mariash, C.N. and Oppneheimer, J.H. 2000. Molecular actions of thyroid hormone. In L.D. Braverman and R.D. Utiger (eds.), *Werner and Ingbar's The Thyroid, Eighth Edition,* pg. 174-195. Lippincott, Philadelphia.

021 Anderson, G.W., Schoonover, C.M. and Jones, S.A. 2003. Control of thyroid hormone action in the developing rat brain. *Thyroid* 13(11):1039-1056.

022 Antle, M.C. and Silver, R. 2005. Orchestrating time: arrangements of the brain circadian clock. *TRENDS in Neuroscience* 28(3):145-151.

023 Anway, M.D., Cupp, A.S., Uzumcu, M., and Skinner, M.K. 2005. Epigenetic transgenerational actions of endocrine disruptors and male fertility. *Science* 308:1466-1469.

024 Arem, R. 1999. *The Thyroid Solution.* Ballantine Books, New York.

025 Arem, R. and Escalante, D. 1996. Subclinical hypothyroidism: epidemiology, diagnosis, and significance. *Advances in Internal Medicine* 41:213-250.

026 Armario, A., Montero, J.L. and Jolin, T. 1987. Chronic food restriction and the circadian rhythms of pituitary-adrenal hormones, growth hormone and thyroid-stimulating hormone. *Annals of Nutrition and Metabolism* 31(2):81-87.

027 Arnason, U., Bodin, K., Gullberg, A., Ledje, C. and Mouchaty, S. 1995. A molecular view of pinniped relationships with particular emphasis on the true seals. *Journal of Molecular Evolution* 40:78-85.

028 Arnason, U., Gullberg, A. and Widegren, B. 1993. Cetacean mitochondrial DNA control region: sequences of all extant baleen whale and two sperm whale species. *Molecular Biology and Evolution* 10:960-970.

029 Arnason, U. and Gullberg, A. 1994. Relationship of baleen whales established by cytochrome b gene sequence comparison. *Nature* 367:726-728.

030 Arnold, M.L. 1997. *Natural Hybridization and Evolution.* Oxford University Press, Oxford.

031 Arnould, A.P., Xu, J., Grisham, W., Chen, X., Kim, Y.-H., et al. 2004. Sex chromosomes and brain sexual differences. *Endocrinology* 145(3):1057-1062.

032 Arons, C.D. and Shoemaker, W.J. 1992. The distribution of catecholamines and beta-endorphin in the brains of three behaviorally distinct breeds of dogs and their F1 hybrids. *Brain Research* 594:31-39.

033 Asdell, S.A. 1964. *Patterns of Mammalian Reproduction.* Cornell University Press, Ithaca.

034 Asfaw, B., Gilbert, W.H., Beyene, Y., Hart, W.K., Renne, P.R., et al. 2002. Remains of *Homo erectus* from Bouri, Middle Awash, Ethiopia. *Nature* 416:317-320.

035 Badyaev, A.V. 2005. Stress-induced variation in evolution: from behavioural plasticity to genetic assimilation. *Proceedings of the Royal Society B* 272:877-886.

036 Badyaev, A.V., Foresman, K.R. and Young, R.L. 2005. Evolution of morphological integration: developmental accommodation of stress-induced variation. *American Naturalist* 166(3):382-395.

037 Balter, M. 2005. Small but smart? Flores hominid shows signs of advanced brain. *Science* 307:1386-1389.

038 Balter, V., Person, A., Labourdette, N., Drucker, D., Renard, M., et al., 2001. Les Néandertaliens étaient-ils essentiellement carnivores? Résultats préliminaires sur les teneurs en Sr et en Ba de la paléobiocénose mammalienne de Saint-Césaire. Comptes Rendus de l'Academie des Sciences de Paris, *Sciences de la Terre et des planétes/Earth and Planetary Sciences* 332:59-65.

039 Bandyopadhyay, A., Roy, P. and Bhattachacharya, S. 1996. Thyroid hormone induces the synthesis of a putative protein in the rat granulose cell which stimulates progesterone release. *Journal of Endocrinology* 150(2):309-318.

040 Banfield, A.W.F. 1974. *Mammals of Canada*. National Museum of Canada, University of Toronto Press, Toronto.

041 Barnard, J.C., Williams, A.J., Harvey, C.B. and Williams, G.R. 2003. Effects of thyroid hormone (T3) on fibroblast growth factor signaling in bone. *Endocrine Abstracts* 5:7.

042 Barres, B.A., Lazar, M.A. and Raff, M.C. 1994. A novel role for thyroid hormone, glucocorticoids and retinoic acid in timing oligodendrocyte development. *Development* 120:1097-1108.

043 Bassett, J.H.D. and Williams, G.R. 2003. The molecular actions of thyroid hormone in bone. *TRENDS in Endocrinology and Metabolism* 14(8):356-364.

044 Baumgartner, A., Dietzel, M., Saletu, B., Wolf, R., Campos-Barros, A., et al. 1993. Influence of partial sleep deprivation on the secretion of thyrotropin, thyroid hormones, growth hormone, prolactin, luteinizing hormone, follicle stimulating hormone, and estradiol in healthy young women. *Psychiatry Research* 48(2):153-178.

045 Baumgartner, A., Hiedra, L., Pinna, G., Eravci, M., Prengel, H., et al. 1998. Rat brain type II 5'-iodothyronine deiodinase activity is extremely sensitive to stress. *Journal of Neurochemistry* 71:817-826.

046 Bede, J.C., Teal, P.E.A., Goodman, W.G. and Tobe, S.S. 2001. Biosynthetic pathway of insect juvenile hormone III in cell suspension cultures of the sedge, *Cyperus iria*. *Plant Physiology* 127:584-593.

047 Behnke, R.J. 1992. *Native Trout of Western North America*. American Fisheries Society Monograph 6, Bethesda.

048 Behrends, J., Prank, K., Dogu, E. and Brabant, G. 1998. Central nervouse system control of thyrotopin secretion during sleep and wakefulness. *Hormone Research* 49:173-177.

049 Bell, M.A. and Travis, M.P., in press. Hybridization, transgressive segregation, genetic covariation, and adaptive radiation. *TRENDS in Ecology and Evolution*

050 Belyaev, D.K. 1979. Destabilizing selection as a factor in domestication. *Journal of Heredity* 70:301-308.

051 Belyaev, D.K. 1984. Foxes. In I.L. Mason (ed.), *Evolution of Domesticated Animals*, pg. 211-214. Longman Co., London.

052 Belyaev, D.K. and Trut, L.N. 1975. Some genetic and endocrine effects of selection for domestication in silver foxes. In M.W. Fox (ed.), *The Wild Canids: Their Systematics, Behavioral Ecology and Evolution*, pg. 416-426. Behavioral Science Series, Van Nostrand Reinhold Co., New York.

053 Belyaev, D.K., Ruvinsky, A.O. and Trut, L.N. 1981. Inherited activation-inactivation of the star gene in foxes: its bearing on the problem of domestication. *Journal of Heredity* 72:267-274.

054 Belyaev, D.K., Trut, L.N. and Ruvinsky, A.O. 1975. Genetics of the W locus in foxes and expression of its lethal effects. *Journal of Heredity* 66:331-338.

055 Benecke, N. 1987. Studies on early dog remains from northern Europe. *Journal of Archaeological Science* 14:31-45.

056 Bennett, D.C. 1991. Color genes, oncogenes and melanocyte differentiation. *Journal of Cell Science* 98:135-139.

057 Berg, O., Gorbman, A. and Kobayashi, H. 1959. The thyroid hormones in invertebrates and lower vertebrates. In A. Gorbman (ed.), *Comparative Endocrinology:Proceedings of the Columbia University Symposium on Comparative Endocrinology, 1958*, pg. 302-319. J. Wiley and Sons, Inc., New York.

058 Berge, C. 2002. Peramorphic processes in the evolution of the hominid pelvis and femur. In N. Minugh-Purvis and K. McNamara (eds.), *Human Evolution Through Developmental Change*, pg. 381-404. Johns Hopkins University Press, Baltimore.

059 Berkeley, S.A., Chapman, C. and Sogard, S.M. 2004. Maternal age as a determinant of larval growth and survival in a marine fish, *Sebastes melanops. Ecology* 85(5):1258-1264.

060 Bernal, J. 2002. Action of thyroid hormone in brain. *Journal of Endocrinological Investigations.* 25(3):268.288.

061 Bernatchez, L., Glemet, H., Wilson, C.C. and Danzmann, R.G. 1995. Introgression and fixation of Arctic char (*Salvelinus alpinus*) mitochondrial genome in an allopatric population of brook trout (*Salvelinus fontinalis*). *Canadian Journal of Fisheries and Aquatic Sciences* 52:179-185.

062 Berry, R.J. 1984. House mouse. In I.L. Mason (ed.), *Evolution of Domesticated Animals*, pg. 273-283. Longman Co., London.

063 Berta, A. and Sumich, J.L. 1999. *Marine Mammals: Evolutionary Biology.* Academic Press, San Diego.

064 Bérubé, M. and Aguilar, A. 1998. A new hybrid between a blue whale, *Balaenoptera musculus*, and a fin whale, *B. physalus:* frequency and implications of hybridization. *Marine Mammal Science* 14(1):82-98.

065 Bhattachacharya, S., Guin, S., Bandyopadhyay, A., Jana, N.R. and Halder, S. 1996. Thyroid hormone induces the generation of a novel putative protein in piscine ovarian follicle that stimulates the conversion of pregnenolone to progesterone. *European Journal of Endocrinology* 134 (1):128-135.

066 Bitman, J., Kahl, S., Wood, D.L. and Lefcourt, A.M. 1994. Circadian and ultradian rhythms of plasma thyroid hormone concentrations in lactating dairy cows. *American Journal of Regulatory, Integrative and Comparative Physiology* 266:1797-1803.

067 Blache, D., Blackberry, M.A., Van Cleeff, J. and Martin, G.B. 2001. Plasma thyroid hormones and growth hormone in embryonic and growing emus (*Dromaius novaehollandiae*). *Reproduction, Fertility and Development.* 13(2-3):125-132.

068 Blackburn-Munro, G. and Blackburn-Munro, R.E. 2001. Chronic pain, chronic stress and depression: coincidence or consequence? *Journal of Neuroendocrinology* 13, 1009-1023.

069 Blűher, S. and Mantzoros, C.S. 2004. The role of leptin in regulating neuro-endocrine function in humans. *Journal of Nutrition* 134:2469S-2474S.

070 Boback, S.M. 2003. Body size evolution in snakes: evidence from island populations. *Copeia* 2003(1):81-94.

071 Bodosi, B., Gardi, J., Hajdu, I., Szentirmai, E., Obal Jr., F., et al. 2004. Rhythms of ghrelin, leptin, and sleep in rats: effects of the normal diurnal cycle, restricted feeding, and sleep deprivation. *American Journal of Physiology – Regulatory, Integrative and Comparative Physiology* 287:R1071-R1079.

072 Boissy, A. 1995. Fear and fearfulness in animals. *Quarterly Revue of Biology* 70:165-191.

073 Boitani, L., Francisci, F., Ciucci, P. and Andreoli, G. 1995. Population biology and ecology of feral dogs in central Italy. In J. Serpell (ed.), *The Domestic Dog: Its Evolution, Behaviour and Interactions with People,* pg. 217-244. Cambridge University Press, Cambridge.

074 Bőkőnyi, S. 1984. Horse. In I.L. Mason (ed.), *Evolution of Domesticated Animals*, pg. 162-173. Longman Co., London.

075 Bortolotti, G.R. 1984. Physical development of nestling bald eagles with emphasis on the timing of growth events. *Wilson Bulletin* 96(4):524-542.

076 Bowling, A.T. 1994. Dominant inheritance of overo spotting in paint horses. *Journal of Heredity* 85:222-224.

077 Bucher, T.L., Bartholomew, G.A., Trivelpiece, W.Z. and Volkman, N.J. 1986. Metabolism, growth, and activity in Adélie and Emperor penguin embryos. *Auk* 103:485-493.

078 Buckley, T.M. and Schatzberg, A.F. 2005. On the interactions of the hypothalamic-pituitary-adrenal (HPA) axis and sleep: normal HPA axis activity and circadian rhythm, exemplary sleep disorders. *Journal of Clinical Endocrinology and Metabolism* 90(5):3106-3114.

079 Budiansky, S. 1992. *The Covenant of the Wild: Why Animals Chose Domestication.* Weidenfeld and Nicolson, London.

080 Bultman, S.J., Michaud, E.J. and Woyckik, R.P. 1992. Molecular characterization of the mouse *agouti* locus. *Cell* 71:1195-1204.

081 Bunce, M., Worthy, T.H., Ford, T., Hoppitt, W., Willerslev, E., et al. 2003. Extreme reversed sexual size dimorphism in the extinct New Zealand moa *Dinornis. Nature* 425:172-175.

082 Bunce, M., Szulkin, M., Lerner, H.R.L., Barnes, I., Shapiro, B., et al. 2005. Ancient DNA provides new insights into the evolutionary history of New Zealand's extinct giant eagle. *Public Library of Science [PloS]* 3(1):e9 (3 pgs).

083 Burchill, S.A., Ito, S. and Thody, A.J. 1993. Effects of melanocyte-stimulating hormone on tyrosinase expression and melanin synthesis in hair follicular melanocytes of the mouse. *Journal of Endocrinology* 137:189-195.

084 Burgner, R.L.1991. Life history of sockeye salmon (*Oncorhynchus nerka).* In C. Groot and L. Margolis (eds.), *Pacific Salmon Life Histories*, pg. 3-117. UBC Press, Vancouver.

085 Burness, G.P., Diamond, J. and Flannery, T. 2001. Dinosaurs, dragons, and dwarfs: the evolution of maximal body size. *Proceedings of the National Academy of Sciences USA* 98:14518-14523.

086 Burrow, G.N. 1997. Editorial: mothers are important! *Endocrinology* 138:3-4.

087 Buyse, J., Tixier-Boichard, M, Berghman, L.R., Huybrechts, L.M. and Decuypere, E. 1990. Growth hormone secretory characteristics of sex-linked dwarf and normal-sized chickens reared on a control or on a 3, 3',5-triiodo-thyronine-supplemented diet. *General and Comparative Endocrinology* 93:406-410.

088 Brabant, G. and Prank, K. 2000. Prediction and significance of the temporal pattern of hormone secretion in disease states. In D.J. Chadwick and J.A. Goode (eds.), *Mechanisms and Biological Significance of Pulsatile Hormone Secretion*, pg.105-118. J. Wiley and Sons, Chichester.

089 Brabant, G., Prank, K., Ranft, U., Schuermeyer, T., Wagner, T.O., et al. 1990. Physiological regulation of circadian and pulsatile thyrotropin secretion in normal man and woman. *Journal of Clinical Endocrinology & Metabolism* 70:403-409.

090 Bradley, D.G., Loftus, R.T., Cunningham, P. and MacHugh, D.E. 1998. Genetics and domestic cattle origins. *Evolutionary Anthropology* 6:79-86.

091 Braverman, L.D. and Utiger, R.D. 1991. *Werner and Ingbar's The Thyroid.* Lippincott, Philadelphia.

092 Breen, S., Rees, S. and Walker, D. 1996. The development of diurnal rhythmicity in fetal suprachiasmatic neurons as demonstrated by fos immunohistochemistry. *Neuroscience* 74(3):917-926.

093 Brent, G.A. 2000. Tissue-specific actions of thyroid hormone: insights from animal models. *Revue of Endocrinology and Metabolic Disorders* 1:27-33.

094 Brent, G.A., Moore, D.D. and Larsen, P.R. 1991. Thyroid hormone regulation of gene expression. *Annual Revue of Physiology* 53:17-35.

095 Brosvic, G.M., Taylor, J.N. and Dihoff, R.E. 2002. Influences of early thyroid hormone manipulations: delays in pup motor and exploratory behavior are evident in adult operant performance. *Physiology and Behavior* 75(5): 697-715.

096 Brown, P., Sutikna, T., Morwood, M.J., Soejono, R.P., Jatmiko, et al. 2004. A new small-bodied homini from the Late Pleistocene of Flores, Indonesia. *Nature* 431:1055-1061.

097 Brucker-Davis, F., Skarulis, M.C., Pikus, A., Ishizawar, D., Mastroianni, M., et al. 1996. Prevalence and mechanisms of hearing loss in patients with resistance to thyroid hormones. *Journal of Clinical Endocrinology and Metabolism* 81:2766-2772.

098 Bruford, M.W., Bradley, D.G. and Luikart, G. 2003. DNA markers reveal the complexity of livestock domestication. *Nature Reviews (Genetics)* 4:900-910.

099 Bunevičius, R., Kažanavičius, G., Žalinkevičius, R. and Prange, Jr., A.J. 1999. Effects of thyroxine as compared with thyroxine plus triiodothyronine in patients with hypothyroidism. *New England Journal of Medicine* 340:424-429.

100 Butler, P. 2000. Pulsatile insulin secretion. In D.J. Chadwick and J.A. Goode, (eds.), *Mechanisms and Biological Significance of Pulsatile Hormone Secretion*, pg.190-205. J. Wiley and Sons, Chichester.

101 Caloi, L., Kotsakis, T., Palombo, M.R. and Petronio, C. 1996. The Pleistocene dwarf elephants of Mediterranean islands. In J. Shoshani and P. Tassy (eds.), *The Proboscidea: The Evolution and Palaeoecology of Elephants and their Relatives,* pg. 234-239. Oxford: Oxford University Press.

102 Camilot, M., Teofoli, F., Gandini, A., Franceschi, R., Rapa, A., et al. 2005. Thyrotropin receptor gene mutations and TSH resistance: variable expressivity in the heterozygotes. *Clinical Endocrinology* 63(2):146-151.

103 Campbell, R.W., Dawe, N.K., McTaggart-Cowan, I., Cooper, J.M., Kaiser, G.W., et al. 1990. *The Birds of British Columbia. Vols I & II.* Royal British Columbia Museum, Victoria.

104 Campos-Barros, A., Musa, A., Flechner, A., Hessenius, C., Gaio, U., et al. 1997. Evidence for circadian variations of of thyroid hormone concentrations and type II 5'-iodothyronine deiodinase activity in the rat central nervous system. *Journal of Neurochemistry* 68:795-803.

105 Canaris, G.J., Manowitz, N.R., Mayor, G. and Ridgeway, E.C. 2000. The Colorado thyroid disease prevalence study. *Archive of Internal Medicine* 160:526-534.

106 Canaris, G.J., Steiner, J.R. and Ridgeway, E.C. 1997. Do traditional symptoms of hypothyroidism correlate with biochemical disease? *Journal of General and Internal Medicine* 12:544-550.

107 Carani, C., Isidori, A.M., Granata, A., Carosa, E., Maggi, M., et al. in press. Multicenter study on the prevalence of sexual symptoms in male hypo- and hyper-thyroid patients. *Journal of Clinical Endocrinology & Metabolism.*

108 Carey, C., Rahn, H. and Parisi, P. 1980. Calories, water, lipid and yolk in avian eggs. *Condor* 82:335-343.

109 Carlquist, S. 1965. *Island Life: A Natural History of the Islands of the World.* The Natural History Press, Garden City.

110 Caro, T. 2005. The adaptive significance of coloration in mammals. *BioScience* 55(2):125-136.

111 Carroll, S.B. 2003. Review: Genetics and the making of *Homo sapiens. Nature* 422:849-857.

112 Carroll, S.B. 2005. *Endless Forms Most Beautiful: The New Science of Evo-Dev and the Making of the Animal Kingdom.* W.W. Norton and Co., New York.

113 Chaix, L. 2000. A preboreal dog from the northern Alps (Savoie, France). In S.J. Crockford (ed.), *Dogs Through Time: An Archaeological Perspective,* pg. 49-59. British Archaeological Reports S889, Oxford.

114 Chan, J., Ogawa, S. and Pfaff, D.W. 2001. Reproduction-related behaviors of Swiss-Webster female mice living in a cold environment. *Proceedings of the National Academy of Sciences USA* 98(2):700-704.

115 Chan, S. and Kilby, M.D. 2000. Review: Thyroid hormone and central nervous system development. *Journal of Endocrinology* 165:1-8.

116 Chan, S. and Rovet, J. 2003. Thyroid hormones in fetal central nervous system development. *Fetal and Maternal Medicine Review* 14:177-208.

117 Chastel, O., Lacroix, A. and Kersten, M. 2003. Pre-breeding energy requirements: thyroid hormone, metabolism and the timing of reproduction in house sparrows *Passer domesticus. Journal of Avian Biology* 34:298-306.

118 Chesemore, D.L. 1975. Ecology of the arctic fox (*Alopex lagopus*) in North America: a review. In M.W. Fox (ed.), *The Wild Canids: Their Systematics, Behavioral Ecology and Evolution,* pg. 143-163. Behavioral Science Series, Van Nostrand Reinhold Co., New York.

119 Chino, Y., Saito, M., Yamasu, K., Suyemitsu, T, and Ishihara, K. 1994. Formation of the adult rudiment of sea urchins is influenced by thyroid hormones. *Developmental Biology* 161:1-11.

120 Chowdhury, D., Ojaman, K., Parnell, V.A., McMahon, C., Sison, C.P., et al. 2001. A prospective randomized clinical study of thyroid hormone treatment after operations for complex congenital heart disease. *Journal of Thoracic and Cardiovascular Surgery* 122:1023-1025.

121 Chubb, S.A.P., Davis, W.A. and Davis, T.M.E. 2005. Interactions among thyroid function, insulin sensitivity, and serum lipid concentrations: the Freemantle diabetes study. *Journal of Clinical Endocrinology & Metabolism* 90(9):5317-5320.

122 Ciucci, P., Lucchini, V., Boitani, L. and Randi, E. 2003. Dewclaws in wolves as evidence of admixed ancestry with dogs. *Canadian Journal of Zoology* 81:2077-2081.

123 Clark, A.G., Glanowski, S., Nielsen, R., Thomas, P.D., Kejariwal, A., et al. 2003. Inferring nonneutral evolution from human-chimp-mouse othologous gene trios. *Science* 302:1960-1963.

124 Clark, B.R. and Price, E.O. 1981. Sexual maturation and fecundity of wild and domestic Norway rats (*Rattus norvegicus*). *Journal of Reproduction and Fertility* 63:215-220.

125 Clark, P.A. and Rogol, A.D. 1996. Growth hormones and sex steroid interactions at puberty. *Endocrinology and Metabolism Clinics of North America.* 25(3):665-681.

126 Clément, K., Viguerie, N., Diehn, M., Alizadeh, A., Barbe, P., et al. 2002. In vivo regulation of human skeletal muscle gene expression by thyroid hormone. *Genome Research* 12:281-291.

127 Clutton-Brock, J. 1981. *Domesticated Animals From Early Times.* British Museum (Natural History), Heinemann.

128 Clutton-Brock, J. 1992. Domestication of animals. In S. Jones, R. Martin, and D. Pilbeam (eds.), *The Cambridge Encyclopedia of Human Evolution,* pg. 380-385. Cambridge University Press, Cambridge.

129 Clutton-Brock, J. 1992. The process of domestication. *Mammal Revue* 22:79-85.

130 Clutton-Brock, J. 1995. Origins of the dog: domestication and early history. In J. Serpell (ed.), *The Domestic Dog: Its Evolution, Behaviour and Interactions with People,* pg. 8-20. Cambridge University Press, Cambridge.

131 Cogburn, L.A. and Freeman, R.M. 1987. Response surface of daily thyroid hormone rhythms in young chickens exposed to constant ambient temperature. *General and Comparative Endocrinology* 68(1):113-123.

132 Collis, K., Roby, D.D., Craig, D.P., Ryan, B.A. and Ledgerwood, R.D. 2001. Avian predation on juvenile salmonids tagged with passive integrated transponders in the Columbia River estuary: vulnerability of different salmonid species, stocks, and rearing types. *Transactions of the American Fisheries Society* 130:385-396.

133 Cooper, A., Lalueza-Fox, C., Anderson, S., Rambaut, A., Austin, J., et al. 2001. Complete mitochondrial genome sequences of two extinct moas clarify ratite evolution. *Nature* 409:704-707.

134 Cooper, A., Mourer-Chauvire, C., Chambers, G.K., von Haeseler, A., Wilson, A.C., et al. 1992. Independent origins of New Zealand moas and kiwis. *Proceedings of the National Academy of Sciences USA* 89:8741-8744.

135 Copinschi, G., Spiegel, K., Leproult, R. and Van Cauter, E. 2000. Pathophysiology of human circadian rhythms. In D.J. Chadwick and J.A Goode (eds.), *Mechanisms and Biological Significance of Pulsatile Hormone Secretion,* pg. 143-162. J. Wiley and Sons, Chichester.

136 Coppinger, R. and Coppinger, L. 2001. *Dogs: A New Understanding of Canine Origins,Behavior and Evolution.* University of Chicago Press, Chicago.

137 Coppinger, R. and Feinstein, M. 1991. 'Hark! Hark! The dogs do bark...' and bark and bark. *Smithsonian* 21:119-129.

138 Coppinger, R. and Schneider, R. 1995. Evolution of working dogs. In J. Serpell (ed.), *The Domestic Dog: Its Evolution, Behaviour and Interactions with People*, pg. 21-47. Cambridge University Press, Cambridge.

139 Corbet, G.B. and Clutton-Brock, J. 1984. Appendix: taxonomy and nomenclature. In I.L. Mason (ed.), *Evolution of Domesticated Animals*, pg. 434-438. Longman Co., London.

140 Cordero, D., Marcucio, R., Hu, D., Gaffield, W., Tapadia, M., et al. 2004. Temporal perturbations in sonic hedgehog signaling elicit the spectrum of holo-prosencephaly phenotypes. *Journal of Clinical Investigation* 114(4):485-495.

141 Cowling, K., Robbins, R.J., Haigh, G.R., Teed, S.K. and Dawson, W.D. 1994. Coat color genetics of *Peromyscus*: IV. Variable white, a new dominant mutation in the deer mouse. *Journal of Heredity* 85:48-52.

142 Crabtree, P.J. 1993. Early animal domestication in the middle East and Europe. In M.B. Schiffer (ed.), *Archaeological Method and Theory*, pg. 201-245. University of Arizona Press, Tucson.

143 Crawford, R.D. 1984. Turkey. In I. L. Mason (ed.), *Evolution of Domesticated Animals*, pg. 325-334. Longman Co., London.

144 Cristofanilli, M., Yamamura, Y., Kau, S-W., Bevers, T., Strom, S., et al. 2005. Primary hypothyroidism is associated with a reduced incidence of primary breast cancer. *Cancer* 103(6):1122-1128.

145 Crockford, S.J. 1997. *Osteometry of Makah and Coast Salish dogs.* Archaeology Press 22, Simon Fraser University, Burnaby.

146 Crockford S.J. 2000. Dog evolution: a role for thyroid hormone physiology in domestication changes. In S.J. Crockford (ed.), *Dogs Through Time: An Archaeological Perspective,* pg. 11-20. British Archaeological Reports S889, Oxford.

147 Crockford, S.J. 2000. A commentary on dog evolution: regional variation, breed development and hybridisation with wolves. In S.J. Crockford (ed.), *Dogs Through Time: An Archaeological Perspective,* pg. 295-312. British Archaeological Reports S889, Oxford.

148 Crockford, S.J. 2002. Animal domestication and heterochronic speciation: the role of thyroid hormone. In N. Minugh-Purvis and K. McNamara (eds.), *Human Evolution Through Developmental Change*, pg. 122-153. Johns Hopkins University Press, Baltimore.

149 Crockford, S.J. 2002. Thyroid hormone in Neandertal evolution: A natural or a pathological role? *Geographical Revue* 92:73-88.

150 Crockford, S.J. 2003. Thyroid hormone rhythms and hominid evolution: a new paradigm implicates pulsatile thyroid hormone secretion in speciation and adaptation changes. *International Journal of Comparative Biochemistry and Physiology Part A – Molecular & Integrative Physiology* 135(1):105-129.

151 Crockford, S.J. 2004. Animal domestication and vertebrate speciation: a paradigm for the origin of species. Ph.D. Dissertation, University of Victoria, Canada.

152 Crockford, S.J. 2005. Native dog types in North America before the arrival of European dogs. Proceedings of the 30th World Congress of World Small Animal Veterinary Association, Mexico City. May 14, 2005. online at: http://www.wsava2005.com/memorias/Diezmo1_2_3_4/04091 CrockfordNativeDogInNorthAmericaBeforeArrivalOfEuropeanDogs.htm

153 Crockford, S.J. and Pye, C.J. 1997. Forensic reconstruction of prehistoric dogs from the Northwest coast. *Canadian Journal of Archaeology* 21:149-153.

154 Crowder, L.B., Rice, J.A., Miller, T.J. and Marschall, E.A. 1992. Empirical and theoretical approaches to size-based interactions and recruitment variability in fishes. In D.L. DeAngelis and L.J.Gross (eds.), *Individual-based Models and Approaches in Ecology*, pg. 237-255. Chapman and Hall, New York.

155 Cubo, J. and Arthur, W. 2001. Patterns of correlated character evolution in flightless birds: a phylogenetic approach. *Evolutionary Ecology* 14:693-702.

156 Cudd, T.A. 2005. Animal model systems for the study of alcohol teratology. *Experimental Biology and Medicine* 230:389-393.

157 Cudd, T.A., Chen, W-J. A. and West, R.W. 2002. Fetal and maternal thyroid hormone responses to ethanol exposure during the third trimester equivalent of gestation in sheep. *Alcoholism:Clinical and Experimental Research* 26(1):53-58.

158 Culotta, E. 2005. Chimp genome catalogs differences with humans. *Science* 309: 1468-1469.

159 Culotta, E. 2005. New "hobbits" bolster species, but origins still a mystery. *Science* 309:208-209.

160 Cunnane, S. and Crawford, M.A. 2003. Survival of the fattest: fat babies were the key to evolution of the large human brain. *Comparative Biochemistry and Physiology A* 136:17-26.

161 Custro, N., Scafidi, V., Gallo, S. and Notarbartolo, A. 1994. Deficient pulsatile thyrotropin secretion in the low-thyoid-hormone state of severe non-thyroidal illness. *European Journal of Endocrinology* 130(2):132-136.

162 Dardente, H., Menet, J.S., Challet, E., Tournier, B.B., Pévet, P., et al. 2004. Daily and circadian expression of neuropeptides in the suprachiasmatic nuclei of nocturnal and diurnal rodents. *Molecular Brain Research* 124:143-151.

163 Darwin, C. 1840. Questions on the breeding of animals. 1968 facsimile, J. Murry, London.

164 Darwin, C. 1859. *On the Origin of Species*. 1966 facsimile, J. Murray, London.

165 Darwin, C. 1868. *The Variation of Animals and Plants Under Domestication*, 1998 facsimile of 1883 edition, Johns Hopkins University Press, Baltimore.

166 Darwin, C. 1871. *The Descent of Man*. J. Murray, London.

167 Darwin, F. (ed.) 1958. *The Autobiography of Charles Darwin and Selected Letters*. Dover Publications, New York.

168 Davis, K.D. and Lazar, M.A. 1992. Selective antagonism of thyroid hormone action by retinoic acid. *Journal of Biological Chemistry* 267:3185-3189.

169 Davis, P.J., Davis, F.B. and Lawrence, W.D. 1989. Thyroid hormone regulation of membrane Ca2(+)-ATPase activity. *Endocrine Research* 15(4):651-682.

170 Davis, S.J.M. 1987. *The Archaeology of Animals*. Yale University Press, New Haven.

171 Davis, S.J.M. and Valla, F.R. 1978. Evidence for domestication of the dog 12,000 years ago in the Natufian of Israel. *Nature* 276:608-610.

172 Dawkins, R. 1976. *The Selfish Gene*. Oxford University Press, New York.

173 Dawson, A., Deeming, D.C., Dick, A.C.K. and Sharp, P.J. 1996. Plasma thyroxine concentrations in farmed ostriches in relation to age, body weight, and growth hormone. *General and Comparative Endocrinology* 103:308-315.

174 Dawson, A., McNaughton, F.J., Goldsmith, A.R. and Degen, A.A. 1994. Ratite-like neotony induced by neonatal thyroidectomy of European starlings, *Sturnus vulgaris*. *Journal of Zoology London* 232:633-639.

175 Dayan, T. 1994. Early domesticated dogs of the Near East. *Journal of Archaeological Science* 21:633-640.

176 Dayan, T. and Simberloff, D. 1994. Character displacement, sexual dimorphism, and morphological variation among British and Irish mustelids. *Ecology* 75:1063-1073.

177 DeAngelis, D.L. and Gross, L.J. 1992. *Individual-based Models and Approaches in Ecology.* Chapman and Hall, New York.

178 De Groef, B., Goris, N., Arckens, L., Kűhn, E.R. and Darras, V.M. 2003. Corticotropin-releasing hormone (CRH)-induced thyrotropin release is directly mediated through CRH receptor type 2 on thyrotropes. *Endocrinology* 144(12):5537-5544.

179 De Groot, L.J. and Hennemann, G. (eds.) 2005. www.thyroidmanager.org Endocrine Education Inc.

180 Dellovad, T.L., Zhu, Y., Krey, L. and Pfaff, D.W. 1996. Thryoid hormone and estrogen interact to regulate behavior. *Proceedings of the National Academy of Sciences USA* 93(22):12581-12586.

181 deMenocal, P.B. 2004. African climate change and faunal evolution during the Pliocene-Pleistocene. *Earth and Planetary Science Letters* 6976:1-22.

182 Demers, L.M. and Spencer, C.A. (eds.) 2002. *Laboratory Medicine Practice Guidelines: Laboratory Support for the Diagnosis and Monitoring of Thyroid Disease.* Monograph published by the National Academy of Clinical Biochemistry, Washington, DC., see www.nacb.org

183 Denver, R.J. 1999. Evolution of the corticotropin-releasing hormone signaling system and its role in stress-induced phenotypic plasticity. *Annals of the New York Academy of Sciences* 897:46-53.

184 De Pablo, F. 1993. Introduction. In Schreibman, M.P., Scanes, C.G. and P.K.T. Pang (eds.), *The Endocrinology of Growth, Development, and Metabolism of Vertebrates,* pg. 1-11. Academic Press, New York.

185 Derégnaucourt, S., Guyomarc'h, J-C. and Belhamra, M. 2005. Comparison of migratory tendency in European quail *Coturnix c. coturnix,* domestic Japanese quail *coturnix c. japonica* and their hybrids. *Ibis* 147:25-36.

186 De Vries, G.J. 2004. Sex differences in adult and developing brains: compensation, compensation, compensation. *Endocrinology* 145(3):1063-1068.

187 Diamond, J. 1999. *Guns, Germs and Steel: The Fates of Human Societies.* W.W. Norton & Co., New York.

188 Diamond, J. 2002. Evolution, consequences and future of plant and animal domestication. *Nature* 418:700-707.

189 Dickhoff, W.W. 1993. Hormones, metamorphosis and smolting. In Schreibman, M.P., Scanes, C.G. and P.K.T. Pang (eds.), *The Endocrinology of Growth, Development, and Metabolism of Vertebrates,* pg. 519-540. Academic Press, New York.

190 Dorus, S., Vallender, E.J., Evans, P.D., Anderson, J.R., Gilbert, S.L., et al. 2004. Accelerated evolution of nervous system genes in the origin of *Homo sapiens. Cell* 119:1027-1040.

191 Doulabi, B.Z., Schiphorst, M.P-T., Kalsbeek, A., Fliers, E., Bakker, O., et al. 2004. Diurnal variation in rat liver thyroid hormone receptor (TR)-α messenger ribonucleic acid (mRNA) is dependent on the biological clock in the suprachiasmatic nucleus, whereas diurnal variation of TRβ1 mRNA is modified by food intake. *Endocrinology* 145(3):1284-1289.

192 Douyon, L. and Schteingart, D.E. 2002. Effect of obesity and starvation on thyroid hormone, growth hormone, and cortisol secretion. *Endocrinology and Metabolism Clinics of North America* 21(1):173-189.

193 Doyle Driedger, S. 2001. Overcoming depression: one woman's terrifying odyssey through a nightmare of despair. *Maclean's* 114(46):34-38.

194 Dubois-Dalcq., M. and Murray, K. 2000. Why are growth factors important in oligodendrocyte physiology? *Pathological Biology* 48:80-86.

195 Duckett, W.M., Manning, J.P. and Weston, P.G. 1989. Thyroid hormone periodicity in healthy adult geldings. *Equine Veterinary Journal* 21(2):123-125.

196 Dufty Jr., A.M., Clobert, J. and Møller, A.P. 2002. Hormones, developmental plasticity and adaptation. *TRENDS in Ecology and Evolution* 17:190-196.

197 Dunn, E.H. 1975. Growth, body components and energy content of nestling Double-crested cormorants. *Condor* 77:431-438.

198 Dunn, J.T. 2002. Editorial: Guarding our nation's thyroid health. *Journal of Clinical Endocrinology & Metabolism* 87:486-488.

199 Eales, J.G. 1997. Iodine metabolism and thyroid-related functions in organisms lacking thyroid follicles: are thyroid hormones also vitamins? *Proceedings of the Society for Experimental Biology and Medicine* 214:302-317.

200 Ebbesson, L.O.E., Björnsson, B.Th., Stefansson, S.O. and Ekström, P. 2000. Free plasma thyroxine levels in coho salmon, *Oncorhynchus kisutch,* during parr-smolt transformation: comparison with total thyroxine, total triiodothyronine, and growth hormone levels. *Fish Physiology and Biochemistry* 22:45-50.

201 Eldredge, N. and Gould, S.J. 1972. Punctuated equilibria: an alternative to phyletic gradualism. In T.J.M. Schopf (ed.), *Models in Paleobiology,* pg. 82-115. Freeman and Cooper, San Francisco.

202 Eliceiri, B.P. and Brown, D.D. 1994. Quantitation of endogenous thyroid hormone receptors α and β during embryogenesis and metamorphosis in *Xenopus laevis. Journal of Biological Chemistry* 269(39):24459-24465.

203 Elinson, R.P. 1987. Change in developmental patterns: embryos of amphibians with large eggs. In R.A. Raff and E.C. Raff (eds.), *Development as an Evolutionary Process*, pg. 1-21. A.R. Liss, Inc., New York.

204 Episkopou, F., Maeda, S., Nishiguchi, S., Shimada, K., Gaitanaris, G.A., et al.1993. Disruption of the transthyretin gene results in mice with depressed levels of plasma retinol and thyroid hormone. *Proceedings of the National Academy of Sciences USA* 90:2375-2379.

205 Epstein, H. 1984. Ass, mule and onager. In I.L. Mason (ed.), *Evolution of Domesticated Animals,* pg. 174-184. Longman Co., London.

206 Epstein, H. and Bichard, M. 1984. Pig. In I.L. Mason (ed.), *Evolution of Domesticated Animals,* pg. 145-161. Longman Co., London.

207 Epstein, H. and Mason, I.L. 1984. Cattle. In I.L. Mason (ed.), *Evolution of Domesticated Animals,* pg. 7-27. Longman Co., London.

208 Eravci, M., Pinna, G., Meinhold, H. and Baumgartner, A. 2000. Effects of pharmacological and nonpharmacological treatments on thyroid hormone metabolism and concentrations in rat brain. *Endocrinology* 141:1027-1040.

209 Evans, P.D., Gilbert, S.L., Mekel-Bobrov, N., Vallender, E.J., Anderson, J.R., et al. 2005. *Microcephalin,* a gene regulating brain size, continues to evolve adaptively in humans. *Science* 309:1717-1720.

210 Evans, R.M. 1988. The steroid and thyroid hormone receptor superfamily. *Science* 240:889-895.

211 Evered, D.C., Young, E.T., Ormston, B.J., Menzies, R., Smith, P.A., et al. 1973. Treatment of hypothyroidism: a reappraisal of thyroxine therapy. *British Medical Journal* 3(5872):131-134.

212 Everson, C.A. and Nowak Jr., T.S. 2002. Hypothalamic thryotropin-releasing hormone mRNA responds to hypothroxinemia induced by sleep deprivation. *American Journal of Physiology:Endcrinology & Metabolism* 283:E85-E93.

213 Ezzel, C. 2003. Ma's eyes, not her ways. *Scientific American* (April):30.

214 Fagan, B.M. 1994. *In the Beginning: An Introduction to Archaeology*. HarperCollins, New York.

215 Falk, D., Hildebolt, C., Smith, K., Morwood, M.J., Sutikna, T., et al. 2005. The brain of LB1, *Homo floresiensis. Science* 308:242-245.

216 Farnsworth, E. 2004. Hormones and shifting ecology throughout plant development. *Ecology* 85(1):5-15.

217 Farwell, A.P. and Leonard, J.L. 2005. Nongenomic actions of thyroid hormone during fetal brain development. *Current Opinion in Endocrinology and Diabetes* 12(1):17-22.

218 Fazio, S., Palmieri, E.A., Lombardi, G. and Biondi, B. 2004. Effects of thyroid hormone on the cardiovascular system. *Recent Progress in Hormone Research* 59:31-50.

219 Fedoseev, G. A. 1975. Ecotypes of the ringed seal (*Pusa hispida* Schreber, 1777) and their reproductive capabilities. In K. Ronald and A. W. Mansfield (eds.), *Biology of the Seal.* Rapports et Proces-verbaux des Reunions, Conseil International Pour L'Exploration de la Mer 169:156-160.

220 Ferguson, D.C. 1994. Update on diagnosis of canine hypothyroidism. In D.C. Ferguson (ed.), Thyroid disorders, pg. 515-539. *Veterinary Clinics of North America, Small Animal Practice* 24, W.B. Saunders Co., Philadelphia.

221 Ferguson, D.C. 1997. Euthyroid sick syndrome. *Canine Practice* 22:49-51.

222 Ferketich, A.K., Ferguson, J.P. and Binkley, P.F. 2005. Depressive symptoms and inflammation among heart failure patients. *American Heart Journal* 150(1):132-136.

223 Finch, C.E. and Rose, M.R. 1995. Hormones and the physiological architecture of life history evolution. *Quarterly Revue of Biology* 70:1-52.

224 Finlayson, C. 2005. Biogeography and evolution of the genus *Homo. TRENDS in Ecology and Evolution* 20(8):457-463.

225 Finley, K. J, Miller, G. W., Davis, R. A. and Koski, W.R. 1983. A distinctive large breeding population of ringed seal (*Phoca hispida*) inhabiting the Baffin Bay pack ice. *Arctic* 36(2):162-173.

226 Fisher, D.C. 1996. Extinction of proboscideans in North America. In J. Shoshani and P. Tassy (eds.), *The Proboscidea: The Evolution and Palaeoecology of Elephants and their Relatives*, pg. 296-315. Oxford: Oxford University Press.

227 Flamant, F. and Samarut, J. 2003. Thyroid hormone receptors: lessons from knockout and knock-in mutant mice. *TRENDS in Endocrinology and Metabolism* 14:85-90.

228 Flatt, T., Tu, M-P. and Tatar, M. 2005. Hormonal pleitropy and the juvenile hormonal regulation of *Drosophila* development and life history. *BioEssays* 27:999-1010.

229 Fleagle, J.G. 1999. *Primate Adaptation and Evolution, Second Edition.* Academic Press, San Diego.

230 Fleming, I.A. 1998. Pattern and variability in the breeding system of Atlantic salmon (*Salmo salar*), with comparisons to other salmonides. *Canadian Journal of Fisheries and Aquatic Science* 55 (Suppl 1):59-76.

231 Fleming, I.A., Agustsson, T., Finstad, B., Johnsson, J.I. and Björnsson, B.T. 2002. Effects of domestication on growth physiology and endocrinology of Atlantic salmon. *Canadian Journal of Fisheries and Aquatic Science* 59:1323-1330.

232 Fleming, I.A., Einum, S., Jonsson, B. and Jonsson, N. 2003. Comment on "Rapid evolution of egg size in captive salmon" (1). *Science* 302: 59b; www.sciencemag.org/cgi/content/full/302/5642/59b.

233 Fondon III, J.W. and Garner, H.R. 2004. Molecular origins of rapid and continuous morphological evolution. *Proceedings of the National Academy of Sciences* 101:18058-18063.

234 Foote, C.J., Brown, G.S. and Hawryshyn, C.W. 2004. Female colour and male choice in sockeye salmon: implications for the phenotypic convergence of anadromous and nonanadromous morphs. *Animal Behaviour* 67:69-83.

235 Forchhammer M.C. 2000. Timing of foetal growth spurts can explain sex ratio variation in polygynous mammals. *Ecology Letters* 3:1-4.

236 Forhead, A.J. and Fowden, A.L. 2002. Effects of thyroid hormones on pulmonary and renal angiotensin-converting enzyme concentrations in fetal sheep near term. *Journal of Endocrinology* 173(1):143-150.

237 Fowden, A.L., Mapstone, J. and Forhead, A.J. 2001. Regulation of glucogenesis by thyroid hormones in fetal sheep during late gestation. *Journal of Endocrinology* 170(2):461-469.

238 Fowler, M. 1989. Physical examination, restraint and handling. In L.W. Johnson (ed.), Llama Medicine. *Veterinary Clinics of North America, Food Animal Practice* 5, pg. 21-35. W.B. Saunders Co., Philadelphia.

239 Fox, H.E., White, S.A., Kao, M.H.F. and Fernald, R.D. 1997. Stress and dominance in a social fish. *Journal of Neuroscience* 17(16):6463-6469.

240 Fox, M.W. 1978. *The Dog: Its Domestication and Behavior.* Garland STPM Press, New York.

241 Friis, L.K. 1985. An investigation of subspecific relationships of the grey wolf, *Canis lupus,*in British Columbia. M.Sc. thesis, University of Victoria, Canada.

242 Franklyn, J.A. 2000. Metabolic changes in thyrotoxicosis. In L.D. Braverman and R.D. Utiger (eds.), *Werner and Ingbar's The Thyroid, Eighth Edition,* pg. 667-672. Lippincott, Philadelphia.

243 Freedman, R.R. and Blacker, C.M. 2002. Estrogen raises the sweating threshold in postmenopausal women with hot flashes. *Fertility and Sterility* 77(3):487-490.

244 Fuge, R. 2002. Transport of iodine in the environment and pathways into the biosphere. *Abstract of a presentation to the 2002 Meeting of the Geological Society of America, October 27-30,* Colorado.

245 Fumihito, A., Miyake, T., Sumi, S-I., Takada, M., Ohno, S., et al. 1994. One subspecies of the red junglefowl (*Gallus gallus gallus*) suggices as the matriarchic ancestor of all domestic breeds. *Proceedings of the National Academy of Sciences USA* 91:12505-12509.

246 Gagneux, P., Amess, B. Diaz, S., Moore, S., Patel, T., et al. 2001. Proteomic omparison of human and great ape blood plasma reveals conserved glycosylation and differences in thyroid hormone metabolism. *American Journal of Physical Anthropology* 115:99-109.

247 Gaines, C.A., Hare, M.P., Beck, S.E. and Rosenbaum, H.C. 2005. Nuclear markers confirm taxonomic status and relationships among highly endangered and closely related right whale species. *Proceedings of the Royal Society B* 272:533-542.

248 Galli, C., Lagutina, I., Crotti, G., Colleoni, S., Turini, R., et al. 2003. A cloned horse born to its dam twin. *Nature* 424:635.

249 Gancedo, B., Alonso-Gomez, A.L., de Pedro, N., Corpas, I., Delgado, M.J., et al. 1995. Seasonal changes in thyroid activity in male and female frog, *Rana perezi. General and Comparative Endocrinology* 97:66-75.

250 Gancedo, B., Alonso-Gomez, A.L., de Pedro, N., Delgado, M.J. and Alonso-Bedate, M. 1996. Daily changes in thyroid activity in the frog *Rana perizi:* variations with season. *Comparative Biochemistry and Physiology C.* 114:79-87.

251 Gancedo, B., Alonso-Gomez, A.L., de Pedro, N., Delgado, M.J. and Alonso-Bedate, M. 1997. Changes in thyroid hormone concentrations and total contents through ontogeny in three anuran species: evidence for daily cycles. *General and Comparative Endocrinology* 107:240-250.

252 Garcia-Segura, L.M. and McCarthy, M.M. 2004. Role of glia in neuroendocrine function. *Endocrinology* 145(3):1082-1086.

253 Gardahaut, M.F., Fontaine-Perus, J., Rouaud, T., Bandman, E. and Ferrand, R. 1992. Developmental modulation of myosin expression by thyroid hormones in avian skeletal muscle. *Development* 115:1121-1131.

254 Gaskell, J. 2000. *Who Killed the Great Auk?* Oxford University Press, Oxford.

255 Gautier, A. 1993. "What's in a name?": a short history of the Latin and other labels proposed for domestic animals. In A. Clasen, S. Payne and H.P. Uerpmann (eds.), *Skeletons in her Cupboard: Festschrift for Juliet Clutton-Brock,* pg. 91-98. Monograph 34, Oxbow Books, Oxford.

256 Gavlik, S., Albina, M. and Specker, J.L. 2002. Metamorphosis in summer flounder: manipulation of thyroid status to synchronize settling behavior, growth, and development. *Aquaculture* 203:359-373.

257 Gavrilets, S. 2003. Models of speciation: what have we learned in 40 years? *Evolution* 57(10):2197-2215.

258 Geelhoed, G.W. 1999. Metabolic maladaptation: individual and social consequences of medical intervention in correcting endemic hypothyroidism. *Nutrition* 15:908-932.

259 Geist, V. 1971. *Mountain Sheep: A Study in Behavior and Evolution.* University of Chicago Press, Chicago.

260 Geist, V. 1986. On speciation in ice age mammals, with special reference to cervids and caprids. *Canadian Journal of Zoology* 65:1067-1084.

261 Geist, V. 1998. *Deer of the World: Their Evolution, Behavior, and Ecology.* Stackpole Books, Mechanicsburg,PA.

262 Gentry, A., Clutton-Brock, J. and Groves, C.P. 2004. The naming of wild animal species and their domestic derivatives. *Journal of Archaeological Science* 31:645-651.

263 Gerhart, J. and Kirschner, M. 1997. *Cells, Embryos and Evolution: Toward a Cellular and Developmental Understanding of Phenotypic Variation and Evolutionary Adaptability.* Blackwell Science, Malden.

264 German, R.Z. and Stewart, S.A. 2002. Sexual dimorphism and ontogeny in primates. In N. Minugh-Purvis and K. McNamara (eds.), *Human Evolution Through Developmental Change*, pg. 207-222. Johns Hopkins University Press, Baltimore.

265 Gibson, G. and Hogness, D.S. 1996. Effect of polymorphism in the Drosophila regulatory gene Ultrabithorax on homeotic stability. *Science* 271:200-203.
266 Gillespie, J.M.A., Chan, B.P.K., Roy, D., Cai, F. and Belsham, D.D. 2003. Expression of circadian rhythm genes in gonadotropin-releasing hormone-secreting GT1-7 neurons. *Endocrinology* 144(12):5285-5292.
267 Gittleman, J.L. 1989. *Carnivore Behavior, Ecology and Evolution*. Cornell University Press, Ithaca.
268 Giuffra, E., Kijas, J.M.H., Amarger, V., Carlborg, Ő., Jeon, J.-T., et al. 2000. The origin of the domestic pig: independent domestication and subsequent introgression. *Genetics* 154:1785-1791.
269 Goldberg, T.L. and Ruvolo, H. 1997. Molecular phylogenetics and historical biogeography of east African chimpanzees. *Biological Journal of the Linnean Society* 61:301-324.
270 Goichot, B., Brandenberger, G., Saini, J., Wittersheim, G. and Follenius, M. 1994. Nycthemeral patterns of thyroid hormones and their relationships with thyrotropin variations and sleep structure. *Journal of Endocrinological Investigation* 17(3):181-187.
271 Goichot, B., Weibel, L., Chapotot, F., Gronfier, C., Piquard, F., et al. 1998. Effect of the shift of the sleep-wake cycle on three robust endocrine markers of the circadian clock. *American Journal of Physiology: Endocrinology & Metabolism* 275(2):E243-E248.
272 Gomez, J.M., Boujard, T., Boeuf, G., Solari, A. and Le Bail, P.Y. 1997. Individual diurnal profiles of thyroid hormones in rainbow trout (*Oncorhynchus myskiss*) in relation to cortisol, growth hormone, and growth rate. *General and Comparative Endocrinology* 107(1):74-83.
273 Goodnight, C.J. and Wade, M.J. 2000. The ongoing synthesis: a reply to Coyne, Barton, and Turelli. *Evolution* 54(1):317-324.
274 Goodwin, D., Bradshaw, J.W.S. and Wickens, S.M. 1997. Paedomorphosis affects agonistic visual signals of domestic dogs. *Animal Behavior* 53:297-304.
275 Gould, S.J. 1977. *Ontogeny and Phylogeny*. Harvard University Press, Cambridge.
276 Gould, S.J. 1983. How the zebra gets its stripes. *Hen's Teeth and Horses Toes*, pg. 366-375. W.W. Norton and Co., New York.
277 Gould, S.J. 1996. *Full House: The Spread of Excellence from Plato to Darwin*. Harmony Books, New York.
278 Gould, S.J. 2002. *The Structure of Evolutionary Theory*. Belknap Press, Harvard.
279 Gould, S.J. and Eldredge, N. 1977. Punctuated equilibria: the tempo and mode of evolution reconsidered. *Paleobiology* 3:115-151.
280 Gould, S.J. and Eldredge, N. 1993. Punctuated equilibrium comes of age. *Nature* 366:223-227.
281 Grandin, T. and Johnson, C. 2005. *Animals in Transition: Using the Mysteries of Autism to Decode Animal Behaviour*. Scribner, New York.
282 Grant, P.R. 1998. Patterns on islands and microevolution. In P.R. Grant (ed.), *Evolution on Islands*, pg. 1-17. Oxford University Press, Oxford.
283 Grant, P.R. 1998. Epilogue and questions. In P.R. Grant (ed.), *Evolution on Islands*, pg. 305-319. Oxford University Press, Oxford.
284 Grant, P.R. and Grant, B.R. 2002. Unpredictable evolution in a 30-year study of Darwin's finches. *Science* 296:707-711.

285 Grau, E.G., Nishioka, R.S., Specker, J.L. and Bern, H.A. 1985. Endocrine involvement in the smoltification of salmon, with specific reference to the role of the thyroid gland. In B. Lofts and W.N. Holmes (eds.), *Current TRENDS in Comparative Endocrinology*, pg. 491-493. Hong Kong University Press, Hong Kong.

286 Grumbach, M.M. 2000. Estrogen, bone, growth and sex: a sea change in conventional wisdom. *Journal of Pediatric Endocrinology and Metabolism* 13 (suppl. 6):1439-1455.

287 Greene, R.T. 1997. Thyroid testing: the full circle. *Canine Practice* 22:10-11.

288 Greenspan, S.L., Klibanski, A., Rowe, J.W. and Elahi, D. 1991. Age-related alterations in pulsatile secretion of TSH: role of dopaminergic regulation. *American Journal of Physiology* 260(3 Pt. 1):486-491.

289 Greenspan, S.L., Klibanski, A., Shoenfeld, D. and Ridgway, E.C. 1986. Pulsatile secretion of thyrotropin in man. *Clinical Endocrinology and Metabolism* 63(3):661-668.

290 Gross, M. R. 1998. One species with two biologies: Atlantic salmon (*Salmo salar*) in the wild and in aquaculture. *Canadian Journal of Fisheries and Aquatic Science* 55 (suppl.):131-144.

291 Guerrant Jr., E.O. 1988. Heterochrony in plants: the intersection of evolution, ecology and ontogeny. In M.L. McKinney (ed.), *Heterochrony in Evolution:A Multidisciplinary Approach,* pg. 111-133. Plenum Press, New York.

292 Gunaratnam, P. 1986. The effects of thyroxine on hair growth in the dog. *Journal of Small Animal Practice* 27:17-29.

293 Gupta, B.B.P. and Premabati, Y. 2002. Differential effects of melatonin on plasma levels of thyroxine and triiodothyronine levels in the air-breathing fish, *Clarias gariepinus*, during breeding and quiescent periods. *General and Comparative Endocrinology* 129:146-151.

294 Gustafsson, A. 1994. Regulation of sexual dimorphism in rat liver. In R.V. Short and E. Balaban (eds.), *The Differences Between the Sexes*, pg. 231-241. Cambridge University Press, Cambridge.

295 Hadley, M.E. 1984. *Endocrinology, First Edition*. Prentice-Hall, Inc., Englewood Cliffs.

296 Hadley, M.E. 2000. *Endocrinology, Fifth Edition*. Prentice-Hall, Inc., Englewood Cliffs.

297 Hafner, J.C. and Hafner, M.S. 1988. Heterochrony in rodents. In M.L. McKinney (ed.), *Heterochrony in Evolution: A Multidisciplinary Approach*, pg. 217-235. Plenum Press, New York.

298 Haisenleder, D.J., Ortolano, G.A., Dalkin, A.C., Yasin, M. and Marshall, J.C. 1992. Differential actions of thyrotropin (TSH)-releasing hormone pulses in the expression of prolactin and TSH subunit messenger ribonucleic acid in rat pituitary genes in vitro. *Endocrinology* 130:2917-2923.

299 Hall, B.K. 1992. *Evolutionary Developmental Biology*. Chapman and Hall, London.

300 Hall, B.K. 2003. Descent with modification: the unity underlying homology and homoplasy as seen through an analysis of development and evolution. *Biological Review* 78:409-433.

301 Hanken, J. and Wake, D.B. 1993. Miniaturization of body size: Organismal consequences and evolutionary significance. *Annual Review of Ecology and Systematics* 24:501-519.

302 Hanna, F.W.F., Lazarus, J.H. and Scanlon, M.F. 1999. Controversial aspects of thyroid disease. *British Medical Journal* 319:894-899.

303 Hansen, P.S., Brix, T.H., Sørensen, T.I.A., Kyvik, K.O. and Hegedüs, L. 2004. Major genetic influence on the regulation of the pituitary-thyroid axis: a study of healthy Danish twins. *Journal of Clinical Endocrinology & Metabolism* 89(3):1181-1187.

304 Hardy, M.P., Sottas, C.M., Ge, R., McKittrick, C.R., Tamashiro, K.L., et al. 2002. Trends of reproductive hormones in male rats during psychosocial stress: role of glucocorticoid metabolism in behavioral dominance. *Biology of Reproduction* 67(6):1750-1755.

305 Hare, B., Plyusnina, I., Ignacio, N., Schepina, O., Stepika, A., et al. 2005. Social cognitive evolution in captive foxes is a correlated by-product of experimental domestication. *Current Biology* 15(3):226-230.

306 Härlid, A. and Arnason, U. 1999. Analysis of mitochondrial DNA nest ratite birds within the Neognathae: supporting a neotenous origin of ratite morphological characters. *Proceedings of the Royal Society of London B.* 266:305-309.

307 Harris, D.R. 1996. The origins and spread of agriculture and pastoralism in Eurasia: an overview. In D.R. Harris (ed.), *The Origins and Spread of Agriculture and Pastoralism in Eurasia.* Smithsonian Institution Press, Washington, DC.

308 Harris, G.W. 1959. Neuroendocrine control of TSH regulation. In A. Gorbman (ed.), *Comparative Endocrinology: Proceedings of the Columbia University Symposium on Comparative Endocrinology 1958,* pg. 202-222. J. Wiley and Sons, Inc., New York.

309 Harvey, S. 1990. Thyrotropin-releasing hormone: a growth hormone-releasing factor. *Journal of Endocrinology* 125:345-358.

310 Hauser, P., McMillin, J.M. and Bhatara, V.S. 1998. Thyroid hormone disruption: dioxins linked to attention deficit, learning problems. *Toxicology & Industrial Health* 34:85-101.

311 Hauser, P. and Rovet, J. 1998. *Thyroid Diseases of Infancy and Childhood.* American Psychiatric Press Inc., Washington, DC.

312 Hauser, P., Soler, R., Brucker-Davis, F. and Weintrub, B.D. 1997. Thyroid hormones correlate with symptoms of hyperactivity but not inattention in attention deficit hyperactivity disorder. *International Journal of Psychological and Neurological Endocrinology* 22:107-114.

313 Hawaleshka, D. 2005. Teflon trouble. *Maclean's* 118(21):70-72.

314 Hawes, R.O. 1984. Pigeons. In I.L. Mason (ed.), *Evolution of Domesticated Animals,* pg. 351-356. Longman Co., London.

315 Hawkins, D.K. and Foote, C.J. 1998. Early survival and development of coastal cutthroat trout (*Oncorhynchus clarki clarki*), steelhead (*Oncorhynchus mykiss*), and reciprocal hybrids. *Canadian Journal of Fisheries and Aquatic Science* 55:2097-2104.

316 Hayes, J.P. and Jenkins, S.H. 1997. Individual variation in mammals. *Journal of Mammalogy* 78:274-293.

317 Hayes, T.B. 1997. Hormonal mechanisms as potential constraints on evolution: examples from the *Anura. American Zoologist* 37:482-490.

318 Hayssen, V. 1998. Effect of transatlantic transport on reproduction of agouti and nonagouti deer mice, *Peromyscus maniculatus. Laboratory Animals* 32:55-64.

319 Hearing, V.J. 1993. Invited editorial: Unraveling the melanocyte. *American Journal of Human Genetics* 52:1-7.

320 Heath, D. D., Heath, J. W., Bryden, C. A., Johnson, R. M. and Fox, C.W. 2003. Rapid evolution of egg size in captive salmon. *Science* 299:1738-1740.

321 Heath, R.B. 1989. Llama anesthetic programs. In L.W. Johnson (ed.), Llama Medicine. *Veterinary Clinics of North America: Food Animal Practice* 5, pg. 71-80. W.B. Saunders Co., Philadelphia.

322 Heaton, T.H. and Grady, F. 2003. The Late Wisconsin vertebrate history of Prince of Wales Island, Southeast Alaska. *In* B.W. Schubert, .I. Mead, and R.W. Graham (eds.), *Ice Age Cave Faunas of North America,* pg. 17-53. Indiana University Press. (also http://www.usd.edu/esci/alaska/index.html]

323 Hedrick, P. W. and McDonald, J.F. 1980. Regulatory gene adaptation: an evolutionary model. *Heredity* 45:83-97.

324 Helbing, C., Werry, K., Crump, D., Domanski, D. Veldhoen, N., et al. 2003. Expression profiles of novel thyroid hormone-responsive genes and proteins in the tail of *Xenopus laevis* tadpoles undergoing precocious metamorphosis. *Molecular Endocrinology* 17(7):1395-1409.

325 Hemmer, H. 1990. *Domestication: The Decline of Environmental Appreciation.* Cambridge University Press, Cambridge.

326 Henley, W.N. and Koehnle, T.J. 1997. Thyroid hormones and the treatment of depression: an example of basic hormonal actions in the mature mammalian brain. *Synapse* 27:36-44.

327 Henshaw, R.E., Lockwood, R., Shideler, R. and Stephenson, R.O. 1979. Experimental release of captive wolves. In E. Klinghammer (ed.), *The Behavior and Ecology of Wolves*, pg. 319-345. Garland STPM Press, New York.

328 Hercbergs, A. A. 1999. Spontaneous remission of cancer: a thyroid hormone-dependent phenomenon? *Anticancer Research* 19:4839-4844.

329 Hercbergs, A. A., Goyal, L. K., Suh, J. H., Lee, S., Reddy, C. A., et al. 2003. Propylthiouracil-induced chemical hypothyroidism with high-dose tamoxifen prolongs survival in recurrent high grade glioma: a phase I/II study. *Anticancer Research* 23:617-626.

330 Hercbergs, A. A., Suh, J. H., Lee, S., Cohen, B. H., Stevens, G. H., et al. 2002. Propylthiouracil-induced chemical hypothyroidism with high-dose tamoxifen prolongs survival with increased reponse rate in recurrent high grade glioma. [Abstract] *Proceedings of the American Association for Cancer Research* 43:490-491.

331 Heyland, A., Hodin, J. and Reitzel, A.M. 2004a. Hormone signaling in evolution and development: a non-model system approach. *BioEssays* 26:1-12.

332 Heyland, A., Reitzel, A.M. and Hodin, J. 2004b. Thyroid hormones determine developmental mode in sand dollars (Echinodermata: Echinoidea). *Evolution and Development* 6(6):382-392.

333 Heyning, J.E. 1997. Sperm whale phylogeny revisited: analysis of the morphological evidence. *Marine Mammal Science* 13:596-613.

334 Hiendleder, S., Kaupe, B., Wassmuth, R. and Janke, A. 2002. Molecular analysis of wild and domestic sheep questions current nomenclature and provides evidence for domestication from two different subspecies. *Proceedings of the Royal Society of London B* 269:893-904.

335 Hinds, D.A., Stuve, L.L., Nilsen, G.B., Halperin, E., Eskin, E., et al., 2005. Whole-genome patterns of common DNA variation in three human populations. *Science* 307:1072-1079.

336 Hoffmann, A.A. and Parsons, P.A. 1991. *Evolutionary Genetics and Environmental Stress.* Oxford University Press, Oxford.

337 Holliday, T.W. 2003. Species concepts, reticulation, and human evolution. *Current Anthropology* 44(5):653-673.

338 Hollowell, J.G., Staehling, N.W., Flanders, W.D., Hannon, W.H., Gunter, E.W., et al. 2002. Serum TSH, T_4 and thyroid antibodies in the United States population (1988 to 1994): National Health and Nutrition Examination Survey (NHANES III). *Journal of Clinical Endocrinology & Metabolism* 87:489-499.

339 Holst, M., Stirling, I. and Calvert, W. 1999. Age structure and reproductive rates of ringed seals (*Phoca hispida*) on the northwest coast of Hudson Bay in 1991 and 1992. *Marine Mammal Science* 15:1357-1364.

340 Horrobin, D. 1997. Fatty acids, phospholipids, and schizophrenia. In S. Yehuda, and D.I. Mostofsky (eds.), *Handbook of Fatty Acid Biology*, pg. 245-256. Humana Press, Totowa, New Jersey.

341 Horrobin, D. 2001. *The Madness of Adam and Eve: How Schizophrenia Shaped Humanity*. Bantam Press, London.

342 Horrobin, D. 2003. Not in the genes: enthusiasts for genomics have corrupted scientific endeavour and undermined hopes of medical progress. *The Guardian* (Comment):February 12.

343 Horrobin, D. and Bennett, C.N. 1999. Depression and bipolar disorder: relationships to impaired fatty acid and phospholipids metabolism and to diabetes, cardiovascular disease, immunological abnormalities, cancer, aging and osteoporosis: possible candidate genes. *Prostaglandins Leukotines & Essential Fatty Acids* 60:111-167.

344 Horvath, T.L. 2005. The hardship of obesity: a soft-wired hypothalamus. *Nature Neuroscience* 8(5):561-565.

345 Hosoda, K., Hammer, R.E., Richardson, J.A., Baynash, A.G., Cheung, J.C., et al. 1994. Targeted and natural (piebald lethal) mutations of endothelin-B receptor gene produce megacolon associated with spotted coat color in mice. *Cell* 79:1267-1276.

346 Hoss, M., Dilling, A., Currant, A. and Paabo, S. 1996. Molecular phylogeny of the extinct ground sloth *Mylodon darwinii*. *Proceedings of the National Academy of Sciences USA* 93:181-185.

347 Huang, H., Cai, L., Remo, B.F. and Brown, D.D. 2001. Timing of metamorphosis and the onset of the negative feedback loop between the thyroid gland and the pituitary is controlled by type II iodothyronine deiodinase in *Xenopus laevis*. *Proceedings of the National Academy of Sciences USA* 98(13):7348-7353.

348 Hulbert, A.J. 2000. Thyroid hormones and their affects: a new perspective. *Biological Revue* 75:519-631.

349 Huynen, L., Miller, C.D., Scofield, R.P. and Lambert, D.M. 2003. Nuclear DNA sequences detect species limits in ancient moa. *Nature* 425:175-178.

350 Iljin, N.A. 1941. Wolf-dog genetics. *Journal of Genetics* 42:359-414.

351 Iltis, H.H. 1983. From teosinte to maize: the catastrophic sexual transmutation. *Science* 222:886-894.

352 Isaac, E. 1970. *Geography of Domestication.* Foundations of Cultural Geography Series. Prentice-Hall Inc., Englewood Cliffs, New Jersey.

353 Ishizuya-Oka, A., Ueda, S., Amano, T., Shimizu, K., Suzuki, K., Ueno, N., and Yoshizato, K. 2001. Thyroid-hormone dependent and fibroblast-specific expression of BMP-4 correlates with adult epithelial development during amphibian intestinal remodeling. *Cell Tissue Research* 303(2):187-195.

354 Jablonski, N.G. and Chaplin, G. 1992. The origin of hominid bipedalism re-examined. *Archaeology in Oceania* 27(3):113-119.

355 Jablonski, N.G. and Chaplin, G. 2000. The evolution of human skin coloration. *Journal of Human Evolution* 39:57-106.

356 Jablonski, N.G. and Chaplin, G. 2003. Skin deep. *Scientific American* 13:72-79.

357 Jana, N.R. and Bhattachacharya, S. 1994. Binding of thyroid hormone to the goat testicular Leydig cell induces the generation of a proteinaceous factor which stimulates androgen release. *Journal of Endocrinology* 143(3):549-556.

358 Jannini, E.A., Ulisse, S. and D'Armiento, M. 1995. Thyroid hormone and male gonadal function. *Endocrine Reviews* 16(4):443-459.

359 Jansen, T., Forster, P., Levine, M. A., Oelke, H., Hurles, M., et al. 2002. Mitochondrial DNA and the origins of the domestic horse. *Proceedings of the National Academy of Sciences USA* 99:10905-10910.

360 Janssen, R. and Janssen, J. 1989. *Egyptian Household Animals.* Shire Publications, Bucks.

361 Jefcoate, C. 2002. High-flux mitochondrial cholesterol trafficking, a specialized function of the adrenal cortex. *Journal of Clinical Investigation.* 110(7):881-890.

362 Jenkins, K., 2001. Not tonight, dear – I'm feeling better: the drugs that relieve depression also sap the libido. *Maclean's* 114(46):40-42.

363 Jensen, P. 2002. Behavior of pigs. In P. Jensen (ed.), *The Ethology of Domestic Animals: An Introductory Text,* pg. 159-171. CABI Publishing, New York.

364 Jobling, M. 1995. *Environmental Biology of Fishes.* Chapman and Hall, London.

365 Johnsgard, P.A. 1970. A summary of intergeneric New World quail hybrids, and a new intergeneric hybrid combination. *The Condor* 72:85-88.

366 Johnson, D. 2002. Dogs throughout time. *American Archaeology* 6(3):32-37.

367 Johnson, L.G. 1997. Thyroxine's evolutionary roots. *Perspectives in Biology and Medicine* 40:529-535.

368 Johnson, L.G. 1998. Stage-dependent thyroxine effects on sea urchin development. *New Zealand Journal of Marine and Freshwater Research* 32:531-536.

369 Johnson, L.G. and Cartwright, C.M. 1996. Thyroxine-accelerated larval development in the crown-of-thorns starfish, *Acanthaster planci. Biological Bulletin* 190:299-301.

370 Jolicoeur, P. 1959. Multivariate geographic variation in the wolf, *Canis lupus. Evolution* 13:283-299.

371 Jolicoeur, P. 1975. Sexual dimorphism and geographic distance as factors of skull variation in the wolf (*Canis lupus*). In M.W. Fox (ed.), *The Wild Canids: Their Systematics,Behavioural Ecology and Evolution,* pg. 54-61. Behavioral Science Series, Van Nostrand Reinhold Co., New York.

372 Jones, S.A., Thoemke, K.R. and Anderson, G.W. 2005. The role of thyroid hormone in fetal and neonatal brain development. *Current Opinion in Endocrinology and Diabetes* 12:10-16.

373 Jonsson, B., Jonsson, N. and Hansen, L.P. 1991. Differences in life history and migratory behavior between wild and hatchery reared Atlantic salmon in nature. *Aquaculture* 98:69-78.

374 Juma, A.H., Ardawi, M.S. Baksh, T.M. and Serafi, A.A. 1991. Alterations in thyroid hormones, cortisol, and catecholamine concentration in patients after orthopedic surgery. *Journal of Surgical Research* 50(2):129-134.

375 Kadwell, M., Fernandez, M., Stanley, H., Baldi, R., Wheeler, J., et al. 2001. Genetic analyses reveals the wild ancestors of the llama and alpaca. *Proceedings of the Royal Society London B* 268:2575-2584.

376 Kahaly, G.J. and Dillmann, W.H. 2005. Thyroid hormone action in the heart. *Endocrine Reviews* 26(5):704-728.

377 Kalsbeek, A., Fliers, E., Franke, A.N., Wortel, J. and Buijs, R.M. 2000. Functional connections between the suprachiasmatic nucleus and the thyroid gland as revealed by lesioning and viral tracing techniques in the rat. *Endocrinology* 141(10):3832-3841.

378 Kalsbeek, A., Buijs, R.M., van Schaik, R., Kaptein, E., Visser, T.J., et al. 2005. Daily variations in type II iodothyronine deiodinase activity in the rat brain as controlled by the biological clock. *Endocrinology* 146(3):1418-1427.

379 Kaptein, E.M., Hays, M.T. and Ferguson, D.C. 1994. Thyroid hormone metabolism: a comparative evaluation. In D.C. Ferguson (ed.), Thyroid disorders, pg. 431-463. *Veterinary Clinics of North America, Small Animal Practice* 24, W.B. Saunders Co., Philadelphia.

380 Kassai, Y., Munne, P., Hotta, Y., Penttila, E., Kavanagh, K., et al. 2005. Regulation of mammalian tooth cusp patterning by ectodin. *Science* 309:2067-2070.

381 Kaufman, L.S., Chapman, L.J. and Chapman, C.A. 1997. Evolution in fast forward: haplochromine fishes of the Lake Victoria region. *Endeavor* 21:23-30.

382 Keeler, C. 1975. Genetics of behaviour variations in colour phases of the red fox. In M.W. Fox (ed.), *The Wild Canids: Their Systematics, Behavioural Ecology and Evolution,* pg. 399-415. Behavioral Science Series, Van Nostrand Reinhold Co., New York.

383 Keenan, D.M., Roelfsema, F., Biermasz, N. and Veldhuis, J.D. 2003. Physiological control of pituitary hormone secretory-burst mass, frequency, and waveform: a statistical formulation and analysis. *American Journal of Physiology – Regulatory, Integrative and Compararative Physiology* 285:R664-R673.

384 Kelly, M.J. and Durant, S.M. 2000. Viability of the Serengeti cheetah population. *Conservation Biology* 14(3):786-797.

385 Kennaway, D.J. 2005. The role of circadian rhythmicity in reproduction. *Human Reproduction Update* 11(1):91-101.

386 Kenyon, C. 1994. If birds can fly, why can't we? Homeotic genes and evolution. *Cell* 78:175-180.

387 Keogh, J.S., Scott, I.A.W. and Hayes, C. 2005. Rapid and repeated origin of insular gigantism and dwarfism in Australian tiger snakes. *Evolution* 59:226-233.

388 Kemppainen, R.J. and Peterson, M.E. 1996. Domestic cats show episodic variation in plasma concentrations of adrenocorticotropin, alpha-melanocyte-stimulating hormone (alpha-MSH), cortisol and thyroxine with circadian variation in plasma alpha-MSH concentrations. *European Journal of Endocrinology* 134(5):602-609.

389 Kerns, J. A., Olivier, M., Lust, G. and Barsh, G.S. 2003. Exclusion of *melanocortin-1 receptor(Mc1r)* and *agouti* as candidates for dominant black in dogs. *Journal of Heredity* 94:75-79.

390 Kershaw, E.E. and Flier, J.S. 2004. Adipose tissue as an endocrine organ. *Journal of Clinical Endocrinology and Metabolism* 89(6):2548-2556.

391 Keynes, R. Darwin (ed.) 1988. *Charles Darwin's Beagle Diary.* Cambridge University Press, Cambridge.

392 Kilby, M.D., Verhaeg, J., Gittoes, N., Somerset, D.A., Clark, P.M.S., et al. 1998. Circulating thyroid hormone concentrations and placental thyroid hormone receptor expression in normal human pregnancy and pregnancy complicated by intrauterine growth restriction (IUGR). *Journal of Clinical Endocrinology & Metabolism* 83(8):2964-2971.

393 Kim, M., McCormick, S., Timmermans, M. and Sinha, N. 2003. The expression of PHANTASTICA determines leaflet placement in compound leaves. *Nature* 424:438-443.

394 Kingdon, J. 2003. *Lowly Origin: Where, When and Why Our Ancestors First Stood Up*. Princeton University Press, Princeton.

395 Kingsley, R.J., Corcoran, M.L., Krider, K.L. and Kriechbaum, K.L. 2001. Thyroxine and vitamin D in the gorgonian, *Leptogorgia virgulata*. *Journal of Comparative Biochemistry and Physiology Part A* 129:897-907.

396 Kirkpatrick, M. 2000. Fish found *in flagrante delicto*. *Nature* 408:298-299.

397 Kitajka, K., Puskas, L.G., Zvara, A., Hackler Jr., L., Barcelo-Coblijn, G., et al. 2002. The role of n-3 polyunsaturated fatty acids in brain: modulation of rat brain gene expression by dietary n-3 fatty acids. *Proceedings of the National Academy of Sciences USA* 99:2619-2624.

398 Klandorf, H. and Sharp, P.J. 1985. Feeding-induced daily rhythms in plasma thyroid hormone levels in chickens. In B. Lofts and W.N. Holmes (eds.), *Current Trends in Comparative Endocrinology*, pg. 517-519. Hong Kong University Press, Hong Kong.

399 Klee, H. 2003. Hormones are in the air. *Proceedings of the National Academy of Sciences USA* 100:10144-10145.

400 Klingenberg, C.P. and Ekau, W. 1996. A combined morphometric and phylogenetic analysis of an ecomorphological trend: pelagization in Antarctic fishes (Perciformes: Nototheniidae). *Biological Journal of the Linnean Society* 59:143-177.

401 Klingenberg, C.P. and Spence, J.R. 1993. Heterochrony and allometry: lessons from the water strider genus Limnoporus. *Evolution* 47:1834-1853.

402 Knap, R. and Moore, M.C. 1996. Male morphs in tree lizards, *Urosaurus ornatus*, have different delayed hormonal responses to aggressive encounters. *Animal Behavior* 52:1045-1055.

403 Knudsen, N., Laurberg, P., Rasmussen, L.B., Bülow, I., Perrild, H., et al. 2005. Small differences in thyroid function may be important for Body Mass Index and the occurrence of obesity in the population. *Journal of Clinical Endocrinology and Metabolism* 90(7):4019-4024.

404 Köhler, M. and Moyà-Solà, S. 1997. Ape-like or hominid-like? The positional behaviour of *Oreopithecus bambolii* reconsidered. *Proceedings of the National Academy of Sciences USA* 94:11747-11750.

405 Köhler, M. and Moyà-Solà, S. 2004. Reduction of brain and sense organs in the fossil insular bovid *Myotragus*. *Brain, Behavior and Evolution* 63:125-140.

406 Kőhrle, J. 2000. Thyroid hormone metabolism and action in the brain and pituitary. *Acta Medica Austriaca*. 27(1):1-7.

407 Koibuchi, N., Natsuzaki, S., Ichimura, K., Ohtake, H. and Yamaoka, S. 1996. Ontogenetic changes in the expression of cytochrome c oxidase subunit I gene in the cerebellar cortex of the perinatal hypothyroid rat. *Endocrinology* 137:5096-5108.

408 Kok, P., Roelfsema, R., Frölich, M., Meinders, A.E. and Hanno, P. in press. Spontaneous diurnal TSH secretion is enhanced in proportion to circulation leptin in obese premenopausal women. *Journal of Clinical Endocrinology & Metabolism*.

409 Kolesnikova, L.A. 1997. Structural and functional features of the pineal gland of silver foxes: changes under the effect of domestication. In L.N. Trut and L.V. Osadchuk (eds.), *Evolutionary-genetic and Genetic-physiological Aspects of Fur Animal Domestication: A Collection of Reports*, pg. 41-54. IFASA/Scientifur, Oslo.

410 Koop B.F., Burbidge M., Byun A., Rink U. and Crockford, S.J. 2000. Ancient DNA evidence of a separate origin for North American indigenous dogs. In S.J. Crockford (ed.), *Dogs Through Time: An Archaeological Perspective*, pg. 271-286. British Archaeological Reports S889, Oxford.

411 Korf, H-W. 1994. The pineal organ as a component of the biological clock: phylogenetic and ontogenetic considerations. In W. Pierpaoli, W. Regelson and N. Fabris (eds.), *The Aging Clock: The Pineal Gland and Other Pacemakers in the Progression of Aging and Carcinogenesis*, pg. 13-42. *Annals of the NewYork Academy of Science* 719, New York.

412 Kovacs, K.M., Lyndersen, C., Hammil, M.O., White, B.N., Wilson, P.J., et al.1997. A harp seal X hooded seal hybrid. *Marine Mammal Science* 13(3):460-468.

413 Kronfol, Z., Nair, M., Zhang, Q., Hill, E.E. and Brown, M.B. 1997. Circadian immune measures in healthy volunteers: relationship to hypothalamic-pituitary-adrenal axis hormones and sympathetic neurotransmitters. *Psychosomatic Medicine* 59(1):42-50.

414 Krumlauf, R. 1994. Hox genes in vertebrate development. *Cell* 78:191-201.

415 Kruska, D.C.T. 2005. On the evolutionary significance of encephalization in some eutherian mammals: effects of adaptive radiation, domestication, and feralization. *Brain, Behavior and Evolution* 65:73-108.

416 Kuhn, E.R., Delmotte, N.M.J. and Darras, V.M. 1983. Persistence of a circadian rhythmicity for thyroid hormones in plasma and thyroid of hibernating male *Rana ridibunda. General and Comparative Endocrinology* 50:838-894.

417 Kuhn, T.S., 1970. *The Structure of Scientific Revolutions, Second Edition.* University of Chicago Press, Chicago.

418 Kuijpens, J.I., Vader, H.L., Drexhage, H.A., Wiersinga, W.M., van Son, M.J., et al. 2001. Thyroid peroxidase antibodies during gestation are a marker for subsequent depression postpartum. *European Journal of Endocrinology* 145:579-584.

419 Künzi, C. and Sachser, N. 1999. The behavioral endocrinology of domestication: A comparison between the domestic guinea pig (*Cavia aperea* f. *porcellus*) and its wild ancestor, the cavy (*Cavia aperea*). *Hormones and Behavior* 35:28-37.

420 Kurtén, B. 1968. *Pleistocene Mammals of Europe*. Weidenfeld and Nicolson, London.

421 Kurtén, B. 1988. *On Evolution and Fossil Mammals*. Columbia University Press, New York.

422 Kurtén, B. and Anderson, E. 1980. *Pleistocene Mammals of North America*. Columbia University Press, New York.

423 Kuzawa, C.W. 1998. Adipose tissue in human infancy and childhood: an evolutionary perspective. *Yearbook of Physical Anthropology* 41:177-209.

424 LaFranchi, S.H., Haddow, J.E. and Hollowell, J.G. 2005. Is thyroid inadequacy during gestation a risk factor for adverse pregnancy and developmental outcomes? *Thyroid* 15(1):60-71.

425 Lagarde, F., Bonnet, X., Henen, B.T., Corbin, J., Nagy, K.A., et al. 2001. Sexual size dimorphism in steppe tortoises (*Testudo horsfieldi*): growth, maturity, and individual variation. *Canadian Journal of Zoology* 79:1433-1441.

426 Lahr, M.M. and Foley, R. 2004. Human evolution writ small. *Nature* 431:1043-1044.

427 Lamont, M.M., Vida, J.T., Hawey, J.T., Jeffries, S., Brown, R., et al. 1996. Genetic substructure of the Pacific harbour seal (*Phoca vitulina richardsoni)* off Washington, Oregon and California. *Marine Mammal Science* 12:402-413.

428 Lamoreux, M.L. and Russell, E.S. 1979. Developmental interaction in the pigmentary system of mice. *Journal of Heredity* 70:31-36.

429 Lampl, M. Velduis, J.D. and M.L. Johnson. 1992. Saltation and statis: a model of human growth. *Science* 258:801-803.

430 Langridge, J. 1991. *Molecular Genetics and Comparative Evolution.* J. Wiley and Sons Inc., New York.

431 Lanier, J. L., Grandin, T., Green, R., Avery, D. and McGee, K. 2001. A note on hair whorl position and cattle temperament in the auction ring. *Applied Animal Behaviour Science* 73:93-101.

432 Lanza, R.P., Cibelli, J.B., Faber, D., Sweeney, R.W., Henderson, B., et al. 2001. Cloned cattle can be healthy and normal. *Science* 294:1893-1894.

433 Lanza, R.P., Dresser, B.L. and Damiani, P. 2000. Cloning Noah's ark. *Scientific American* (November):84-89.

434 Lapseritis, J.M. and Hayssen, V. 2001. Thyroxine levels in agouti and non-agouti deer mice (*Peromyscus maniculatus*). *Comparative Biochemistry and Physiology A* 130:295-299.

435 Larson, G., Dobney, K. Albarella, U., Fang, M., Matisoo-Smith, E., et al. 2005. Worldwide phylogeography of wild boar reveals multiple centers of pig domestication. *Science* 307:1618-1621.

436 Larsson, H.J., Eaton, W.W., Madsen, K.M., Vestergaard, M., Olesen, A.V., et al. 2005. Risk factors for autism: perinatal factors, parental psychiatric history, and socioeconomic status. *American Journal of Epidemiology* 161(10):916-925.

437 Lavado-Autric, R., Ausó, E., Garcia-Velasco, J.V., Carmen del Arufe, M., Escobar del Rey, F., et al. 2003. Early maternal hypothyroxinemia alters histogenesis and cerebral cortex cytoarchitecture of the progeny. *Journal of Clinical Investigation* 111:1073-1082.

438 Lawrence, P.A. and Morata, G. 1994. Homeobox genes: their function in *Drosophila* segmentation and pattern formation. *Cell* 78:181-189.

439 Leach, H.M. 2003. Human domestication reconsidered. *Current Anthropology* 44:349-368.

440 Leakey, M. and Walker, A. 2003. Early hominid fossils from Africa. *Scientific American* 13:14-19.

441 Lee, B.C., Kim, M.K., Jang, G., Oh, H.J., Yuda, F., et al. 2005. Dogs cloned from adult somatic cells. *Nature* 436:641.

442 Lehman, N., Eisenhawer, A., Hansen, K., Mech, L.D., Peterson, R.O., et al. 1990. Introgression of coyote mitochondrial DNA into sympatric North American gray wolf populations. *Evolution* 45:104-119.

443 Lento, G.M., Haddon, M., Chambers, G.K. and Baker, C.S. 1997. Genetic variation in southern hemisphere fur seals (*Arctocephalus* spg.): investigation of population structure and species identity. *Journal of Heredity* 88:202-208.

444 Leonard, J.A., Wayne, R.K., Wheeler, J., Valadez, R., Guillén, S., et al. 2002. Ancient DNA evidence for Old World origin of New World dogs. *Science* 298:1613-1616.

445 Leonard, W.R. 2003. Food for thought. *Scientific American* 13:62-71.

446 Leonard, W.R. and Robertson, M.L. 1997. Rethinking the energetics of bipedality. *Current Anthropology* 38:304-309.

447 Levin, D.A. 2002. Hybridization and extinction. *American Scientist* 90:254-261.

448 Lewinson, D., Harel, Z., Shenzer, P., Silbermann, M. and Hochberg, Z. 1989. Effect of thyroid hormone and growth hormone on recovery from hypothyroidism of epiphyseal growth plate cartilage and its adjacent bone. *Endocrinology* 124(2):937-945.

449 Li, X.F., Bowe, J.E., Lightman, S.L. and O'Bryne, K.T. 2005. Role of cortictropin-releasing factor receptor-2 in stress-induced suppression of pulsatile luteinizing hormone secretion in the rat. *Endocrinology* 146(1):318-322.

450 Licinio, J., Negrão, A.B., Mantzoros, C., Kaklamani, V., Wong, M-L., et al. 1998. Sex differences in circulating human leptin pulse amplitude: clinical implications. *Journal of Clinical Endocrinology and Metabolism* 83:4140-4147.

451 Lieberman, D.E. 2001. Another face in our family tree. *Nature* 410:419-420.

452 Lieberman, D.E. 2005. Further fossil finds from Flores. *Nature* 437:957-958.

453 Lien, R.J. and Siopes, T.D. 1990. The relationship of plasma thyroid hormone and prolactin concentrations to egg laying, incubation behavior, and molting by female turkeys exposed to a one-year natural daylength cycle. *General and Comparative Endocrinology* 90:205-213.

454 Lightman, S.L., Windle, R.J., Julian, M.D., Harbuz, M.S., Shanks, N., et al. 2000. Significance of pulsatility in the HPA axis. In D.J. Chadwick and J.A. Goode (eds.), *Mechanisms and Biological Significance of Pulsatile Hormone Secretion*, pg. 244-260. J. Wiley and Sons, Chichester.

455 Lister, A.M. 1989. Rapid dwarfing of red deer on Jersey in the last Interglacial. *Nature* 342:539-542.

456 Lister, AM. 2004. The impact of Quaternary Ice Ages on mammalian evolution. *Philisophical Transactions of the Royal Society of London B*. 359:221-241.

457 Little, C.C. 1958. Coat color genes in rodents and carnivores. *Quarterly Revue of Biology* 33:103-137.

458 Liu, Y-W. and Chan, W-K. 2002. Thryoid hormones are important for embryonic to larval transitory phase in zebrafish. *Differentiation* 70(1): 1432-1436.

459 Livezey, B.C. 1988. Morphometrics of flightlessness in the Alcidae. *The Auk* 105: 681-698.

460 Livezey, B.C. 1989. Phylogenetic relationships and incipient flightlessness of the extinct Auckland Islands merganser. *Wilson Bulletin* 101(3):410-435.

461 Livezey, B.C. 1990. Evolutionary morphology of flightlessness in the Auckland Islands teal. *Condor* 92:639-673.

462 Loftus, R.T., MacHugh, D.E., Bradley, D.G., Sharp, P.M. and Cunningham, P. 1994. Evidence for two independent domestications of cattle. *Proceedings of the National Academy of Sciences USA* 91:2757-2761.

463 Longcope, C. 2000. The male and female reproductive systems in thyrotoxicosis. In L.D. Braverman and R.D. Utiger (eds.), *Werner and Ingbar's The Thyroid, Eighth Edition,* pg. 653-6658. Lippincott, Philadelphia.

464 Lord, E.M. and Hill, J.P. 1987. Evidence for heterochrony in the evolution of plant form. In R.A. Raff and E.C. Raff (eds.), *Development as an Evolutionary Process,* pg. 47-70. A.R. Liss Inc., New York.

465 Lovejoy, C.O., Cohn, M.J. and White, T.D. 1999. Morphological analysis of the mammalian postcranium: a developmental perspective. *Proceedings of the National Academy of Sciences USA* 96:13247-13252.

466 Lucke, C., Hehrmann, R., von Mayersbach, K. and von zur Muhlen, A. 1977. Studies on circadian variations of plasma TSH, thyroxine and triiodothyronine in man. *Acta Endocrinologica* 86(1):81-88.

467 Luikart, G., Gielly, L., Excoffier, L., Vigne, J.-D., Bouvet, J., et al. 2001. Multiple maternal origins and weak phylogeographic structure in domestic goats. *Proceedings of the National Academy of Sciences USA* 98:5927-5932.

468 Luo, Z.C., Albertsson-Wikland, K. and Karlberg, J. 1998. Length and body mass index at birth and target height influences on patterns of postnatal growth in children born small for gestational age. *Pediatrics* 102:72-78.

469 Lynch, W. 1993. *Bears: Monarchs of the Northern Wilderness.* Douglas and McIntyre, Vancouver.

470 Lyons, D.M. and Mosher, J.A. 1987. Morphological growth, behavioral development, and parental care of broad-winged hawks. *Journal of Field Ornithology* 58(3):334-344.

471 McConkey, E.H. and Varki, A. 2005. Thoughts on the future of great ape research. *Science* 309:1499-1501.

472 McGinnity, P., Prodőhl, P., Ferguson, A., Hynes, R., Ó Maoiléidigh, N., et al. 2003. Fitness reduction and potential extinction of wild populations of Atlantic salmon, *Salmo salar*, as a result of interactions with escaped farm salmon. *Proceedings of the Royal Society London B* 270(1532):2443-2450.

473 McKinney, M.L. 1988. *Heterochrony in Evolution: A Multidisciplinary Approach.* Plenum Press, New York.

474 McKinney, M.L. 1998. The juvenilzed ape myth-our "overdeveloped" brain. *BioScience* 48:109-116.

475 McKinney, M.L. 1999. Heterochrony: beyond words. *Paleobiology* 25:149-153.

476 McKinney, M.L. 2002. Brain evolution by stretching the global mitotic clock of development. In N. Minugh-Purvis and K. McNamara, K. (eds.), *Human Evolution Through Developmental Change*, pg. 173-188. Johns Hopkins University Press, Baltimore.

477 McKinney, M.L. and Gittleman, J.L. 1995. Ontogeny and phylogeny: tinkering with covariation in life history, morphology, and behavior. In K.J. McNamara (ed.), *Evolutionary Change and Heterochrony*, pg. 21-47. J. Wiley and Sons, Chichester.

478 McKinney, M.L. and McNamara, K.J. 1991. *Heterochrony – The Evolution of Ontogeny.* Plenum Press, New York.

479 McKinnon, J.S. and Rundle, H.D. 2002. Speciation in nature: the threespine stickleback model systems. *TRENDS in Ecology and Evolution* 17:480-488.

480 McNabb, A.F.M. and King, D.B. 1993. Thyroid hormone effects on growth, development, and metabolism. In M.P. Schreibman, C.G. Scanes and P.K.T. Pang (eds.), *The Endocrinology of Growth, Development,and Metabolism of Vertebrates,* pg. 393-417. Academic Press, New York.

481 McNamara, J.M. and Houston, A.I. 1996. State-dependent life histories. *Nature* 380:215-221.

482 McNamara, K.J. 1995. *Evolutionary Change and Heterochrony.* J. Wiley & Sons, Chichester.

483 McNamara, K.J. 1997. *Shapes of Time.* Johns Hopkins University Press, Baltimore.

484 McNamara, K.J. 1999. Embryos and evolution. *New Scientist* 16 (October) *Inside Science* 4 pg. pullout.

485 MacHugh, D.E. and Bradley, D.G. 2001. Livestock genetic origins: Goats buck the trend. *Proceedings of the National Academy of Sciences USA* 98:5382-5384.

486 Madson, T. and Shine, R. 2000. Silver spoons and snake body sizes: prey availability early in life influences long-term growth rates of free-ranging pythons. *Journal of Animal Ecology* 69:952-958.

487 Mallet, J. 1995. A species definition for the modern synthesis. *TRENDS in Ecology and Evolution* 10:294-299.

488 Manna, P.R., Tena-Sempere, M. and Huhtaneimi, I.T. 1999. Molecular mechanisms of thyroid hormone-stimulated steroidogenesis in mouse Leydig tumor cells. *Journal of Biological Chemistry* 272(9):5909-5918.

489 Mannen, H., Tsuji, S., Loftus R.T. and Bradley, D.G. 1998. Mitochondrial DNA variation and evolution of Japanese black cattle (*Bos taurus*). *Genetics* 150:1169-1175.

490 Mantzoros, C.S., Ozata, M., Negrao, A.B., Suchard, M.A., Ziotopoulou, M., et al. 2001. Synchronicity of frequently sampled thyrotropin (TSH) and leptin concentrations in healthy adults and leptin-deficient subjects: evidence for possible partial TSH regulation by leptin in humans. *Journal of Clinical Endocrinology & Metabolism* 86(7):3284-3291.

491 Manzano, J., Morte, B., Scanlan, T.S. and Bernal, J. 2003. Differential effects of triiodothyronine and the thyroid hormone receptor β-specific agonist GC-1 on thyroid hormone target genes in the brain. *Endocrinology* 144(12):5480-5487.

492 Marchetti, M.P. and Nevitt, G.A. 2003. Effects of hatchery rearing on brain structures of rainbow trout, *Oncorhynchus mykiss. Environmental Biology of Fishes* 66:9-14.

493 Marino, L. 2005. Commentary: Big brains do matter in new environments. *Proceedings of the National Academy of Sciences USA* 102(15):5306-5307.

494 Marino, L., McShea, D. and Uhen, M.D. 2004. Origin and evolution of large brains in toothed whales. *The Anatomical Record Part A* 281A:1247-1255.

495 Marlow, N., Wolke, D., Bracewell, M.A. and Samara, M. 2005. Neurologic and developmental disability at six years of age after extremely preterm birth. *New England Journal of Medicine* 352:9-19.

496 Martin, B.C., Warram, J.H., Krolewski, A.S., Bergman, R.N., Soeldner, J.S., et al. 1992. Role of glucose and insulin resistance in development of type 2 diabetes mellitus: results of a 25-year follow-up study. *Lancet* 340:925-929.

497 Martin, R.D., Willner, L.A. and Dettling, A. 1994. The evolution of sexual size dimorphism in primates. In R.V. Short and E. Balaban (eds.), *The Differences Between the Sexes,* pg. 159-200. University Press, Cambridge.

498 Martinez, R. and Gomes, F.C.A. 2005. Proliferation of cerebellar neurons induced by astrocytes treated with thyroid hormone is mediated by a cooperation between cell contact and soluble factors and involves the epidermal growth factor-protein kinase A pathway. *Journal of Neuroscience Research* 80:341-349.

499 Mason, I.L. 1984a. *Evolution of Domesticated Animals*, Longman Co., London.

500 Mason, I.L. 1984b. Goat. In I.L. Mason (ed.), *Evolution of Domesticated Animals*, pg. 85-99. Longman Co., London.

501 Mason, I.L. 1984c. Camels. In I.L. Mason (ed.), *Evolution of Domesticated Animals*, pg. 106-115. Longman Co., London.

502 Mathew, S.J., Coplan, J.D., Goetz, R.R., Feder, A., Greenwald, S., et al. 2003. Differentiating depressed adolescent 24h cortisol secretion in light of their adult clinical outcome. *Neuropsycholpharmacology* 28:1336-1343.

503 Matochik, J.A., London, E.D., Yildiz, B.O., Ozata, M., Caglayan, S., et al. 2005. Effect of leptin replacement on brain structure in genetically leptin-deficient adults. *Journal of Clinical Endocrinology & Metabolism* 90(5):2851-2854.

504 Mayr, E. 1942. *Systematics and the Origin of Species.* Columbia University Press, New York.

505 Mayr, E. 1963. *Animal Species and Evolution.* Harvard University Press, Cambridge.

506 Mayr, E. 1982. *The Growth of Biological Thought: Diversity, Evolution and Inheritance*. The Belknap Press of University of Harvard Press, Cambridge.

507 Mayr, E. 1988. *Towards a New Philosophy of Biology: Observations of an Evolutionist*. Harvard University Press, Cambridge.

508 Mayr, E. 1991. *One Long Argument: Charles Darwin and the Genesis of Modern Evolutionary Thought*. Harvard University Press, Cambridge, Mass.

509 Mayr, E. 1994. Recapitulation reinterpreted: the somatic program. *Quarterly Review of Biology* 69:223-232.

510 Mayr, E. 1996. The modern evolutionary theory. *Journal of Mammalogy* 77:1-7.

511 Mayr, E. 2001. *What Evolution Is*. Basic Book, New York.

512 Mech, L.D. 1970. *The Wolf.* Natural History Press, Garden City.

513 Mech, L.D. 1999. Alpha status, dominance, and division of labor in wolf packs. *Canadian Journal of Zoology* 77:1196-1203.

514 Meddle, S.L. and Follett, B.K. 1997. Photoperiodically driven changes in fos expression within the basal tuberal hypothalamus and median eminence of Japanese quail.*Journal of Neuroscience* 17(22):8909-8918.

515 Mekel-Bobrov, N., Gilbert, S.L., Evans, P.D., Vallender, E.J., Anderson, J.R., et al. 2005. Ongoing adaptive evolution of *ASPM,* a brain size determinant in *Homo sapiens*. *Science* 309:1720-1722.

516 Meldrum, D.R., Defazio, J.D., Erlik, Y., Lu, J.K., Wolfsen, A.F., et al. 1984. Pituitary hormones during the menopausal hot flash. *Obstetrics and Gynecology* 64:752-756.

517 Melinda, A. 2000. The initial domestication of goats. *Science* 287:2254-2257.

518 Mengel, R.M. 1971. A study of dog-coyote hybrids and implications concerning hybridization in Canis. *Journal of Mammalogy* 52:316-336.

519 Metcalfe, N.B. 1998. The interaction between behaviour and physiology in determining life history patterns in Atlantic salmon (*Salmo salar*). *Canadian Journal of Fisheries and Aquatic Science* 55 (Suppl. 1):93-103.

520 Meyer, A., Kocher, T.D., Basasibwaki, P. and Wilson, A.C. 1990. Monophyletic origin of Lake Victoria cichlid fishes suggested by mitochondrial DNA sequences. *Nature* 347:550-553.

521 Michalopoulou, G., Alevizaki, M., Piperingos, G., Mitsibounas, D., Mantzos, E., et al. 1998. High serum cholesterol levels in persons with "high-normal" TSH levels: should one extend the definition of subclinical hypothyroidism? *European Journal of Endocrinology* 138:141-145.

522 Miller, E.H., Ponce de Leon, A. and DeLong, R.L. 1996. Violent interspecific sexual behavior by male sea lions (Otariidae):evolutionary and phylogenetic implications. *Marine Mammal Science* 12:468-476.

523 Miller, R.A., Harper, J.M., Dysko, R.C., Durkee, S.J. and Austad, S.N. 2002. Longer life spans and delayed maturation in wild-derived mice. *Experimental Biology and Medicine* 227:500-508.

524 Miller, V.M., Clarkson, T.B., Harman, S.M., Brinton, E.A., Cedars, M., et al. 2005. Women, hormones, and clinical trials: a beginning, not an end. *Journal of Applied Physiology* 99:381-383.

525 Miller, V.M., Tindall, D.J. and Liu, P.Y. 2004. Of mice, men, and hormones. *Arteriosclerosis, Thrombosis and Vascular Biology* 24:995-997.

526 Milne, J.A., Loudon, A.S.I., Sibbald, A.M., Curlewis, J.D. and McNeilly, A.S. 1990. Effects of melatonin and a dopamine agonist and antagonis on seasonal changes in voluntary intake, reproductive activity and plasma concentrations of prolactin and tri-iodothyronine in red deer hinds. *Journal of Endocrinology* 125:241-249.

527 Minugh-Purvis, N. 2002. Heterochronic change in the neurocranium and the emergence of modern humans. In N. Minugh-Purvis and K. McNamara (eds.), *Human Evolution Through Developmental Change*, pg. 479-498. Johns Hopkins University Press, Baltimore.

528 Minugh-Purvis, N. and Crockford, S.J. 2002. Pulsatile hormone secretion, episodic growth patterning, heterochrony and punctuated equilibrium: A unifying model. Meeting of the American Association of Physical Anthropology, Buffalo.

529 Minugh-Purvis, N. and McNamara, K. 2002. *Human Evolution Through Developmental Change*, Johns Hopkins University Press, Baltimore.

530 Morell, V. 1997. The origin of dogs: running with the wolves. *Science* 276: 1647-1648.

531 Morey, D.F. 1990. Cranial allometry and the evolution of the domestic dog. Ph.D. dissertation, University of Tennessee, Knoxville.

532 Morey, D.F. 1992. Size, shape and development in the evolution of the domestic dog. *Journal of Archaeological Science* 13:119-145.

533 Morey, D.F. 1994. The early evolution of the domestic dog. *American Scientist (*July/August):336-347.

534 Morey, D.F. and Wiant, M.D. 1992. Early Holocene domestic dog burials from the North American midwest. *Current Anthropology* 33:224-229.

535 Morgan, B.A. and Tabin, C. 1994. Hox genes and growth: early and late roles in limb bud morphogenesis. *Development (Suppl.)*:181-186.

536 Morgan, M.A., Dellovade, T.L. and Pfaff, D.W. 2000. Effect of thyroid hormone administration on estrogen-induced sex behavior in female mice. *Hormones and Behavior* 37:15-22.

537 Morita, S., Matsuo, K., Tsuruta, M., Leng, S., Yamashita, S., et al. 1990. Stimulatory effects of retinoic acid and triiodothyronine in rat pituitary cells. *Journal of Endocrinology* 125:251-256.

538 Morreale de Escobar, G., Pastor, R., Obregon, M.J. and Escobar del Rey, F. 1985. Effects of maternal hypothyroidism on the weight and thyroid hormone content of rat embryonic tissues, before and after onset of fetal thyroid function. *Endocrinology* 117:1890-1900.

539 Morris, R. and Morris, D. 1966. *Men and Pandas*. McGraw-Hill, New York.

540 Morton, A.J., Wood, N.I., Hastings, M.H., Hurelbrink, C., Barker, R.A., et al. 2005. Disintegration of the sleep-wake cycle and circadian timing in Huntington's disease. *Journal of Neuroscience* 25(1):157-163.

541 Morwood M.J., R.P. Soejono, R.G. Roberts, T. Suktina, C.S.M. Turney, K.E., et al. 2004. Archaeology and age of a new hominin from Flores in eastern Indonesia. *Nature* 431:1087-1091.

542 Morwood, M.J., Brown, P. Jatmiko, Sutikna, T., Wahyu Saptomo, E., et al. 2005. Further evidence for small-bodied hominins from the late Pleistocene of Flores, Indonesia. *Nature* 437:1012-1017.

543 Moum, T., Arnason, U. and Árnason, E. 2002. Mitochondrial DNA sequence evolution and phylogeny of the Atlantic Alcidae, including the extinct great auk (*Pinguinus impennis*). *Molecular Biology and Evolution* 19:1434-1439.

544 Moyle, P.B. 1969. Comparative behavior of young brook trout of domestic and wild origin. *Progressive Fish Culturist* 31:51-59.

545 Müller, G.B. 1990. Developmental mechanisms at the origin of morphological novelty: A side-effect hypothesis. In. M.H. Nitecki (ed.), *Evolutionary Innovations,* pg. 99-130. University of Chicago Press, Chicago.

546 Murakami, N., Hayafuji, C. and Takahashi, K. 1984. Thyroid hormone maintains normal circadian rhythm of blood coricosterone levels in the rat by restoring the release and synthesis of ACTH after thyroidectomy. *Acta Endocrinologia* 107(4):519-524.

547 Musil, R. 2000. Domestication of wolves in central European Magdalenian sites. In Crockford, S.J. (ed.) *Dogs Through Time: An Archaeological Perspective*, pg. 271-286. British Archaeological Reports S889, Oxford.

548 Ng, L., Goodyear, R.J., Woods, C.A., Schneider, M.J., Diamond, E., et al. 2004. Hearing loss and retarded cochlear development in mice lacking type 2 iodothyronine deiodinase. *Proceedings of the National Academy of Sciences USA* 101(10):3474-3479.

549 Nielsen, R. 2005. Disclosure of variation. *Nature* 434:288-289.

550 Nijhout, H.F. 1999. Control mechanisms of polyphenic development in insects. *BioScience* 49(3):181-192.

551 Noren, J.G. and Alm, J. 1983. Congenital hypothyroidism and changes in the enamel of deciduous teeth. *Acta Paediatrica Scandinavica* 72(4):485-489.

552 Nova 2004. *Dogs, Dogs, and More Dogs.* A documentary available on tape and CD [includes interviews with Ray Coppinger on domestication and rare original footage of the Russian fox experiment]. WGBH Boston Video www.shop.wgbh.org

553 Novacek, M.J. 1992. Mammalian phylogeny: shaking the tree. *Nature* 356:121-125.

554 Novacek, M.J. 1993. Reflections on higher mammalian phylogenetics. *Journal of Mammalian Evolution* 1:3-30.

555 Nowak, R.M. 1979. *North American Quaternary Canis.* Museum of Natural History Monograph Number 6, University of Kansas, Lawrence.

556 Nunez, E.A., Becker, D.V., Furth, E.D., Belshaw, B.E. and Scott, J.P. 1970. Breed differences and similarities in thyroid function in purebred dogs. *American Journal of Physiology* 218:1337-1341.

557 Nunez, J. 1985. Microtubules and brain development: the effects of thyroid hormone. *Neurochemistry International* 7:959-968.

558 O'Connor, T.M., O'Halloran, D.J. and Shanahan, F. 2000. The stress response and the hypothalamic-pituitary-adrenal axis: from molecule to melancholia.*Quarterly Journal of Medicine* 93:323-333.

559 O'Connor, T.P. 1997. Working at relationships: another look at animal domestication. *Antiquity* 71:149-156.

560 Ogilvy-Stuart, A.L. and Shalet, S.M. 1992. Growth hormone and puberty. *Journal of Endocrinology* 135:405-406.

561 Olivier, R.C.D. 1984. Asian elephant. In I.L. Mason (ed.), *Evolution of Domesticated Animals*, pg. 185-192. Longman Co., London.

562 Olsen, S.J. 1985. *Origins of the Domestic Dog: The Fossil Record.* University of Arizona Press, Tucson.

563 Olsen, S.J. 1993. Evidence of early domestication of the water buffalo in China. In A. Clasen, S. Payne and H.P. Uerpmann (eds.), *Skeletons in her cupboard: festschrift for Juliet Clutton-Brock*, pg. 151-156. Monograph 34, Oxbow Books, Oxford.

564 Olson, T.A. 1981. The genetic basis for piebald patterns in cattle. *Journal of Heredity* 72:113-116.

565 O'Malley, B.P., Cook, N., Richardson, A, Barnett, D.B. and Rosenthal, F.D. 1984. Circulating catecholamine, thyrotropin, thyroid hormone and prolactin responses of normal subjects to acute cold exposure. *Clinical Endocrinology* 21(3):285-291.

566 Oppenheimer, J.H. and Schwartz, H.L. 1997. Molecular basis of thyroid hormone dependent brain development. *Endocrine Review* 18:462-475.

567 O'Reilly, D.S., 2000. Thyroid function tests---time for a reassessment. *British Medical Journal* 320:1332-1334.

568 Osadchuk, L.V. 1997. Effects of domestication on the adrenal cortisol production in silver foxes during embryonic development. In L.N. Trut and L.V. Osadchuk (eds.), *Evolutionary-genetic and Genetic-physiological Aspects of Fur Animal Domestication: A Collection of Reports,* pg. 73-81. IFASA/Scientifur, Oslo.

569 O'Steen, S. and Janzen, F.J. 1999. Embryonic temperature effects metabolic compensation and thyroid hormones in hatchling snapping turtles. *Physiology, Biochemistry and Zoology* 72(5):520-533.

570 Othman, S., Philips, D.I.W., Parkes, A.B., Richards, C., Jr., Harris, B., et al. 1990. A long term follow up of postpartum thyroiditis. *Clinical Endocrinology* 32:559-564.

571 Otte, D. and Endler, J.A. (eds.) 1989. *Speciation and Its Consequences.* Sinauer, Sunderland.

572 Owen, R.B., Crossley, R. and Johnson, T.C. 1990. Major low levels of Lake Malawi and their implications for speciation rates in cichlid fishes. *Proceedings of the Royal Society of London B* 240:519-53.

573 Ozcan, O., Cakir, E., Yaman, H., Akgul, E.O., Erturk, K., et al. 2005. The effects of thyroxine replacement on the levels of serum asymmetric dimethylarginine (ADMA) and other biochemical cardiovascular risk markers in patients with subclinical hypothyroidism. *Clinical Endocrinology* 63(2):203-206.

574 Palkovacs E.P. 2003. Explaining adaptive shifts in body size on islands: a life history approach. *Oikos* 103:37-44.

575 Palombo, M.R. 2001a. Paedomorphic features and allometric growth in the skull of *Elephas falconeri* from Spinagallo (Middle Pleistocene, Sicily). In G. Cavarretta, P. Gioia, P. Mussi, and M. Palombo (eds.), *The World of Elephants: Proceedings of the First International Congress,* pg. 492-496. CNR, Rome.

576 Palombo, M.R. 2001b. Endemic elephants of the Mediterranean Islands: knowledge, problems and perspectives. In G. Cavarretta, P. Gioia, P. Mussi, and M. Palombo (eds.), *The World of Elephants: Proceedings of the First International Congress,* pg. 486-491. CNR, Rome.

577 Palombo, M.R. and Ferretti, M.P. 2005. Elephant fossil record from Italy: knowledge, problems and perspectives. *Quaternary International* 126:107-136.

578 Park, S.K., Solomon, D. and Vartanian, T. 2001. Growth factor control of CNS myelination. *Developmental Neuroscience* 23:327-37.

579 Parker, S.T. 2002. Evolutionary relationships between molar eruption and cognitive development in anthropoid apes. In N. Minugh-Purvis and K. McNamara (eds.), *Human Evolution Through Developmental Change,* pg. 305-316. Johns Hopkins University Press, Baltimore.

580 Parker, S.T. and McKinney, M. 1999. *Origins of Intelligence: The Evolution of Cognitive Development in Monkeys, Apes, and Humans.* Johns Hopkins University Press, Baltimore.

581 Pavan, W.J., Mac, S., Cheng, M. and Tilghman, S.M. 1995. Quantitative trait loci that modify the severity of spotting in piebald mice. *Genome Research* 5:29-41.

582 Pavan, W.J. and Tilghman, S.M. 1994. Piebald lethal (sl) acts early to disrupt the development of neural crest-derived melanocytes. *Proceedings of the National Academy of Sciences USA* 91:7159-7163.

583 Paxinos, E.E., James, H.F., Olson, S.L, Sorenson, M.D., Jackson, J., et al. 2002. mtDNA from fossils reveals a radiation of Hawaiian geese recently derived from the Canada goose (*Branta canadensis*). *Proceedings of the National Academy of Sciences USA* 99:1399-1404.

584 Pedraza, P., Calvo, R., Obregon, M.J., Asuncion, M., Escobar del Rey, F., et al. 1996. Displacement of T4 from transthyretin by the synthetic flavonoid EMD 21388 results in increased production of T3 from T4 in rat dams and fetuses. *Endocrinology* 137:4902-4914.

585 Peeters, R.P., Wouters, P.J., van Toor, H., Kaptein, E., Visser, T.J., et al. in press. Serum rT3 and T3/rT3 are prognostic markers in critically ill patients and are associated with post-mortem tissue deiodinase activities. *Journal of Clinical Endocrinology and Metabolism*

586 Peeters, R.P., van der Geyten, S., Wouters, P.J., Darras, V.M. and van Toor, H., et al. in press. Tissue thyroid hormone levels in critical illness. *Journal of Clinical Endocrinology and Metabolism.*

587 Pelly, D.F. 2001. *Sacred Hunt: A Portrait of the Relationship Between Seals and Inuit.* University of Washington Press, Seattle.

588 Pennisi, E. 2002. A shaggy dog history (editorial). *Science* 298:1540-1542.

589 Pennisi, E. 2004. Bonemaking protein shapes beaks of Darwin's finches. *Science* 305:1383.

590 Penny, D. and Phillips, M.J. 2004. The rise of birds and mammals: are micro-evolutionary processes sufficient for macroevolution? *TRENDS in Ecology and Evolution* 19(10):516-522.

591 Pfaff, D.W., Vasudevan, N., Kia, H.K., Zhu, Y-S., Chan, J., et al. 2000. Estrogens, brain and behavior: studies in fundamental neurobiology and observations related to women's health. *Journal of Steroid Biochemistry and Molecular Biology* 74(5):365-373.

592 Philp, H.A. 2003. Hot flashes – a review of the literature on alternative and complementary treatment approaches. *Alternative Medicine Review* 8(3):284-302.

593 Pickard, M.R., Sinha, A.K., Ogilvie, L. and Ekins, R.P. 1993. The influence of the maternal thyroid hormone environment during pregnancy on the ontogenesis of brain and placental ornithine decarboxylase activity in the rat. *Journal of Endocrinology* 139:205-212.

594 Pincus, S.M. 2000. Orderliness of hormone release. In D.J. Chadwick and J.A. Goode, (eds.), *Mechanisms and Biological Significance of Pulsatile Hormone Secretion*, pg. 82-104. J. Wiley and Sons, Chichester.

595 Piosik, P.A., van Groenigen, M., van Doorn, J., Baas, F. and de Vijlder, J.J.M. 1997. Effects of maternal thryoid status on thyroid hormones and growth in congenitally hypothyroid goat fetuses during the second half of gestation. *Endocrinology* 138:5-11.

596 Pirinen, S. 1995. Endocrine regulation of craniofacial growth. *Acta Odontologia Scandinavica* 53(3):179-185.

597 Polziehn, R.O., Strobeck, C., Sheraton, J. and Beech, R. 1995. Bovine mtDNA discovered in North American bison populations. *Conservation Biology* 9:1638-1643.

598 Porterfield, S.P. 2000. Thyroidal dysfunction and environmental chemicals – potential impact on brain development. *Environmental Health Perspectives* 108 (Suppl. 3):433-438.

599 Potts, R. 1998. Environmental hypothesis of hominin evolution. *Yearbook of Physical Anthropology* 41:93-136.

600 Powell, K. 2003. Eat your veg. *Nature* 426:378-379.

601 Power, D.M, Llewellyn, L., Faustino, M., Nowell, M.A., Björnsson, B.Th., et al. 2001. Thyroid hormones in growth and development of fish. *Comparative Biochemistryand Physiology Part C* 130:447-459.

602 Prasolova, L.A., Trut, L.N. and Vsevolodov, E.B. 1997. Morphology of mottling hairs in domesticated silver foxes (*Vulpes vulpes*) and relation between the expression of the star and the mottling mutation. In L.N. Trut and L.V. Osadchuk (eds.), *Evolutionary-genetic and Genetic-physiological Aspects of Fur Animal Domestication: A Collection of Reports,* pg.31-40. IFASA/Scientifur, Oslo.

603 Prestrud, P. and Nilssen, K. 1995. Growth, size and sexual dimorphism in arctic foxes. *Journal of Mammalogy* 76:522-530.

604 Price, E.O. 1984. Behavioral aspects of animal domestication. *Quarterly Revue of Biology* 59:1-32.

605 Pringle, H. 1998. Reading the signs of ancient animal domestication. *Science* 282:1448.

606 Prinz, P. 2004. Sleep, appetite and obesity: what is the link? *Public Library of Science Medicine (www.plosmedicine.org)* 1(3):e61.

607 Provine, W.B. 1986. *Sewell Wright and Evolutionary Biology*. University of Chicago Press, Chicago.

608 Provine, W.B. 2005. Ernst Mayr, a retrospective. *TRENDS in Ecology and Evolution* 20(8):411-413.

609 Quammen, D. 1996. *The Song of the Dodo: Island Biogeography in an Age of Extinctions*. New York: Scribner.

610 Quinn, T.P., Graynoth, E., Wood, C.C. and Foote, C.J. 1998. Genotypic and phenotypic divergence of sockeye salmon in New Zealand from their ancestral British Columbia populations. *Transations of the American Fisheries Society* 127:517-534.

611 Raff, R.A. 1996. *The Shape of Life: Genes, Development, and the Evolution of Animal Form.* University of Chicago Press, Chicago.

612 Raff, R.A., Anstrom, J.A., Chin, J.E., Field, K.G., Ghiselin, M.T., et al. 1987. Molecular and developmental correlates of macroevolution. In R.A. Raff and E.C. Raff (eds.), *Development as an Evolutionary Process*, pg. 109-138. A.R. Liss Inc., New York.

613 Raia, P., Barbera, C. and Conte, M. 2003. The fast life of a dwarfed giant. *Evolutionary Ecology* 17:293-312.

614 Ramirez Rozzi, F. 2002. Enamel microstructure in hominids. In: N. Minugh-Purvis and K. McNamara (eds.), *Human Evolution Through Developmental Change*, pg. 319-348. Johns Hopkins University Press, Baltimore.

615 Rand-Weaver, M., Kawauchi, H. and Ono, M. 1993. Evolution of the structure of the growth hormone and prolactin family. In M.P. Schreibman, C.G. Scanes and P.K.T. Pang (eds.), *The Endocrinology of Growth, Development, and Metabolism of Vertebrates,* pg. 13-42. Academic Press, New York.

616 Raser, J.M. and O'Shea, E.K. 2005. Noise in gene expression: origins, consequences, and control. *Science* 309:2010-2013.

617 Reed, C.A. 1984. The beginnings of animal domestication. In I.L. Mason (ed.), *Evolution of Domesticated Animals*, pg. 1-6. Longman Co., London.

618 Reed, K.E. 1997. Early hominid evolution and ecological change through the African Plio-Pleistocene. *Journal of Human Evolution* 32:289-322.

619 Reeves, R.R, Stewart, B.S., Clapham, P.J. and Powell, J.A. 2002. *National Audobon Society's Guide to Marine Mammals of the World.* Alfred A. Knopf, New York.

620 Refetoff, S., Weiss, R.E. and Usak, S.J., 1993. The syndromes of resistance to thyroid hormones. *Endocrine Review* 14:348-399.

621 Reilly, S.M. 1994. The ecological morphology of metamorphosis: heterochrony and the evolution of feeding mechanisms in salamanders. In P.C. Wainwright and S.M Reilly (eds.), *Ecological Morphology: Integrative Organismal Biology,* pg. 319-338. University of Chicago Press, Chicago.

622 Reilly, S.M., Wiley, E.O. and Meinhardt, D.J. 1997. An integrative approach to heterochrony: the distinction between interspecific and intraspecific phenomena. *Biological Journal of the Linnaean Society* 60:119-143.

623 Rennie, J. 2002. Fifteen answers to creationist nonsense. *Scientific American* (July):78-85.

624 Reppert, S.M. and Schwartz, W.J. 1984. Functional activity of the suprachiasmatic nuclei in the fetal primate. *Journal of Neuroscience Letters* 46(2):145-149.

625 Reppert, S.M. and Schwartz, W.J. 1986. Maternal suprachiasmatic nuclei are necessary for maternal coordination of the developing circadian systerm. *Journal of Neuroscience* 6:2724-2729.

626 Reppert, S.M. and Weaver, D.R. 2002. Coordination of circadian timing in mammals. *Nature* 418:935-941.

627 Rice, W.R. and Hostert, E.E. 1993. Laboratory experiments on speciation: what have we learned in 40 years? *Evolution* 47:1637-1653.

628 Richardson, G. and Tate, B. 2000. Hormonal and pharmacological manipulation of the circadian clock: recent developments and future strategies. *Sleep* 23 (Suppl. 3):S77-85.

629 Richardson, M.K. 1995. Heterochrony and the phylotypic period. *Developmental Biology* 172:412-421.

630 Richardson, S.J., Monk, J.A., Shepherdly, C.A., Ebbesson, L.O.E., Sin, F., et al. 2005. Developmentally regulated thyroid hormone distributor proteins in marsupials, a reptile, and fish. *American Journal of Physiology: Regulatory and Integrative Comparative Physiology* 288:R1264-R1272.

631 Ricklefs, R.E. and Marks, H.L. 1985. Anatomical response to selection for four-week body mass in Japanese quail. *Auk* 102:323-333.

632 Riddle R.D. and Tabin, C.J. 1999. How limbs develop. *Scientific American* (February):74-70.

633 Rindos, D. 1984. *The Origins of Agriculture: An Evolutionary Perspective.* Academic Press, Orlando.

634 Rivkees, S.A., Fox, C.A., Jacobson, C.D. and Reppert, S.M. 1988. Anatomic and functional development of the suprachiasmatic nuclei in the gray short-tailed opossum. *Journal of Neuroscience* 8:4269-4276.

635 Robbins, L.S., Nadeau, J.H., Johnson, K.R., Kelly, M.A., Roseli-Rehfuss, L., et al. 1993. Pigmentation phenol-types of variant extension locus alleles result from point mutations that alter MSH receptor function. *Cell* 72:827-834.

636 Roberts, C.G.P. and Ladenson, P.W. 2004. Hypothyroidism. *Lancet* 363:793-803.

637 Robins, A.H. 1991. *Biological Perspectives on Human Pigmentation.* Cambridge University Press, Cambridge.

638 Robinson, I.C.F. 2000. Control of growth hormone (GH) release by GH secretagogues. In D.J. Chadwick and J.A. Goode (eds.), *Mechanisms and Biological Significance of Pulsatile Hormone Secretion*, pg. 206-224. J. Wiley and Sons, Chichester.

639 Robinson, R. 1984a. Cat. In I.L. Mason (ed.), *Evolution of Domesticated Animals*, pg. 217-224. Longman Co., London.

640 Robinson, R. 1984b. Syrian hamster. In I.L. Mason (ed.), *Evolution of Domesticated Animals*, pg. 263-266. Longman Co., London.

641 Robinson, R. 1984c. Norway rat. In I.L. Mason (ed.), *Evolution of Domesticated Animals*, pg. 284-290. Longman Co., London.

642 Roff D.A. 2000. Trade-offs between growth and reproduction: an analysis of the quantitative genetic evidence. *Journal of Evolutionary Biology* 13:434-445.

643 Rolland, N. and Crockford, S. 2005. Late Pleistocene Dwarf *Stegodon* from Flores, Indonesia? *Antiquity* 79(304), online. http://antiquity.ac.uk/ProjGall/rolland/index.html

644 Rol'nik, V.V. 1970. *Bird Embryology*. Israel Program for Scientific Translations, Jerusalem.

645 Rook, L. 1999. *Oreopithecus* was a bipedal ape after all: evidence from the iliac cancellous architecture. *Proceedings of the National Academy of Sciences USA* 96:8795-8799.

646 Root-Bernstein, R.S. 1989. *Discovering: Inventing and Solving Problems at the Frontiers of Scientific Knowledge.* Harvard University Press, Cambridge.

647 Rose, C.S. 2005. Integrating ecology and developmental biology to explain the timing of frog metamorphosis. *TRENDS in Ecology and Evolution* 20(3): 129-135.

648 Rosenbaum, H.C., Brownell Jr., R.L., Brown, M.W., Schaeff, C., Portway, V., et al. 2000. World-wide genetic differentiation of *Eubalaena:* questioning the number of right whale species. *Molecular Ecology* 9:1793-1802.

649 Ross Cockrill, W. 1984. Water buffalo. In I.L. Mason (ed.), *Evolution of Domesticated Animals,* pg. 52-62. Longman Co., London.

650 Roth, V.L. 1996. Pleistocene dwarf elephants of the California islands. In J. Shoshani and P. Tassy (eds.), *The Proboscidea: The Evolution and Palaeoecology of Elephants and their Relatives,* pg 249-253. Oxford: Oxford University Press.

651 Rovet, J.F. 2004. Neurodevelopmental consequences of maternal hypothyroidism during pregnancy. 76th Annual meeting of the American Thyroid Association, Vancouver. Abstract 88. *Thyroid* 14(9):710.

652 Roy, M.S., Geffen, E., Smith, D., Ostrander, E.A. and Wayne, R.K. 1995. Patterns of differentiation and hybridization in North American wolf-like canids revealed by analysis of microsatellite loci. *Molecular Biology and Evolution* 11:553-570.

653 Rũber, L. and Adams, D.C. 2001. Evolutionary convergence of body shape and trophic morphology in cichlids from Lake Tanganyika. *Journal of Evolutionary Biology* 14:325-332.

654 Russell-Aulet, M., Dimaraki, E.V., Jaffe, C.A., DeMott-Friberg, R. and Barkan, A.L. 2001. Aging-related growth hormone (GH) decrease is a selective hypothalamic GH-releasing hormone pulse amplitude mediated phenomenon. *Journal of Gerontology A Biological Sciences and Medical Sciences* 56(2):M124-M129.

655 Ryder, M.L. 1984. Sheep. In I.L. Mason (ed.), *Evolution of Domesticated Animals,* pg. 63-84. Longman Co., London.

656 St. Aubin, D.J., Ridgway, S.H., Wells, R.W. and Rhinehardt, H. 1996. Dolphin thyroid and adrenal hormones: circulating levels in wild and semi-domesticated *Tursiops truncatus*, and influences of sex, age, and season. *Marine Mammal Science* 12:1-13.

657 Sablin, M.V. and Khlopachev, G.A. 2002. The earliest ice age dogs: evidence from Eliseevichi I. *Current Anthropology* 43(5):795-799.

658 Safer, J.D., Crawford, T.M. and Holick, M.F. 2005. Topical thyroid hormone accelerates wound healing in mice. *Endocrinology* 146:4425-4430.

659 Saper, C.B., Lu, J., Chou, T.C. and Gooley, J. 2005. The hypothalamic integrator for circadian rhythms. *TRENDS in Neuroscience* 28(3):152-157.

660 Satinoff, E., Li, H., Tcheng, T.K., Liu, C., McArthur, A.J., et al. 1993. Do the suprachiasmatic nuclei oscillate in old rats as they do in young ones? *American Journal of Physiology* 265(5 Pt 2):R1216-R1222.

661 Satoh, Y. and Sairenji, T. 1997. Regulation of the expression of epidermal growth factor receptor mRNA with thyroid hormone L-3,5,3'-triiodothyronine in rat hepatoma cells. *Yonaga Acta Medica* 40:133-136.

662 Scanlon, M.F. and Toft, A.D. 2000. Regulation of thyrotropin secretion. In L.D. Braverman and R.D. Utiger (eds.), *Werner and Ingbar's The Thyroid, Eighth Edition,* pg. 235-253. Lippincott, Philadelphia.

663 Schaefer, F. 2000. Pulsatile parathyroid hormone secretion in health and disease. In D.J. Chadwick and J.A. Goode, (eds.), *Mechanisms and Biological Significance of Pulsatile Hormone Secretion,* pg. 224-243. J. Wiley and Sons, Chichester.

664 Scheer, F.A.J.L., Ter Horst, G.J., van der Vliet, J. and Buijs, R.M. 2001. Physiological and anatomic evidence for regulation of the heart by suprachiasmatic nucleus in rats. *American Journal of Physiology Heart and Circulatory Physiology* 280:H1391-H1399.

665 Schew, W.A., McNabb, F.M.A. and Scanes, C.G. 1996. Comparison of the ontogenesis of thyroid hormones, growth hormone, and insulin-like growth factor-I in ad Libitum and food-restricted (altricial) European starlings and (precocial) Japanese quail. *General and Comparative Endocrinology* 101:304-316.

666 Schilthuis, J.G., Gann, A.A.F. and Brockes, J.P. 1995. Chimeric retinoic acid/ thyroid hormone receptors implicate RAR-à1 as mediating growth inhibition by retinoic acid. *EMBO* 12:3459-3466.

667 Schmutz, S.M., Berryere, T.G., Ellinwood, N.M., Kerns, J.A. and Barsh, G.S. 2003. *MC1R* studies in dogs with melanistic mask or brindle patterns. *Journal of Heredity* 94:69-73.

668 Schreiber, A.M., Das, B., Huang, H., Marsh-Armstrong, N. and Brown, D.D. 2001. Diverse developmental programs of *Xenopus laevis* metamorphosis are inhibited by a dominant negative thyroid hormone receptor. *Proceeding of the National Academy of Sciences USA* 98(19):10739-10744.

669 Schreibman, M.P., Scanes, C.G. and Pang, P.K.T. 1993. *The Endocrinology of Growth, Development, and Metabolism in Vertebrates.* Academic Press, San Diego.

670 Schwartz, J.H. 1999. *Sudden Origins: Fossils, Genes, and the Emergence of Species.* J. Wiley and Sons, New York.

671 Scott, A.J., Harvey, C.B., O'Shea, P.J., Stevens, D.A., Samarut, J., et al. 2002. Thyroid hormone activates fibroblast growth factor receptor-1 in bone. *Endocrine Abstracts* 3:OC19.

672 Scott, J.P., Fuller, J.L. and King, J.A. 1959. The inheritance of annual breeding cycles in hybrid basenji-cocker spaniel dogs. *Journal of Heredity* 50:254-261.

673 Scribner, K. 1993. Hybrid zone dynamics are influenced by genotype-specific variation in life-history traits: experimental evidence form hybridizing *Gambusia* species. *Evolution* 47:632-646.

674 Searle, A.G. 1968. *Comparative Genetics of Coat Colour in Mammals.* Logos Press Ltd., London.

675 Sedinger, J.S. 1986. Growth and development of Canada goose goslings. *Condor* 88:169-180.

676 Segal, J. and Ingbar, S.H. 1989. Evidence that an increase in cytoplasmic calcium is an initiating event in certain plasma membrane-mediated responses to 3,5,3'-triiodothyronine in rat thymocytes. *Endocrinology* 124:1949-1955.

677 Shanafelt, T.D., Barton, D.L., Adjei, A.A. and Loprinzi, C.L. 2002. Pathophysiology and treatment of hot flashes. *Mayo Clinic Proceedings* 77(11):1207-1218.

678 Shapiro, B., Sibthorpe, D., Rambaut, A., Austin, J., Wragg, G.M., et al. 2002. Flight of the dodo. *Science* 295:1683.

679 Shepherdley, C.A., Daniels, C.B., Orgeig, S., Richardson, S.J., Evans, B.K., et al. 2002. Glucocorticoids, thyroid hormones, and iodothyronine deiodinases in embryonic saltwater crocodiles. *American Journal of Physiology-Regulatory, Integrative and Comparative Physiology* 283:R1155-R1163.

680 Shi, Z.H. and Barrel, B.K. 1992. Requirement of thyroid function for the expression of seasonal reproductive and related changes in the red deer (*Cervus elaphus*) stags. *Journal of Reproduction and Fertility* 94:251-260.

681 Shibata, S. and Moore, R.Y. 1988. Development of a fetal circadian rhythm after disruption of the maternal circadian system. *Brain Research* 469:313-317.

682 Shin, D-J. and Osborne, T.F. 2003. Thyroid hormone regulation and cholesterol metabolism are connected through sterol regulatory element-binding protein-2 (SREBP-2). *Journal of Biological Chemistry* 278(36):34114-34118.

683 Shin, T., Kraemer, D., Pryor, J., Liu, L., Rugila, J., et al. 2002. A cat cloned by nuclear transplantation. *Nature* 415:859.

684 Silva, J.E. 1995. Thyroid hormone control of thermogenesis and energy balance. *Thyroid* 5(6):481-492.

685 Silva, J.E. 2000. Catecholamines and the sypathoadrenal system in thyroxicosis. In L.D. Braverman and R.D. Utiger (eds.), *Werner and Ingbar's The Thyroid, Eighth Edition,* pg. 642-651. Lippincott, Philadelphia.

686 Silva, J.E. and Larsen, P.R. 1983. Adregenic activation of triiodothyronine production in brown adipose tissue. *Nature* 305:712-713.

687 Silva, J.E. and Larsen, P.R. 1985. Potential of brown adipose tissue type II thyronine 5'-deiodinase as a local and systemic source of triiodothyronine in rats. *Journal of Clinical Investigation* 76:2296-2305.

688 Silvers, W.K. 1961. Genes and the pigment cells of mammals. *Science* 134:368-373.

689 Silvers, W.K. 1979. *The Coat Colors of Mice: A Model for Mammalian Gene Action and Interaction.* Springer-Verlag, New York.

690 Simmons, A.H. 1999. *Faunal Extinction in an Island Society: Pygmy Hippopotamus Hunters of Cyprus.* Kluwer Academic, New York.

691 Simoons, F.J. 1984. Gayal or mithan. In I.L. Mason (ed.), *Evolution of Domesticated Animals,* pg. 34-39. Longman Co., London.

692 Siracus, L.D., Washburn, L.L., Swing, D.A., Argeson, A.C., Jenkins, N.A., et al. 1995. Hypervariable yellow (Ahvy), a new murine agouti mutation: (Ahvy) displays the largest variation in coat color phenotypes of all known agouti alleles. *Journal of Heredity* 86:121-128.

693 Skibola, C.F., Curry, J.D., VandeVoort, C., Conley, A. and Smith, M.T. 2005. Brown kelp modulates endocrine hormones in female Sprague-Dawley rats and in human luteinized granulose cells. *Journal of Nutrition* 135: 296-300.

694 Skjenneberg, S. 1984. Reindeer. In I.L. Mason (ed.), *Evolution of Domesticated Animals,* pg. 128-137. Longman Co., London.

695 Slattery, J.P. and O'Brien, S.J. 1995. Molecular phylogeny of the red panda (*Ailurus fulgens*). *Journal of Heredity* 86:413-422.

696 Smallridge, R.C. and Ladenson, P.W. 2001. Hypothyroidism in pregnancy: consequences to neonatal health. *Journal of Clinical Endocrinology and Metabolism* 86:2349-2353.

697 Smith, F.A., Betancourt, J.L. and Brown, J.H. 1995. Evolution of body size in the woodrat over the past 25,000 years of climate change. *Science* 270:2012-2014.

698 Smith, J.A. 1989. Noninfectious diseases, metabolic diseases, toxicities and neoplastic diseases of South American camelids. In L.W. Johnson (ed.), *Llama Medicine, Veterinary Clinics of North America: Food Animal Practice* 5, pg. 101-143. W.B. Saunders Co., Philadelphia.

699 Smith, R.G., Betancourt, L. and Sun, Y. 2005. Molecular endocrinology and physiology of the aging central nervous system. *Endocrine Reviews* 26(2):203-250.

700 Smith, T.B. and Skúlason, S. 1996. Evolutionary significance of resource polymorphisms in fishes, amphibians, and birds. *Annual Review of Ecology and Systematics* 27:111-133.

701 Sol, D., Duncan, R.P., Blackburn, T.M., Cassey, P. and Lefebvre, L. 2005. Big brains, enhanced cognition, and response of birds to novel environments. *Proceedings of the National Academy of Sciences USA* 102(15):5460-5465.

702 Solomon, G.M. and Schettler, T. 2000. Environment and health: 6. Endocrine disruption and potential human health implications. *Canadian Medical Association Journal* 163(11):1471-1476.

703 Sondaar, P.Y. 1977. Insularity and its effect on mammal evolution. In M.K. Hecht, P.C. Goody and B.M. Hecht (eds.), *Major Patterns in Vertebrate Evolution,* pg. 671-707. Plenum, New York.

704 Song, C., Hiipakka, R.A., Kokontis, J.M. and Liao, S. 1995. Ubiquitous receptor: structures, immunocytochemical localization, and modulation of gene activation by receptors for retinoic acids and thyroid hormones. In D. Henderson, D. Philibert, A.K. Roy and G. Teutsch (eds.), *Steroid Receptors and Antihormones,* pg. 38-49. Annals of the New York Academy of Science 761, New York.

705 Sotherland, P.R. and Rahn, H. 1986. On the composition of bird eggs. *Condor* 89:48-65.

706 Southren, A.C., Olivo, J., Gordon, G.G., Vittek, J., Briner, J., et al. 1974. The conversion of androgens to estrogen in hyperthyroidism. *Journal of Clinical Endocrinology and Metabolism* 38:207-214.

707 Specker, J.L., Eales, J.G., Tagawa, M. and Tyler III, W.A. 2000. Parr-smolt transformation in Atlantic salmon: thyroid hormone deiodination in liver and brain and endocrine correlates of change in rheotactic behavior. *Canadian Journal of Zoology* 78:696-705.

708 Sponheimer, M. and Lee-Thorpe, J.A. 1999. Isotopic evidence for the diet of an early hominid, *Australopithecus africanus. Science* 283:368-369.

709 Sponheimer, M. and Lee-Thorp, J.A. 2003. Differential resource utilization by extant great apes and Australopithecines: towards solving the C4 conundrum. *Comparative Biochemistry and Physiology A*136:27-34.

710 Stagnaro-Green, A., Chen, X., Bogden, J., Davies, T.F. and Scholl, T.O. 2004. The thyroid and pregnancy: a novel risk factor for very preterm delivery. 76th Annual meeting of the American Thyroid Association, Vancouver. Abstract 84. *Thyroid* 14(9):708.

711 Staub, J. 1998. Minimal thyroid failure: effects on lipid metabolism and peripheral target tissues. *European Journal of Endocrinology* 138:137-138.

712 Stedman, H.H., Kozyak, B.W., Nelson, A., Thesier, D.M., Su, L.T., et al. 2004. Myosin gene mutation correlates with anatomical changes in the human lineage. *Nature* 428:415-418.

713 Steinhart, P. 1996. *The Company of Wolves.* Vintage Books, New York.

714 Stephanou, A. and Handwerger, S. 1995. Retinoic acid and thyroid hormone regulate placental lactogen expression in human trophoblast cells. *Endocrinology* 136:933-938.

715 Stewart, A.G., Carter, J., Parker, A. and Alloway, B.J. 2003. The illusion of environmental iodine deficiency. *Environmental Geochemistry and Health* 25:165-170.

716 Stock, J.M., Surks, M.I. and Oppenheimer, J.H. 1974. Replacement dosage of L-thyroxine in hypothyroidism: a re-evaluation. *New England Journal of Medicine* 290(10):529-533.

717 Stockard, C.R. 1941. *The Genetic and Endocrinic Basis for Differences in Form and Behavior (as elucidated by studies of contrasted pure-line dog breeds and their hybrids).* American Anatomical Memoirs 19. Wistar Institute, Philadelphia.

718 Stockigt, J.R. 2000. Serum thyrotropin and thyroid hormone measurements and assessment of thyroid hormone transport. In L.D. Braverman and R.D. Utiger (eds.), *Werner and Ingbar's The Thyroid, Eighth Edition,* pg. 376-392. Lippincott, Philadelphia.

719 Stolow, M.A. and Shi, Y-B. 1995. Xenopus sonic hedgehog as a potential morphogen during embryogenesis and thyroid hormone-dependent metamorphosis. *Nucleic Acids Research* 23(13):2555-2562.

720 Streelman, J.T. and Danley, P.D. 2003. The stages of vertebrate evolutionary radiation. *TRENDS in Ecology and Evolution* 18:126-131.

721 Stringer, C. 2002. Modern human origins: progress and prospects. *Philosophical Transactions of the Royal Society of London B* 357, 563-579.

722 Stryer, L. 1988. *Biochemistry.* W.H. Freeman and Co., New York.

723 Stuermer, I.W., Plotz, K., Leybold, A., Zinke, O., Kalberlah, O., et al. 2003. Intraspecific allometric comparison of laboratory gerbils with Mongolian gerbils trapped in the wild indicates domestication in *Meriones unguiculatus* (Milne-Edwards, 1867) (Rodentia: Gerbillinae). *Zoologischer Anzeiger* 242:249-266.

724 Struzik, E. 2003. Grizzlies on ice. *Canadian Geographic* 123(6):38-48.

725 Sullivan, P.F. 2005. The genetics of schizophrenia. *Public Library of Science Medicine (www.plosmedicine.org)* 2(7):e212.

726 Surks, M.I., Chopra, I.J., Mariash, C.N., Nicoloff, J.T. and Solomon, D.H. 1990. American Thyroid Association guidelines for use of laboratory tests in thyroid disorders. *Journal of the American Thyroid Association* 263:1529-1532.

727 Sussman, M.A. 2001. When the thyroid speaks, the heart listens. *Circulation Research* 89:557-559.

728 Sweeney, B.M. 1969. *Rhythmic Phenomena in Plants*. Academic Press, London.

729 Talbot, S.L. and Shields, G.F. 1996a. Phylogeography of brown bears (*Ursus arctos*) of Alaska and paraphyly within the Ursidae. *Molecular Phylogenetics and Evolution* 5:477-494.

730 Talbot, S.L. and Shields, G.F. 1996b. A phylogeny of the bears (Ursidae) inferred from complete sequences of three mitochondrial genes. *Molecular Phylogenetics and Evolution* 5:567-575.

731 Tautz, D. 1996. Selector genes, polymorphisms and evolution. *Science* 271: 160-161.

732 Taylor, E.B., Foote, C.J. and Wood, C.C. 1996. Molecular genetic evidence for parallel life-history evolution within a Pacific salmon (sockeye salmon and kokanee, *Oncorhynchus nerka*). *Evolution* 50(1):401-416.

733 Tchernov, E. 1993a. The impact of sedentism on animal exploitation in the southern Levant. In H. Buitenhuis and A.T. Clason (eds.), *Archaeozoology of the Near East: Proceedings of the First International Symposium on the Archaeozoology of South-western Asia and Adjacent Areas,* pg. 10-26. Universal Book Services, Leiden.

734 Tchernov, E. 1993b. From sedentism to domestication-a preliminary review for the southern Levant. In A. Clasen, S. Payne and H.P. Uerpmann (eds.), *Skeletons in Her Cupboard: Festschrift for Juliet Clutton-Brock*, pg. 189-233, Monograph 34, Oxbow Books, Oxford.

735 Tchernov, E. and Horwitz, L.K. 1991. Body size diminution under domestication: unconscious selection in primeval domesticates. *Journal of Anthropological Archaeology* 10:54-75.

736 Tchernov, E. and Valla, F.F. 1997. Two new dogs, and other Natufian dogs, from the Southern Levant. *Journal of Archaeological Science* 24:65-95.

737 Tegelstrom, H. 1987. Transfer of mitochondrial DNA from the northern red-backed vole (*Cleithrionomys rutilus*) to the bank vole (*C. glareolus). Journal of Molecular Evolution* 24:218-227.

738 Teichert, M. 1993. Size and utilization of the most important domesticated animals in Central Europe from the beginning of domestication until the late Middle Ages. In A. Clasen, S. Payne and H.P. Uerpmann (eds.), *Skeletons in her Cupboard: Festschrift for Juliet Clutton-Brock,* pg. 235-238. Monograph 34, Oxbow Books, Oxford.

739 Templeton, A.R. 1989. The meaning of species and speciation: a genetic perspective. In D. Otte and J.A. Endler (eds.), *Speciation and Its Consequences,* pg. 3-27. Sinauer, Sunderland.

740 Terasawa, E. 2001. Luteinizing hormone-releasing hormone (LHRH) neurons: mechanism of pulsatile LHRH release. *Vitamins and Hormones* 63:91-129.

741 Thayer, K. and Houlihan, J. 2002. Perfluorinated chemicals: justification for inclusion of this chemical class in the National Report on Human Exposure to Environmental Chemicals. *Nomination Report by the Environmental Working Group* (Washington, DC) to the US Center for Disease Control (Atlanta, GA).

742 Thewissen, J.G.M. 1998. *The Emergence of Whales: Evolutionary Patterns in the Origin of Cetacea.* Plenum Press, New York.

743 Thommes, R.C. and Woods, J.E. 1993. Endocrine regulation of the growth/development of warm-blooded vertebrate embryo/fetuses. In M.P. Schreibman, C.G. Scanes and P.K.T. Pang (eds.), *The Endocrinology of Growth, Development, and Metabolism of Vertebrates,* pg. 495-518. Academic Press, New York.

744 Thompson, Jr., D.L., Rahmanian, M.S., DePew, C.L., Burleigh, D.W., DeSouza, C.J., et al. 1992. Growth hormone in mares and stallions: pulsatile secretion, response to growth hormone-releasing hormone, and effects of exercise, sexual stimulation, and pharmacological agents. *Journal of Animal Science* 70:1201-1207.

745 Thomson, K.S. 1996. The fall and rise of the English bulldog. *American Scientist* 84:220-223.

746 Tickle, C. 1991. Retinoic acid and chick limb bud development. *Development* (Suppl.1):113-121.

747 Tikhonov, A. and Vartanyan, S. 2001. Populations of wooly mammoth in North-East Siberia: dwarfing in isolation or last stage of extinction? In G. Cavarretta, P. Gioia, P. Mussi, and M. Palombo (eds.), *The World of Elephants: Proceedings of the First International Congress*, pg. 519. CNR, Rome.

748 Timmer, D.C., Bakker, O. and Wiersinga, W.M. 2003. Triiodothyronine affects the alternative splicing of thyroid hormone receptor alpha mRNA. *Journal of Endocrinology* 179(2):217-225.

749 Tomasi, T.E., Hellgren, E.C. and Tucker, T.J. 1998. Thyroid hormone concentrations in black bears (*Ursus americanus*): hibernation and pregnancy effects. *General and Comparative Endocrinology* 109:192-199.

750 Tomasi, T.E. and Mitchell, D.A. 1994. Seasonal shifts in thyroid function in the cotton rat (*Sigmodon hispidus*). *Journal of Mammalogy* 75:520-528.

751 Troy, C. S., MacHugh, D. E., Bailey, J. F., Magee, D. A., Loftus, R. T., et al. 2001. Genetic evidence for Near-Eastern origins of European cattle. *Nature* 410:1088-1091.

752 Trut, L.N. 1988. The variable rates of evolutionary transformations and their parallelism in terms of destabilizing selection. *Journal of Animal Breeding and Genetics* 105:81-90.

753 Trut, L.N. 1991. The intercranial allometry and morphological changes in silver foxes (*Vulpes vulpes* Desm.) under domestication (in Russian, English abstract). *Genetika* 27:1605-1611.

754 Trut, L.N. 1997. Domestication of the fox: roots and effects. In L.N. Trut and L.V. Osadchuk (eds.) *Evolutionary-genetic and Genetic-physiological Aspects of Fur Animal Domestication: A Collection of Reports*, pg. 7-14. IFASA/Scientifur, Oslo.

755 Trut, L.N. 1999. Early canid domestication: the farm-fox experiment. *American Scientist* 87:160-169.

756 Trut, L.N., Dzerzhinsky, F.Ya. and Nikolsky, V.S. 1991. A principal component analysis of changes in cranial characteristics appearing in silver foxes (*Vulpes vulpes* Desm.) under domestication (in Russian, English abstract). *Genetika* 27:1440-1449.

757 Trut, L.N., Naumenko, E.V. and Belyaev, D.K. 1972. Change in the pituitary-adrenal function of silver foxes during selection according to behavior. *Soviet Genetics* 8:35-40.

758 Trut, L.N. and Osadchuk, L.V. 1997. *Evolutionary- genetic and Genetic-physiological Aspects of Fur Animal Domestication: A Collection of Reports*. IFASA/Scientifur, Oslo.

759 Turek, F.W., Joshua, C., Kohsaka, A., Lin, E., Ivanova, G., et al. 2005. Obesity and metabolic syndrome in circadian *Clock* mutant mice. *Science* 308:1043-1045.

760 Turner, S.E. 2003. Behavioural aspects of maternal investment and disability in mother and infant Japanese macaques *(Macaca fuscata)* with congenital limb malformations. M. A. thesis, University of Victoria, Canada.

761 Uddin, M., Wildman, D.E., Liu, G., Xu, W., Johnson, R.M., et al. 2004. Sister grouping of chimpanzees and humans as revealed by genome-wide phylogenetic analysis of brain gene expression profiles. *Proceedings of the National Academy of Sciences USA* 101(9):2957-2962.

762 Uerpmann, H.P. 1993. Proposal for a separate nomenclature of domestic animals. In A. Clasen, S. Payne and H.P. Uerpmann (eds.), *Skeletons in her Cupboard: Festschrift for Juliet Clutton-Brock*, pg. 239-241. Monograph 34, Oxbow Books, Oxford.

763 Unwin, M.J. and Glova, G.J. 1997. Changes in life history parameters in a naturally spawning population of Chinook salmon (*Oncorhynchus tshawytscha*) associated with releases of hatchery reared fish. *Canadian Journal of Fisheries and Aquatic Science* 54:1235-1245.

764 van Coevorden, A., Mockel, J., Laurent, E., Kerkhofs, M., L'Hermite-Baleriaux, M., et al. 1991. Neuroendocrine rhythms and sleep in aging men. *American Journal of Physiology: Endocrinology & Metabolism* 260(4):E651-E661.

765 van den Beld, A.W., Visser, T.J., Feelders, R.A., Grobbee, D.E. and Lamberts, S.W. in press. Thyroid hormone concentrations, disease, physical function and mortality in elderly men. *Journal of Clinical Endocrinology & Metabolism.*

766 van den Bergh, G.D. 1999. The Late Neogene elephantoid-bearing faunas of Indonesia and their paleozoogeographic implications. *Scripta Geologica* 117.

767 van den Bergh, G.D., de Vos, J., Aziz, F. and Morwood, M.J. 2001. Elephantoidea in the Indonesian region: new *Stegodon* findings from Flores. In G. Cavarretta, P. Gioia, P. Mussi, and M. Palombo (eds.), *The World of Elephants: Proceedings of the First International Congress*, pg. 623-627. Rome: CNR.

768 van den Berghe, G., Wouters, P., Weekers, F., Mohan, S., Baxter, R.C., et al. 1999. Reactivation of pituitary hormone release and metabolic improvement by infusion of growth hormone-releasing peptide and thyrotropin-releasing hormone in patients with protracted critical illness. *Journal of Clinical Endocrinology & Metabolism* 84(4):1311-1323.

769 Vanderpump, M.P.J., Turnbridge, W.M.G., French, J.M., Appleton, D., Bates, M., et al. 1995. The incidence of thyroid disorders in the community: a twenty-year follow-up of the Whickham survey. *Clinical Endocrinology* 43:55-68.

770 van Tuinen, M., Sibley, C.G. and Hedges, S.B. 1998. Phylogeny and biogeography of ratite birds inferred from DNA sequences of the mitochondrial ribosomal genes. *Molecular Biology and Evolution* 15(4):370-376.

771 Vartanyan, S.L., Garutt, V.E. and Sher, A.V. 1993. Holocene dwarf mammoths from Wrangel Island in the Siberian Arctic. *Nature* 362:337-340.

772 Veldhuis, J.D. 2000. Nature of altered pulsatile hormone release and neuro-endocrine network signaling in human ageing: clinical studies of the somatotropic, gonadotropic, corticotropic and insulin axes. In D.J. Chadwick and J.A. Goode (eds.), *Mechanisms and Biological Significance of Pulsatile Hormone Secretion*, pg. 163-189. J. Wiley and Sons, Chichester.

773 Veldhuis, J.D., Anderson, S.M., Shah, N., Bray, M., Vick, T., et al. 2001. Neuro-physiological regulation and target-tissue impact of the pulsatile mode of growth hormone secretion in the human. *Growth Hormones and IGF Research* 11 (Suppl. A):S25-S37.

774 Veldhuis, J.D. and Bowers, C.Y. 2003. Sex-steroid modulation of growth hormone (GH) secretory control: three-peptide ensemble regulation under dual feedback restraint by GH and IGF-I. *Endocrine* 22(1):25-40.

775 Veldhuis, J.D., Iranmanesh, A. and Bowers, C.Y. 2005. Joint mechanisms of impaired growth-hormone pulse renewal in aging men. *Journal of Clinical Endocrinology and Metabolism* 90(7):4177-4183.

776 Vendramin-Gallo, M., Pessutti, C., Pezzato, A.C. and Vicentini-Paulino, M.L.M. 2001. Effect of age on seed digestion in parrots (*Amazona aestiva*). *Physiological and Biochemical Zoology* 74(3):398-403.

777 Vermiglio, F., Lo Presti, V.P. Moleti, M., Sidoti, M., Tortorella, G., et al. 2004. Attention deficit and hyperactivity disorders in offspring of mothers exposed to mild-moderate iodine deficiency: a possible novel iodine deficiency disorder in developed countries. *Journal of Clinical Endocrinology & Metabolism* 89(12):6054-6060.

778 Vidal-Rioja, L., Zambelli, A. and Semorile, L. 1994. An assessment of the relationships among species of Camelidae by satellite DNA comparisons. *Herediatas* 121:283-290.

779 Vilá, C., Savolainen, P., Maldonado, J.E., Amorim, I.R., Rice, J.E., et al.1997. Multiple and ancient origins of the domestic dog. *Science* 276:1687-1689.

780 Vincent, R. E. 1960. Some influences of domestication upon three stocks of brook trout (*Salvelinis fontinalis* Mitchill). *Transactions of the American Fisheries Society* 89:35-52.

781 Vohr, B.R. 2005. Editorial: Extreme prematurity – the continuing dilemma. *New England Journal of Medicine* 352:71-71.

782 Voss, S.R. 1995. Genetic basis of paedomorphosis in the axolotl, *Ambystoma mexicanum:* a test of the single-gene hypothesis. *Journal of Heredity* 86:441-447.

783 Voss, S.R. and Shaffer, H.B. 1997. Adaptive evolution via a major gene effect: paedomorphosis in the Mexican axolotl. *Proceedings of the National Academy of Science USA* 94:14185-14189.

784 Wallace, A.R. 1880. *Island Life: Or, The Phenomenon and Causes of Insular Faunas and Floras, Including a Revision and Attempted Solution of the Problem of Geological Climates.* London: MacMillan. [Reprint addition, a facsimile of the revised third edition of 1911: AMS Press, New York, 1975)

785 Ward, C.V., Leakey, M.G. and Walker, A. 2001. The earliest known *Australopithecus, A. anamensis*. *Journal of Human Evolution* 41:255-368.

786 Wartofsky, L. and Dickey, R.A. 2005. The evidence for a narrower thyrotropin reference range is compelling. *Journal of Clinical Endocrinology & Metabolism* 90(9):5483-5488.

787 Watanobe, T., Ishiguro, N. and Nakano, M. 2003. Phylogeography and population structure of the Japanese wild boar *Sus scrofa leucomystax:* mitochondrial DNA variation. *Zoological Science* 20:1477-1489.

788 Watanobe, T., Ishiguro, N., Nakano, M., Matsui, A., Hongo, H., et al. 2004. Prehistoric Sado Island populations of *Sus scrofa* distinguished from contemporary Japanese wild boar by ancient mitochondrial DNA. *Zoological Science* 21:219-228.

789 Watts, A.G., Tanimura, S. and Sanchez-Watts, G. 2004. Corticotropin-releasing hormone and arginine vasopressin gene transcription in the hypothalamic paraventricular nucleus of unstressed rats: daily rhythms and their interactions with corticosterone. *Endocrinology* 145(2):529-540.

790 Wayne, R.K. 1986. Limb morphology of domestic and wild canids: the influence of development on morphological change. *Journal of Morphology* 187:301-319.

791 Wayne, R.K. 1986. Developmental constraints on limb growth in domestic and some wild canids. *Journal of Zoology London A* 210:381-399.

792 Wayne, R.K. 1986. Cranial morphology of domestic and wild canids: the influence of development on morphological change. *Evolution* 40:243-261.

793 Wayne, R.K. and Jenks, S.M. 1991. Mitochondrial DNA analysis implying extensive hybridization of the endangered red wolf *Canis rufus. Nature* 351:565-568.

794 Wayne, R.K. and Gittleman, J.L. 1995. The problematic red wolf. *Scientific American* 273:36-39.

795 Wayne, R.K., Meyer, A., Leyman, N., Van Valkenburgh, B., Kat, P.W., et al. 1990. Large sequence divergence among mitochondrial DNA genotypes within populations of eastern black-backed jackals. *Proceedings of the National Academy of Science USA* 87(5):1772-1776.

796 Weber, D.S., Stewart, B.S., Garza, J.C. and Lehman, N. 2000. An empirical genetic assessment of the severity of the northern elephant seal population bottleneck. *Current Biology* 10(20):1287-1290.

797 Weetman, A.P. 1997. Hypothyroidism: screening and subclinical disease. *British Medical Journal* 314:1175-1178.

798 Weidensaul, S. 1999. Tracking America's first dogs. *Smithsonian Magazine* Vol. 29 (March):44-57.

799 Weiss, R.E. and Refetoff, S. 1996. Effect of thyroid hormone on growth: lessons from the syndrome of resistance to thyroid hormone. *Endocrinology and Metabolism Clinics of North America* 25(3):719-730.

800 Weissman, M.M. 2001. *Treatment of Depression: Bridging the 21st Century.* American Psychiatric Press Inc., Washington, DC.

801 West-Eberhard, M.J. 2003. *Developmental Plasticity and Evolution.* Oxford University Press, Oxford.

802 West-Eberhard, M.J. 2005. Developmental plasticity and the origin of species differences. *Proceedings of the National Academy of Science USA* 102 (Suppl.1):6543-6549.

803 Wheeler, P.E. 1984. The evolution of bipedality and loss of functional body hair in humans. *Journal of Human Evolution* 13:91-98.

804 Wheeler, P.E. 1991. The influence of bipedalism on the energy and water budgets of early hominids. *Journal of Human Evolution* 21:117-136.

805 White, T., Suwa, G. and Asfaw, B. 1994. *Australopithecus ramidus,* a new species of early hominid from Aramis, Ethiopia. *Nature* 371:306-312.

806 Whiting, M.F., Bradler, S. and Maxwell, T. 2003. Loss and recovery of wings in stick insects. *Nature* 421:264-267.

807 Wichman, H.A. and Lynch, C.B. 1991. Genetic variation for seasonal adaptation in *Peromyscus leucopus*: nonreciprocal breakdown in a population cross. *Journal of Heredity* 82:197-204.

808 Wickelgren, W. 2005. Autistic brains out of synch? *Science* 308:1856-1858.

809 Wiig, O., Derocher, A. E. and Belikov, S.E. 1999. Ringed seal (*Phoca hispida*) breeding in the drifting pack ice of the Barents Sea. *Marine Mammal Science* 15(2):595-598.

810 Wilcox, B.W. and Walkowicz, C. 1989. *The Atlas of Dog Breeds of the World.* T.F.H Publications, Neptune City.

811 Williamson, M. 1981. *Island Populations.* Oxford: Oxford University Press.

812 Willis, P.M., Crespi, B.J., Dill, L.M., Baird, R.W. and Hanson, M.B. 2004. Natural hybridization between Dall's porpoises (*Phocoenoides dalli*) and harbour porpoises (*Phocoena phocoena). Canadian Journal of Zoology* 82:828-834.

813 Wills, C. 1998. *Children of Promethius: The Accelerating Pace of Human Evolution.* Persius Books, Reading.

814 Wilson, C.M. and McNabb, F.M.A. 1997. Maternal thyroid hormones in Japanese quail eggs and their influence on embryonic development. *General and Comparative Endocrinology* 107:153-165.

815 Windle, R.J., Wood, S.A., Lightman, S.L. and Ingram, C.D. 1998. The pulsatile characteristics of hypothalamo-pituitary-adrenal activity in female Lewis and Fischer 344 rats and its relationship to differential stress responses. *Endocrinology* 139(10):4044-4052.

816 Windle, R.J., Wood, S.A., Shanks, N., Lightman, S.L. and Ingram, C.D. 1998. Ultradian rhythm of basal corticosterone release in the female rat: dynamic interaction with the response to acute stress. *Endocrinology* 139(2):443-450.

817 Wintle, A.G. 1996. Archaeologically-relevant dating techniques for the next century. *Journal of Archaeological Science* 23:123-138.

818 Wise, P.M. 1998. Menopause and the brain. *Scientific American* Women's Health: A Lifelong Guide (Special Issue):78-81.

819 Wise, P.M., Smith, M.J., Dubal, D.B., Wilson, M.E., Krajnak, K.M., et al. 1999. Neuroendocrine influences and repercussions of the menopause. *Endocrine Reviews* 20(3):243-248.

820 Witte, O.N. 1990. Steel locus defines new multipotent growth factor. *Cell* 63:5-6.

821 Wolf, M., Ingbar, S.H. and Moses, A.C. 1989. Thyroid hormone and growth hormone interact to regulate insulin-like growth factor-I messenger ribonucleic acid and circulating levels in the rat. *Endocrinology* 125(6):2905-2914.

822 Woller, M.J., Everson-Binotto, G., Nichols, E., Acheson, A., Keen, K.L., et al. 2002. Aging-related changes in release of growth hormone and luteinizing hormone in female rhesus monkeys. *Journal of Clinical Endocrinology and Metabolism* 87(11):5160-5167.

823 Wong, K. 2001. The search for autism's roots. *Nature* 411:882-884.

824 Wong, K. 2005. The littlest human. *Scientific American* (February):56-65.

825 Wood, B. 2002. Palaeoanthropology: Hominid revelations from Chad. *Nature* 418: 133-135.

826 Wood, C.C. and Foote, C.J. 1996. Evidence for sympatric genetic divergence of anadromous and nonanadromous morphs of sockeye salmon (*Oncorhynchus nerka*). *Evolution* 50(3):1265-1279.

827 Woods, G.L., White, K.L., Vanderwall, D.K., Li, G-P., Aston, K.I., et al. 2003. A mule cloned from fetal cells by nuclear transfer. *Science* 301:1063.

828 Woolf, C.M. 1995. Influence of stochastic events on the phenotype variation of common white leg markings in the Arabian horse: implications for various genetic disorders in humans. *Journal of Heredity* 86:129-135.

829 Wrangham, R. and Conklin-Brittain, N. 2003. Cooking as a biological trait. *Journal of Comparative Biochemistry and Physiology Part A* 136:35-46.

830 Wright, K. 2002. Times of our lives. *Scientific American* (September):58-65.

831 Wright, M. L. 2002. Melatonin, diel rhythms, and metamorphosis in anuran amphibians. *General and Comparative Endocrinology* 126:251-254.

832 Wright, M.L., Duffy, J.L., Guertin, C.J., Alves, C.D., Szatkowski, M.C., et al. 2003. Developmental and diel changes in plasma thyroxine and plasma and ocular melatonin in the larval and juvenile bullfrog, *Rana catesbeiana. General and Comparative Endocrinology* 130:120-128.

833 Wright, M.L., Guertin, C.J., Duffy, J.L., Szatkowski, M.C., Visconti, R.F., et al. 2003. Developmental and diel profiles of plasma corticosteroids in the bullfrog, *Rana catesbeiana. Comparative Biochemistry and Physiology Part A* 135:585-595.

834 Wrutniak, C., Casa, F. and Cabello, G. 2001. Thyroid hormone action in mitochondria. *Journal of Molecular Endocrinology* 26:67-77.

835 Wu, P., Jiang, T-X., Suksaweang, S., Widelitz, R.B. and Chuong, C-M. 2004. Molecular shaping of the beak. *Science* 305:1465-1466.

836 Wu, R., Ma, C-X., Lou, X-Y. and Casella, G. 2003. Molecular dissection of allometry, ontogeny, and plasticity: a genomic view of developmental biology. *BioScience* 53(11):1041-1047.

837 Yaoita, Y., Shi, Y.-B. and Brown, D.D. 1990. *Xenopus laevis* α and β thyroid hormone receptors. *Proceedings of the National Academy of Sciences USA* 87:7090-7094.

838 Yeh, J. 2002. The effect of miniaturized body size on skeletal morphology in frogs. *Evolution* 56:628-641.

839 Yen, P.M. 2001. Physiological and molecular basis of thyroid hormone action. *Physiological Reviews* 81(3):1097-1142.

840 Yen, P.M. 2003. Molecular basis of resistance to thyroid hormone. *TRENDS in Endocrinology and Metabolism* 14 (7):327-333.

841 Yoshida, T., Monkawa, T., Hayashi, M. and Saruta, T. 1997. Regulation of expression of leptin mRNA and secretion of leptin by thyroid hormone in 3T3-L1 adipocytes. *Biochemical and Biophysical Research Communictions* 232(3):822-826.

842 Yoshimura, T., Yasuo, S., Watanabe, M., Iigo, M., Yamamura, T., et al. 2003. Light-induced hormone conversion of T_4 to T_3 regulates photoperiodic response of gonads in birds. *Nature* 426:178-181.

843 Young, J.B., Bürgi-Saville, M.E., Bürgi, U. and Landsberg, L. 2005. Sympathetic nervous system activity in rat thyroid: potential role in goitrogenesis. *American Journal of Physiology: Endocrinology & Metabolism* 288:E861-E867.

844 Young, S.P. 1951. *The Clever Coyote*. Wildlife Management, Washington,DC.

845 Young, S.P and Goldman, E.A. 1944. *The Wolves of North America*. Dover Publications Inc., New York.

846 Zeuner, F.E. 1963. *A History of Domesticated Animals*. Hutchinson, London.

847 Zhao, X., Lorenc, H., Stephenson, H., Wang, Y.J., Witherspoon, D., et al. 2005. Thyroid hormone can increase estrogen-mediated transcription from a consensus estrogen response element in neuroblastoma cells. *Proceedings of the National Academy of Sciences USA* 102(13):4890-4895.

848 Zhou, Q., Renard, J-P., Le Friec, G., Brochard, V., Beaujean, N., et al. 2003. Generation of fertile cloned rats by regulating oocyte activation. *Science* 302:1179.

849 Zhu, Y-G., Huang, Y., Hu, Y., Liu, Y. and Christie, P. 2004. Interactions between selenium and iodine uptake by spinach (*Spinacia oleracea* L.) in solution culture. *Plant and Soil* 261:99-105.

850 Zoeller, R.T. 2003. Transplacental thyroxine and fetal brain development. *Journal of Clinical Investigations* 111(7):954-957.

851 Zoeller, R.T., Bansal, R. and Parris, C. 2005. Bisphenol-A, an environmental contaminant that acts as a thyroid hormone receptor antagonist *in vitro,* increases serum thyroxine, and alters RC3/neurogranin expression in the developing rat brain. *Endocrinology* 146(2):607-612.

852 Zoeller, R.T., Dowling, A.L.S., Herzig, C.T.A., Iannacone, E.A., Gauger, K.J., et al. 2002. Thyroid hormone, brain development, and the environment. *Environmental Health Perspectives* 110(Suppl.3):355-361.

853 Zornetzer, H.R. and Duffield, D.A. 2003. Captive-born bottlenose dolphin x common dolphin (*Tursiops truncates x Delphinus capensis*) intergeneric hybrids. *Canadian Journal of Zoology* 81:1755-1762.

Index